Molecular Modeling and Simulation

Applications and Perspectives

Series editor

Edward Maginn, University of Notre Dame, Notre Dame, IN, USA

This series aims at providing a comprehensive collection of works on developments in molecular modeling and simulation, particularly as applied to the various research fields of engineering. The Series covers a broad range of topics related to modeling matter at the atomistic level. The series provides timely and detailed treatment of advanced methods and their application in a broad range of interrelated fields such as biomedical and biochemical engineering, chemical engineering, chemistry, molecular biology, mechanical engineering and materials science. The Series accepts both edited and authored works, including textbooks, monographs, reference works, and professional books. The series welcomes manuscripts concerned with developments in and applications of molecular modeling and simulation to contemporary research in myriad technological fields, including, but not limited to:

- New Materials Development
- Process Engineering
- Fuel Engineering
- Combustion
- Polymer Engineering
- Biomechanics
- Biomaterials
- Fluid Flow and Modeling
- Nano and Micro Fluidics
- Nano and Micro Technology
- Thin Films
- Phase Equilibria
- Transport Properties
- Computational Biology

More information about this series at http://www.springer.com/series/13829

Gabriele Raabe

Molecular Simulation Studies on Thermophysical Properties

With Application to Working Fluids

Springer

Gabriele Raabe
Institut für Thermodynamik
TU Braunschweig
Braunschweig
Germany

ISSN 2364-5083 ISSN 2364-5091 (electronic)
Molecular Modeling and Simulation
ISBN 978-981-10-3544-9 ISBN 978-981-10-3545-6 (eBook)
DOI 10.1007/978-981-10-3545-6

Library of Congress Control Number: 2016961267

Printed on acid-free paper

This Springer imprint is published by Springer Nature
The registered company is Springer Nature Singapore Pte Ltd.
The registered company address is: 152 Beach Road, #22-06/08 Gateway East, Singapore 189721, Singapore

Für Jörg

Series Editor Preface

The series *Molecular Modeling and Simulation – Application and Perspectives* publishes volumes from leading authors on a wide range of topics related to molecular modeling and simulation. The format of the series allows authors to treat their subject in a deep and comprehensive manner—something that is not always possible to do in an archival journal article. With this volume entitled "Molecular Simulation Studies on Thermophysical Properties", Gabriele Raabe of the Institut für Thermodynamik at TU Braunschweig has provided us with a detailed and informative treatment of how atomistic molecular simulations are used to determine thermophysical and structural properties of industrially relevant fluids. She covers many of the fundamentals of molecular simulation, including foundational topics in statistical mechanics. She also provides an introduction to Monte Carlo and molecular dynamics simulation techniques, gives details on how force fields are parameterized and shows how properties are computed from simulation trajectories. A comprehensive summary of the latest simulation techniques is provided, along with examples from her own work on simulating working fluids for different technical applications.

This volume is appropriate both for the expert and the novice looking to learn more about molecular modeling and simulation. In addition to its value as a reference, it could be used as a textbook for a graduate course in molecular modeling and simulation.

I am very grateful to Dr. Gabriele Raabe for taking on this project and for writing such a comprehensive and illuminating volume. I believe it will be an outstanding resource for the molecular modeling community for many years to come.

Edward Maginn
University of Notre Dame

Preface

This volume is based on my habilitation thesis (Habilitationsschrift) that was accepted from the Faculty of Mechanical Engineering of the Technische Universität Braunschweig for granting the *venia legendi* for the field of teaching "Molecular Thermodynamics".

While my early research activities involved traditional experimental studies on thermophysical properties in the lab, I had the chance to perform first "computational experiments" during a research visit in the group of Richard Sadus. Since then I have focused my research on molecular simulations due to my fascination for its ability to allow for predictive studies and to gain insights into the systems on a molecular level that helps to interpret their properties. Shortly after I also started to teach a master course on molecular simulation. This course not only covers the fundamentals of molecular simulation, but also includes an introduction to the statistical mechanics that provide its theoretical framework, and details on force field models and their parametrization. This has motivated me to compose a comprehensive textbook that covers all these topics.

Molecular simulations have a wide range of application, but my research has always focused on studies on thermophysical properties. This is therefore also the scope of this volume, and it was important for me to devote a special chapter on the analysis of simulation outputs to derive different thermophysical properties. A main field of my research for the last years has been the molecular modeling and simulation for new HFO workings fluids. These simulation studies are mainly described at the end of this textbook as illustrative example to demonstrate the ability of molecular simulations to provide predictions on the thermophysical properties of pure components and mixtures for which only limited experimental data are available.

Most of the simulation studies were performed at the Laboratory for Thermodynamics (IfT) of the Technische Universität Braunschweig, and I express my sincere gratitude to the head of the IfT, Jürgen Köhler, for enabling me to work in the fascinating research field of molecular simulation, and for his encouragement to accomplish this habilitation thesis (Habilitationsschrift). I'd also like to express my gratitude to Jadran Vrabec (ThEt, University Paderborn, Germany) and Hans

Hasse (LTD, TU Kaiserslautern, Germany) for agreeing to review this thesis, and I thank the members of the habilitation committee (Habilitationskommission) for their commitment.

I am grateful to all current and former colleagues, PhD students and graduate students from the IfT and TLK, who have somehow contributed to this thesis, for instance, by being involved in some molecular or system simulation studies, by reading versions of the text, by contributing figures for thermodynamics cycles etc.

As mentioned before, I had my first contact with molecular simulation during my stay at the Centre for Molecular Simulation of the Swinburne University in Hawthorn, Australia. I'm indepted to Richard Sadus and Billy Todd for introducing me to this research field, and for many fruitful discussions we have had since then. I especially want to thank Richard Sadus for his comments on this manuscript and for helpful suggestions.

The first stage of the force field for HFO/HCFO compounds, i.e. the model for fluorinated propenes, was developed during may stay in the group of Ed Maginn at the University of Notre Dame, USA. This stay was founded by a fellowship of the Deutsche Forschungsgemeinschaft (DFG). I appreciate this fellowship as it offered me the chance to spend some time in the group of Ed Maginn. I am grateful for his mentoring during my stay and since then. He influenced my work in various respects, and he deserves special thanks for many helpful and inspiring discussions, his support and encouragement. I'd also like to thank him for his valuable comments and helpful suggestions on several chapters of this manuscript. I am also indepted to Eric Lemmon, Arno Läsecke and Marcia Huber from the National Institute of Standard and Technology (NIST) in Boulder, USA, for providing the EOS and ECS models for the HFO and HCFO compounds that enabled the comparative REFPROP simulations, for helpful discussions and encouragement.

I'd also like to thank Ryo Akasaka (Kyushu Sangyo University, Fukuoka, Japan) and Katsuyuki Tanaka (Nihon University, Japan) for sharing their experimental data for the HFO and HCFO compounds with me, often prior to publication. Their data and EOS models allowed for a validation of the molecular model and the simulation results.

I reserve my deepest gratitude for my partner Jörg, for his support, encouragement and forbearance.

23rd November 2016 Gabriele Raabe
Braunschweig, Germany

Contents

About the Author

PD Dr.-Ing. Gabriele Raabe graduated in Mechanical Engineering. She received her Ph.D. working on experimental studies on vapor–liquid phase equilibria at low temperatures and their modeling by equations of state. She continued to work as thermodynamicist and senior scientist at the Laboratory for Thermodynamics, TU Braunschweig, and her research activities involve the modeling and prediction of thermophysical properties, focusing on force-field modeling and molecular simulation studies with a wide range of applications that cover, for instance, predicting the thermophysical properties of working fluids and refrigerants, studies on ionic liquids and simulations of drug solubilities. She also has many years of experience in teaching master's courses on molecular simulations and thermodynamics of mixtures.

Nomenclature

Latin Letters

a	Lattice constant (Å)
a_i	Activity (mol mol^{-1})
$\mathbf{a}$	Acceleration vector (ms^{-2})
acc	Acceptance probability (–)
A	Thermodynamic property (–)
A	Adjustable parameter in EAM, MEAM (–)
$\mathbf{A}$	Transition matrix (–)
b	Third derivative of coordinate vector (ms^{-3})
B	2. Virial coefficient (m^{3}kmol^{-1})
c	Cohesive energy density (kJm^{-3})
c	Reduced correlation coefficient (–)
C	Specific heat capacity (Jg^{-1}K^{-1})
C	Parameter to estimate force constants in MMFF94 (–)
C	Parameter in screening function of EAM, MEAM (–)
C_i	Weights in histogram reweighting (–)
C_i	Corrector coefficient (Gear) (–)
C_{lg}	Dispersion coefficient in ReaxFF-lg (–)
$C(t)$	Time correlation coefficient (–)
$C(I,J)$	Collection matrix (–)
C_m	Molar heat capacity (Jmol^{-1}K^{-1})
d	Distance (m)
D	Self-diffusion coefficient (m^{2}s^{-1})
D_e	Well-depth of the Morse potential (J)
e_i	Energy of an atom (a.u.)
E	Internal energy of a microstate (J)
E_c	Sublimation energy (J)
E_i	Electric field on site i (Vm^{-1})
f	Fugacity (MPa)

$f(), F()$	Function (–)
$\mathbf{f}$	Force vector (N)
F	Embedding function in EAM, MEAM (–)
F	Free energy, Helmholtz energy (J)
g	Degeneracy of an energy level, density of state (–)
g	Constraint force (N)
g_K	Infinite system Kirkwood factor (–)
G_K	Finite system Kirkwood factor (–)
G_s	Van Hove self-correlation function (–)
G	Gibbs free energy (J)
ΔG_{sim}	Gibbs free energy change from free energy methods (Jmol^{-1})
$G(\Gamma)$	Function in EAM, MEAM (–)
H	Enthalpy (Jmol^{-1})
$H(x, y)$	Histogram of x and y values (–)
$H_{i(j)}$	Henry coefficient from i in j (MPa)
$\mathcal{H}$	Hamiltonian (J)
$\mathbf{H}$	Cell matrix (nm)
I	Inertia tensor (kgm^2)
I	Ionization energy (Jmol^{-1})
j	Quantum state for rotation (–)
j	Angular momentum (Nms)
J	Hardness of an atom (Jmol^{-1})
$\dot{J}$	Flux (mass, heat, current) (–)
$\mathbf{k}$	Reciprocal space vector (Ewald sum) (–)
k_r	Force constant for bond stretching (kJ mol^{-1}Å^{-2})
k_θ	Force constant for angle bending (kJ mol^{-1}rad^{-2})
k_ϕ	Force constant for torsion (kJ mol^{-1})
k_ω	Force constant for improper torsion (kJ mol^{-1}rad^{-2})
K_i	Grid points in SPME method (–)
l	Bond length (Å)
L	(Simulation box) length (Å)
$\mathcal{L}$	Lagrangian (J)
n	Multiplicity (dihedral angle) (–)
n	Number of ensemble systems (–)
n	Preceding state in MC simulations (–)
$\mathbf{n}$	Cell vector (–)
n_i	Quantum number (–)
N	Number of particles, atoms (–)
m	Mass (g)
m	State of attempted move in MC simulations (–)
m	Exponent (–)
M	Molar mass (g mol^{-1})
M	System's total dipole moment (Cm)
$\mathbf{M}$	Torque (Nm)

M_n	B-spline functions (–)
p	Pressure (MPa)
p	Parameter (–)
P	Momentum vector (Ns)
p	(Un-normalized) probability (–)
$\wp$	Probability distribution (–)
$\wp$	Instantaneous pressure (MPa)
$\mathbf{q}$	Generalized coordinate (m)
q_i	Partial charge (e)
q_m	Quaternion (–)
q	Molecular partition function (NVE ensemble) (–)
Q	Partition function (–)
Q_T	Quadrupole moment (Cm^2)
Q_i	Effective charge in EAM (–)
$Q_{T/p}$	Fictive mass in thermostat/barostat (–)
$Q(k_1, k_2, k_3)$	Charge array (SPME method) (–)
$\mathbf{r}$	Position vector (Å)
r_{ij}	Interatomic distance (Å)
r_{max}	Cutoff distance in the Buckingham exp-6 potential (Å)
R_{andom}	Random number (–)
$\mathbf{R}$	Vector describing the orientation of a molecule (–)
R_G	Radius of gyration (m)
R_{min}	vdW interaction parameter, distance at the minimum (Å)
s	Reduced coordinate r/L (–)
s	Scaling factor Nosé–Hoover thermostat (–)
S	Entropy (JK^{-1})
S	Screening function in EAM or MEAM (–)
$S(k)$	Reciprocal structure factor (–)
$\mathbf{T}_{ij}$	Dipole tensor (V C^{-1}m^{-2})
t	Time (s)
$t^{(h)}$	Adjustable parameter in EAM, MEAM (–)
T	Thermodynamic temperature (K)
$\mathcal{T}$	Instantaneous temperature (K)
u	Contribution to the total potential energy (J)
u	Scaled coordinates in SPME method (–)
U_{pot}	Potential (configurational) energy (Jmol^{-1})
U	Internal energy (J)
v	Velocity vector (ms^{-1})
v_s	Speed of sound (ms^{-1})
V	Volume (m^3)
w_i	Statistical weights in CBMC methods (–)
W	Thermodynamic probability (–)
W	Virial (Nm)
W	Rosenbluth weight (–)

x_i Mol fraction ($mol\ mol^{-1}$)
Z Compressibility factor (–)
Z Configurational integral (–)
Z Parameter to estimate force constants in MMFF94 (–)
Z Number of nearest neighbors (–)

Greek Letters and Other Symbols

α Polarizability tensor ($Cm^2\ V^{-1}$)
α Parameter in the Buckingham exp-6 potential (–)
α_p Thermal expansivity (K^{-1})
α Parameter in Ewald sum (–)
α Heat transfer coefficient ($WK^{-1}m^{-2}$)
$\alpha(t)$ Non-Gaussian parameter (–)
$\alpha(n \rightarrow m)$ Proposal probability from MC trial move (–)
β Reciprocal temperature ($k_B^{-1}T^{-1}$) (J^{-1})
β Parameter in the Morse potential (–)
β Parameter to estimate force constants in MMFF94 (–)
$\beta^{(h)}$ Adjustable parameter in EAM, MEAM (–)
γ_i Activity coefficient (–)
Γ Phase space (–)
δ Phase of a dihedral angle (rad)
δ_i Partial charge increment (e)
δ_{ij} Kronecker delta (–)
Δ Deviation (–)
Δt Time step size (s)
ε_i Energy level, quantum state (a.u)
ε_{ij} VdW interaction parameter, depth of the potential well (J)
ε_r Dielectric constant (Fm^{-1})
ζ Friction parameter thermostat (–)
η Shear viscosity (mPas)
$\eta(N)$ Weighting factor in flat histogram method (–)
θ Bond angle (rad)
κ_T Isothermal compressibility (Pa^{-1})
λ Coupling parameter in free energy methods (–)
λ Wavelength (m)
λ Velocity scaling (–)
λ_T Thermal conductivity ($Wm^{-1}\ K^{-1}$)
Λ Thermal de Broglie wavelength (m)
μ Chemical potential ($J\ mol^{-1}$)
μ Dipole moment (Cm)
μ_{JT} Joule–Thomson coefficient (KPa^{-1})

$\nu_{\mu\mu\mu}$	Non-additive coefficient (–)
ν	Frequency, i.e. of vibration (s^{-1})
ξ	Fugacity fraction (–)
π	Microstate transition matrix in Metropolis MC simulations (–)
Π	Macrostate transition matrix (–)
ρ	Electron density in EAM, MEAM (–)
ϱ	Density (kmolm^{-3})
ϱ	Charge distribution (–)
σ	Lennard–Jones parameter (Å)
σ^{elec}	Electric conductivity (Sm^{-1})
σ	Symmetry number (–)
σ	Stress tensor (Pa)
$\sigma(A)$	Standard error, variance of property A (–)
τ	Correlation time (s)
τ	Relaxation time (s)
$\tau(n \rightarrow m)$	Transition probability (–)
φ	Fugacity coefficient (–)
Θ	Characteristic temperature (K)
θ	Bond angle (rad)
Θ, Φ, Ψ	Euler angles (rad)
ϕ	Dihedral angle (rad)
ϕ_{ij}	Pair-potential in EAM, MEAM (–)
ϕ	Electrostatic potential (JC^{-1})
χ	Electronegativity (J mol^{-1})
χ	Scaling factor in barostat (–)
χ	Correction factor (–)
Ψ	Thermodynamic potential (–)
ω	Acentric factor (–)
ω	Improper torsion angle (rad)
ω_{ij}	Charge increment between two atoms i and j (e)
ω	Angular velocity (rads^{-1})
Ω_{mn}	Phase space function (–)

Subscript

α, β	Referring to atoms within molecules
bond	Bond stretching potential
B	Block
c	Critical point
cal	Calculated data
Cell	Cell list
conf	Referring to the configurational energy

cut	Cutoff (radius)
C	Component
CoM	Center of mass
D	Drude particle
exp	Experimental data
f	Degree of freedom
Gauss	Gaussian distribution
i, j, k	Index (molecules)
I, II	Referring to two phases at equilibrium
kin	Kinetic
l	Linear
List	Verlet list
m	Molar
NVE	*NVE* (microcanonical) ensemble
NVT	*NVT* (canonical) ensemble
NpT	*NpT* ensemble
over	Overcoordinated
p	Isobaric, at constant pressure
pm	Partial molar
μVT	μVT (grand canonical) ensemble
sim	Simulated data
s	Saturation
S	Isotropic, at constant entropy
tors	Torsion
under	Undercoordinated (ReaxFF)
val	Valence (angle)
vap	Vapor, vaporization
vdW	Van der Waals interactions
V	At constant volume
x, y, z	Referring to the dimensions in space
0	Reference value, equilibrium
$0i$	Pure fluid *i*
2	Referring to pairs, dimers
3	Referring to triples
$N+1$	Trial particle insertion in MC
$N-1$	Trial particle deletion in MC
Δr	Trial random displacement in MC
ΔV	Trial volume change in MC

Superscript

a	Atomic
b	Body-fixed frame
C	Corrected
Coul	Coulombic interaction
el	Electrostatic
elec	Electronic
eff	Effective
E	Excess property
ext	External (intermolecular) contributions
id	Ideal gas
ind	Induced (dipole)
inter	Referring to intermolecular terms or contributions
intra	Referring to intramolecular terms or contributions
kin	Referring to the kinetic energy
L	Liquid
LJ	Lennard–Jones potential
LRC	Long-range correction
nucl	Nuclear
(*n*)	Old state in Metropolis sampling
(*m*)	New (trial) state in Metropolis sampling
or	Orientation
p	Predicted
pos	Position
ref	Reference value
res	Residual, deviation from the ideal gas
rot	Contribution of rotational motion
s	Scape-fixed frame
self	Self energy
trans	Contribution of translational motion
vib	Contribution of vibrational motion
V	Vapor
*	Reduced quantity
*	Virtual variable
′	Saturated liquid phase
″	Saturated vapor phase
—	Averaged value
0	Reference state

Abbreviations

AA	All Atoms (force field)
AIMD	Ab initio molecular dynamics
a.u.	Atomic unit
AMBER	Assisted Model Building and Energy Refinement (force field)
amim	1-alkyl-3-methyl-imidazolium-based IL
BO	Bond order (ReaxFF)
BSE	Running estimate of standard error in block average technique
CBMC	Configurational-bias Monte Carlo
CHARMM	Chemistry at HARvard Molecular Mechanics (force field)
CHELPG	CHarges from ELectrostatic Potential, Grid method
CFC	Chlorofluorocarbon
COM	Center of mass
COS	Charge-on-Spring
DFT	Density functional theory
EAM	Embedded atom model
ECS	Extended corresponding state
EOS	Equation of state
ESP	Electrostatical potential
FF	Force field
GAFF	General AMBER Force Field
GCMC	Grand canonical Monte Carlo simulation
GEMC	Gibbs Ensemble Monte Carlo simulation
GROMOS	GROningen Molecular Simulation package (also force field)
GWP	Global warming potential
HCFC	Hydrochlorofluorocarbon
HCFO	Hydrochlorofluoroolefine
HFO	Hydrofluoroolefine
HTHP	High-temperature heat pump
HVAC	Heating ventilation and air conditioning systems
IL	Ionic Liquid
LJ	Lennard–Jones potential
LRC	Long-range correction
MC	Monte Carlo
MD	Molecular dynamics
MEAM	Modified embedded atom model
MEP	Molecular electrostatic potential
MIC	Minimum image convention
MM_n	Mechanical Mechanics force field
MMFF	Merck Molecular Force Field
MSD	Mean square displacement
ODP	Ozone depletion potential
OPLS	Optimized Potential for Liquid Simulations (force field)

ORC	Organic Rankine Cycle
PBC	Periodic boundary conditions
QM	Quantum Mechanical, ab initio
ReaxFF	Reactive force field
RDF	Radial (pair) distribution function
RESP	Restrained Electrostatical potential
RMS	Root mean square (deviation)
SCF	Self-consistent field
SPMC	Smooth Particle Mesh Ewald
UA	United-Atoms (force field)
UFF	Universal Force Field
VACF	Velocity autocorrelation function
VLCC	Vapor–liquid coexistence curve
VLE	Vapor–liquid phase equilibrium

Fundamental Physical Constants

Quantity	Symbol	Numerical value
Dielectric constant (vacuum permittivity)	ϵ_0	$8.854187817.10^{-12}$ C V^{-1} m^{-1}
Planck's constant	h	$6.6260689633 \cdot 10^{-34}$ Js
Avogadro constant	N_a	$6.02214199 \cdot 10^{23} mol^{-1}$
Boltzmann constant	k_B	$1.3806503 \cdot 10^{-23} J K^{-1}$
Molar gas constant	R	$8.314472 J mol^{-1} K^{-1}$

Chapter 1
Introduction

Molecular simulation techniques are increasingly used in chemical engineering as complement to experimental studies to provide thermophysical properties of pure components and mixtures over wide temperature and pressure ranges. Molecular simulations do not only yield data for thermophysical properties, but at the same time allow gaining an insight into the systems on a molecular level. The analysis of local structures in mixtures by molecular simulations, for instance, provides information on preferred interactions and enables a molecular interpretation of the properties. This thesis covers the fundamentals of molecular simulation, beginning with the fundamentals in statistical mechanics and providing an introduction to the Monte Carlo and molecular dynamics simulation techniques. It then summarizes the available know-how for using molecular simulations to derive information on thermophysical and structural properties. The determination of these properties from molecular simulations are thereby illustrated with examples from studies on different systems. However, molecular simulation studies require an adequate molecular model for the components of interest. Thus, this thesis also provides an introduction to, and discussion on specific aspects of the molecular modeling in so-called force fields.

A benefit of molecular simulation is its ability to provide information for 'not-so-well-described' fluids, i.e. for components for which only limited experimental data are available. Limitation of experimental data may have different reasons: the components are difficult to handle in experiments (toxic, hazardous, corrosive compounds), data are needed at extreme conditions of temperature and pressure, or the compounds are newly synthesized, so that they are not available in sufficient quantities for elaborated experimental studies. This is the case for the hydrofluoroolefine (HFO) and hydrochlorofluoroolefine (HCFO) working fluids covered in this thesis. Due to their high Global Warming Potential (GWP), currently used hydrofluorocarbon (HFC) refrigerants are affected by phase-out regulations. HFO and also HCFO compounds have been proposed as new class of low GWP refrigerants, either as pure compounds or as components in blends. However, limited experimental data on

G. Raabe, *Molecular Simulation Studies on Thermophysical Properties*,
Molecular Modeling and Simulation, DOI 10.1007/978-981-10-3545-6_1

their thermophysical properties hampers the exploration of the performance of these new working fluids in technical applications. Based on a newly developed all-atoms force field for HFO and HCFO compounds, simulation results on the pure components and also potential refrigerant blends as working fluids for different applications are presented. These simulation studies illustrate the value of molecular simulations to provide predictions for poorly known compounds and mixtures, and to gain a molecular level understanding of their properties.

The structure of this thesis is as follows:

Statistical mechanics provides the theoretical framework for molecular simulations that allow for determining macroscopic properties from studying the properties of the microstate. Therefore, important terms of statistical mechanics are explained in Chap. 2. This covers explanations of different kinds of ensembles, the derivations of their probability densities and partition functions, and the connection of the partition functions to thermodynamic state variables. Additionally, the molecular partition function of the ideal gas is introduced that serves as a reference state for simulations on real components. Finally, the chapter deals with the Maxwell-Boltzmann velocity distribution that provides important relations between temperature, velocities and kinetic energy.

Chapter 3 is concerned with the basic ideas of Monte Carlo (MC) simulations. It provides an introduction to the importance sampling technique and its application to simulations in various ensembles. Furthermore, biased sampling techniques for complex molecules and reweighting methods are discussed. A special focus is dedicated to Monte Carlo simulation techniques for fluid phase equilibria, i.e. the Gibbs Ensemble, the Gibbs Duhem, and the Histogram Reweighting technique.

Chapter 4 provides an introduction to the basics of molecular dynamics (MD) simulations. It deals with the fundamental equations that describe the dynamics of the system, and with different finite-difference methods for the numerical integration of the equation of motions. Then different thermostatting and barostatting techniques to allow for simulations at constant temperatures and pressures are explained and discussed. The last section of Chap. 4 treats the "handling" of (partly) rigid molecules in MD simulations. It thereby introduces the constraint dynamics for rigid bonds, and the rotational equations of motion for rigid bodies.

Chapter 5 deals with the practical aspects of molecular simulation studies. These comprise the setting up of simulations, the application of boundary conditions, the speed up of simulations by using neighbor lists, and the quantifications of uncertainties of simulation results. Another important aspect that is discussed in this chapter is the proper treatment of long-range electrostatic interactions. Thereby the widely used Ewald summation and its modifications are explained, and additionally the Wolf damped coulomb potential is introduced.

Chapter 6 provides an overview on force field models, i.e. analytical functions that are used to determine the configurational energy of the simulation system as function of the molecular configuration. The introduction to molecular models and their parametrization thereby mainly focuses on so-called class 1 force fields that are widely used in studies on fluid systems. The parametrization of these molecular models is exemplified by the introduction of an all-atom force field for hydrofluoro-(HFO) and hydrochlorofluoroolefines (HCFO). Classical force fields employ fixed partial charges to account for electrostatic interactions. However, in Sect. 6.2, approaches to introduce polarization into the molecular models are also briefly discussed. To complete the overview, this chapter also provides a short overview on other force field types, i.e. potentials for metals, and the reactive force field ReaxFF—both illustrated by examples from own simulation studies. Chapter 6 is complemented with a discussion on specific aspects in the molecular modeling. These are approaches to account for many-body effects when ab initio models as purely two-body potentials are used in molecular simulation studies. In this context, simulation studies on mercury serve as example for using a semi-empirical approach. Furthermore, the influence of intermolecular flexibility on the description of thermophysical properties of water by SPC type models is discussed in detail. This discussion is based on extensive studies on the flexible SPC/Fw water model.

Chapter 7 provides an extensive compilation of the analysis of molecular simulation outputs to derive thermophysical and structural properties of pure compounds and mixtures. Thereby, the determination of pVT, transport and dielectric properties as well as second-derivative properties in single-phase simulations are explained. Additionally, this chapter deals with the evaluation of phase equilibria simulations to deduce saturation properties of both pure compounds and mixtures. A special focus is dedicated to free energy techniques, and thereby to approaches to derive the chemical potential of pure compounds and components in mixtures. Based on this, the chapter also provides detailed deductions to obtain fugacity and activity coefficients from molecular simulations. The different aspects regarding the determination of thermophysical properties from molecular simulations are illustrated with examples from own simulations studies on different systems such as water, refrigerants or ionic liquids. Beyond this, molecular simulation not only allows for the determination of thermodynamic and transport properties, but also enables the analysis of microscopic structures. Chapter 7 is therefore complemented by the discussion of structural analysis by distribution functions. Molecular simulation studies on ionic liquids and their mixtures with alcohols then exemplify the value of molecular simulations to provide insight into structural features of the systems.

Chapter 8 is concerned with the application of molecular simulation techniques to derive information on the thermophysical properties of working fluids. The molecular simulations are based on the all-atom force field for hydrofluoro-(HFO) and hydrochlorofluoroelefines (HCFO) introduced in Chap. 6. The simulation studies on this new class of working fluids well demonstrate the ability of molecular simulations to provide predictions on the thermophysical properties of pure components and mixtures for which limited experimental data are available. Prior to the discussion of the molecular simulation results, Chap. 8 provides an overview on the historical development of working fluids and current regulations that affect their progression today. Molecular simulation results for different HFO and HCFO compounds and their mixtures with "conventional" refrigerants such as R-32, CO_2 or R-134a are presented. The simulation results for the thermophysical properties of the working fluids are thereby discussed in the context of their potential technical application. This covers studies on working fluids for (mobile) air-conditioning, Organic Rankine cycles and heat pumps, chillers, and blends for different stationary refrigeration applications.

Chapter 9 concludes the thesis with a summary of the most important findings of the simulation studies and gives an outlook on future research trends in both the field of molecular simulation and of working fluids. The thesis is supplemented by an appendix that lists the force field parameters for the molecular modeling of HFO and HCFO compounds.

Chapter 2
Introduction to Statistical Mechanics

This chapter gives an introduction to the theoretical framework of molecular simulation studies provided by statistical mechanics Molecular simulations are aimed at predicting macroscopic thermophysical properties based on the behavior of the particles contained in the system. The detailed configuration of the system on the molecular level is called a microstate, and statistical mechanics is the theoretical framework that allows for determining macroscopic properties from studying the properties of the microstate.

Imagine for instance a macroscopic system with rigid and adiabatic walls that contains a constant amount of particles (N), and has a constant volume (V) and internal energy (E). Although these macroscopic properties are fixed, i.e. constrained, we observe a permanent rearrangement of the configuration on the microscopic level as the particles move, exchange energy through collision and redistribute their energies among the different available energy modes (translation, rotation, vibration). Thus, for each macroscopic state there exists a large number of different configurations on the molecular level, the microstates. This collection of microstates that satisfies the fixed properties of the macroscopic system (constraints) is called an ensemble. The choice of constrained properties characterizes different kinds of ensembles that will be discussed in detail in Sect. 2.1.

In molecular simulations, a large number of microstates corresponding to the macroscopic system of interest is generated. In order to perform simulations that are comparable to experimental studies, the simulation ensemble is generally selected in such a way that the constraints correspond to the properties controlled in the experiment. The properties that are not constrained are fluctuating, and predictions for these properties can be provided by analyzing the simulation results by statistical mechanics.

There are two methods to generate a large number of microstates for a given macrostate. In molecular dynamics (MD) simulations, the time evolution of a

G. Raabe, *Molecular Simulation Studies on Thermophysical Properties*,
Molecular Modeling and Simulation, DOI 10.1007/978-981-10-3545-6_2

microstate is followed by numerically integrating the Newton's equations of motion of its particles with time. The statistical average of a property of interest A is then determined as time average, i.e. from the values $A_i(t)$ of the property at different time steps Δt

$$A_{macro} = \langle A \rangle_{time} = \frac{1}{N_{\Delta t}} \sum_{i=1}^{N_{\Delta t}} A_i(t). \tag{2.1}$$

In Monte Carlo (MC) simulations, different microstates are generated stochastically by elementary changes to the previous configuration. The statistical averages of the thermodynamic properties are then calculated as so called ensemble averages

$$A_{macro} = \langle A \rangle_{ensemble} = \int_{\Gamma} A(\Gamma)\wp(\Gamma)d\Gamma \approx \sum^{N_{microstate}} A(\Gamma)\wp(\Gamma)d\Gamma, \tag{2.2}$$

where Γ is the phase space of possible states, and $\wp$ is the probability distribution, i.e. the probability to observe a particular microstate. A basic postulate of statistical mechanics, and therefore of molecular simulation studies, is the "ergodic hypothesis". It states that ensemble averages are identical with time averages

$$A_{macro} = \langle A \rangle_{time} = \langle A \rangle_{ensemble}, \tag{2.3}$$

i.e. both molecular dynamics and Monte Carlo simulation methods yield the same results for equilibrium properties within the given statistical uncertainties [3]. For the calculation of an ensemble average of a property of interest, the value of this property in a microstate is weighted by the probability that this state occurs. The probability to observe a particular state depends on the constraints imposed on the system. Thus, each kind of ensemble is characterized by its probability distribution or probability density $\wp$. Another important property of an ensemble is its partition function Q that is the sum of all available states under the given constraints. The partition function contains all the thermodynamic information of the system, i.e. it enables the calculation of all its thermodynamic properties. The partition function therefore is a fundamental quantity for statistical mechanics, comparable to the fundamental equation of state in macroscopic thermodynamics.

Thus, the following Sect. 2.1 will provide an introduction to different kinds of ensembles and the derivations of their probability densities and partition functions. Section 2.2 then connects the partition functions to thermodynamic state variables. Section 2.3 introduces the molecular partition function of the ideal gas as a reference state for simulations on real components. The final section of this chapter deals with the Maxwell-Boltzmann velocity distribution that provides important relations between temperature, velocities and kinetic energy required in MD simulations.

2.1 Ensembles

The concept of ensembles plays a key role in statistical mechanics. As described above, an ensemble is a large number of microstates that all have the same constrained properties as the macroscopic system of interest. Figure 2.1 shows a comparison of the classical ensemble types used in statistical mechanics and the corresponding macroscopic thermodynamic systems.

The microcanonical (NVE) ensemble comprises the collection of microstates that all have the same constant volume V, internal energy E and occupation with N particles. It corresponds to an isolated thermodynamic system with rigid, adiabatic and impermeable walls, which disable the exchange of energy and mass with the environment. The microcanonical ensemble, however, is of little importance for molecular simulation studies on thermophysical properties as the constraints do not correspond to those of real experimental set-ups.

In the canonical (NVT) ensemble, the imposed variables are the volume V, the temperature T, and the number of particles N. Thus, each microstate can be regarded as a closed system with rigid and impermeable walls. Though now the walls are diathermic to allow energy exchange with an isothermal heat bath to ensure constant temperature. This ensemble is used in simulation studies for fluids with known density to predict properties such as pressure, chemical potential, internal energy or transport properties.

The grand canonical (μVT) ensemble is a large number of microstates that all have the same constant volume V and temperature T. Additionally, the chemical potential

ensemble		imposed variables (constraints)	fluctuating properties	corresponding thermodynamic system
microcanonical (NVE)	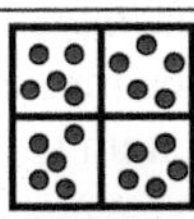	number of particles N volume V internal energy E	energy (quantum state) of each particle ε_i	isolated
canonical (NVT)	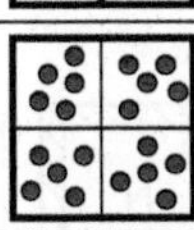	number of particles N volume V temperature T	internal energy E_j	closed
grand canonical (μVT)	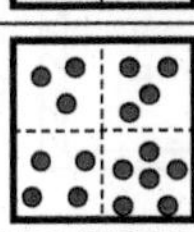	chemical potential μ volume V temperature T	number of particles N_i internal energy E_j	open
isothermal-isobaric (NPT)	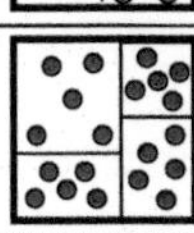	number of particles N pressure p temperature T	volume V_i internal energy E_j	-

Fig. 2.1 Comparison of the classical ensemble types: the constraints properties by which they are characterized and the properties that fluctuate. Also given is the macroscopic thermodynamic system type it can be compared with

μ is imposed, whereas the amount of particles in the system can fluctuate. The grand canonical ensemble corresponds to an open system with rigid but diathermic and permeable walls that enable both mass and energy exchange with the environment. The grand canonical ensemble is the only ensemble in which the number of particles in the system is not an imposed variable. It is therefore used in molecular simulation studies on adsorption or chemical reactions.

The imposed variables of the isothermal-isobaric (*NpT*) ensemble are pressure p, temperature T and the amount of particles N. Although it does not correspond to a classical thermodynamic system, the *NpT* ensemble is widely used in molecular simulation studies as its constraints match usual experimental conditions. Simulations in the *NpT* ensemble therefore enable the prediction of thermophysical properties as a function of temperature and pressure in comparison with experimental results.

Following this short introduction to the classical ensemble types and their application in molecular simulation studies, the next sections provide more details on their theoretical description in statistical mechanics and the derivation of their probability density and partition function. A more detailed description is provided for instance in the textbooks by Van Carey [9], Hill [4], Weingärtner [10], Sears and Salinger [8] or Gasser and Richards [2].

2.1.1 The Microcanonical Ensemble

The microcanonical ensemble can be visualized as n copies of an isolated system that all have the same volume V, internal energy E and constant number of particles N, as shown in Fig. 2.2.

Although the total internal energy of the system E is constant, it is continuously redistributed between the particles, so that the energy of each particle ε_i—or its quantum state—fluctuates. When N_i is the number of particles that occupy the quantum state i with the energy level ε_i, the total energy of the system is given by

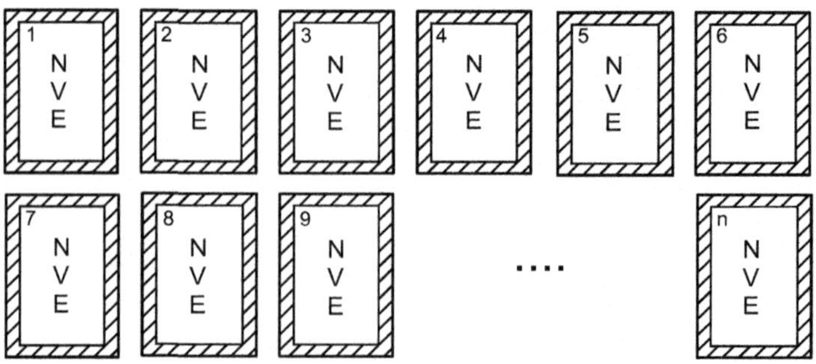

Fig. 2.2 Depiction of the microcanonical (*NVE*) ensemble: n copies of an isolated system with constant volume V and internal energy E, consisting of N equal particles

$$E = \sum_i \varepsilon_i N_i \tag{2.4}$$

The given constraint that the total energy has to be constant yields

$$dE = \sum_i \varepsilon_i dN_i = 0. \tag{2.5}$$

Additionally, the total number of particles is constant, i.e.

$$N = \sum_i N_i = const. \quad \text{and} \quad dN = \sum_i dN_i = 0. \tag{2.6}$$

However, for a given number of particles N and total energy E, a number of possibilities still exist to distribute the particles on the different available quantum states ε_i, i.e. different distribution functions $N_1, N_2, ...N_i$. Figure 2.3 exemplarily shows some feasible distributions of $N = 8$ particles on four different quantum states ε_0, ε_1, ε_2, ε_3 for a constraint total energy of $E = 16\,meV$. Provided that the different particles are distinguishable (in the example of Fig. 2.3 by the numbering), a number of possibilities (microstates) exist to realize a specific distribution. The number of different microstates for a given distribution can be obtained from combinatorial analysis

$$W_{NVE} = \frac{N!}{\Pi_{i=0}^{\infty} N_i!}. \tag{2.7}$$

W is called the thermodynamic probability and gives the total number of microstates consistent with the given distribution—it can therefore be regarded as degree of disorder of the system. As exemplification, Fig. 2.3 also provides the thermodynamic probabilities W of the given distributions. The most famous equation in statistical mechanics is Boltzmann's entropy hypothesis

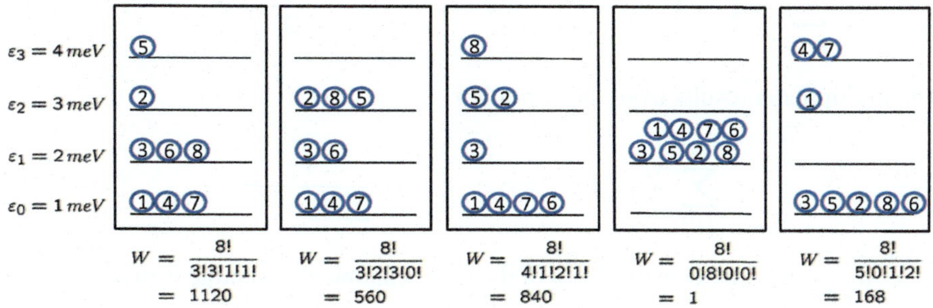

Fig. 2.3 Example for different potential distributions of $N = 8 = const.$ particles on four quantum states $\varepsilon_0 = 1\,\text{meV}$, $\varepsilon_1 = 2\,\text{meV}$, $\varepsilon_2 = 3\,\text{meV}$, $\varepsilon_3 = 4\,meV$ for a constrained total energy of $E = 16\,\text{meV}$. Also given is the number of different ways W to realize a particular distribution for distinguishable particles, employing Eq. 2.7

$$S = k_B \ln(W_{NVE}) \tag{2.8}$$

that relates the entropy to the logarithm of the thermodynamic probability W, and with this links statistical mechanics to classical thermodynamics.

Thus, the equilibrium state is the state with maximum entropy and according to Eq. 2.8, the state with the highest probability, i.e. the distribution with the highest number of different microstates. The condition of equilibrium can be formulated as

$$S \rightarrow max, \quad W_{NVE} \rightarrow max, \quad \ln(W_{NVE}) \rightarrow max. \tag{2.9}$$

$$\Rightarrow d(\ln(W_{NVE})) = 0 \quad \rightarrow dS = 0. \tag{2.10}$$

Applying the Sterling formula

$$lnx! = x \cdot \ln x - x$$

yields

$$\ln(W_{NVE}) = \ln(N!) - \sum_{i=0}^{\infty} \ln(N_i!) \tag{2.11}$$

$$= -\sum_{i=0}^{\infty} N_i \ln\left(\frac{N_i}{N}\right). \tag{2.12}$$

When determining the most probable distribution by minimizing Eq. 2.11, the existing constraints of the ensemble have to be taken into account. This is usually done by employing Lagrange's method of undetermined multipliers [7], i.e. by multiplying the differentials of the equations of constraint in Eqs. 2.5 and 2.6 with Lagrangian multipliers β and α to incorporate them into Eq. 2.11. Thus, the resulting equation to be solved is

$$\sum_{i=0}^{\infty} \left[\ln\left(\frac{N_i}{N}\right) + \alpha + \beta\varepsilon_i\right] dN_i = 0 \quad \Rightarrow \quad \frac{N_i}{N} = e^{-\alpha} e^{-\beta\varepsilon_i}. \tag{2.13}$$

The multiplier α results from the condition

$$\sum_{i=0}^{\infty} \frac{N_i}{N} = 1 \quad \Rightarrow \quad e^{-\alpha} = \frac{1}{\sum_{i=0}^{\infty} e^{-\beta\varepsilon_i}}. \tag{2.14}$$

Thus, the probability density $\wp_{NVE}$ of the microcanonical ensemble is then given by

$$\wp_{NVE} = \frac{N_i}{N} = \frac{e^{-\beta\varepsilon_i}}{\sum_{j=0}^{\infty} e^{-\beta\varepsilon_j}} \quad \text{with} \quad \beta = \frac{1}{k_B T}, \tag{2.15}$$

and expresses the probability that particles occupy the quantum state i with an energy level ε_i. The denominator in Eq. 2.15 is the so called molecular partition function q

$$q = \sum_{j=0}^{\infty} e^{-\beta\varepsilon_j}, \tag{2.16}$$

and is the sum of all available energy levels (quantum states) of the particles. When the energy levels ε_i are g_i times degenerated, Eq. 2.16 can also be expressed by

$$q = \sum_{energy\ levels\ i} g_i e^{-\beta\varepsilon_i}. \tag{2.17}$$

The molecular partition function q has to be distinguished from the partition function Q_{NVE} of the microcanonical ensemble. Q_{NVE} gives the number of accessible state associated with the system energy E at given constant values of N and V. It can be written as [1]

$$Q_{NVE} = \sum_{\Gamma} \delta_i(E_i - E) \quad \text{with} \tag{2.18}$$

$$\delta_i(E_i - E) = 1, \ \text{for}\, E - \Delta E \leq E_i < E$$
$$\delta_i(E_i - E) = 0, \ \text{else.}$$

Therein δ_i is the Kronecker delta, and Δ the uncertainty with which the energy E can be determined [4].

The term $exp(-\beta\varepsilon_i)$ in Eq. 2.16 is the so called Boltzmann factor. The resulting distribution of the particles among the different energy levels (quantum states) ε_i,

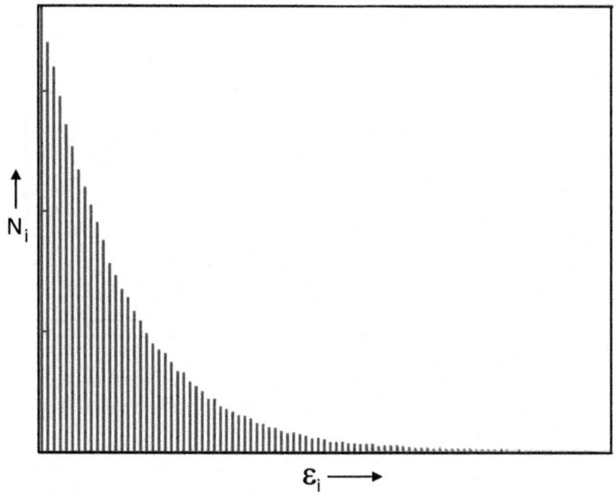

Fig. 2.4 Boltzmann distribution as most probable distribution of $N = const.$ particles on different energy levels for a constrained total energy $E = const$

is the Boltzmann distribution shown in Fig. 2.4 that represents the most probable distribution of a thermodynamic system with $N, E = const.$

2.1.2 The Canonical Ensemble

The canonical ensemble represents n copies of a closed system that all have the same volume V, temperature T and constant number of particles N. All copies are in thermal contact by being arranged in a large isothermal heat bath. Thus, the energy E_i of each copy, i.e. ensemble member, may fluctuate. However, the entire collection of the n copies is adiabatic, as shown in Fig. 2.5.

When n_i stands for the number of ensemble members with the same energy E_i, the total energy of the system is given by

$$U = \sum_i n_i E_i. \tag{2.19}$$

As the total system is thermally isolated with regard to the environment, its total energy U has to be constant, which yields the constraint

$$dU = \sum_i E_i dn_i = 0. \tag{2.20}$$

Additionally, the total number n of ensemble members is constant that gives the constraints

$$n = \sum_i n_i = const. \quad \text{and} \quad dn = \sum_i dn_i = 0. \tag{2.21}$$

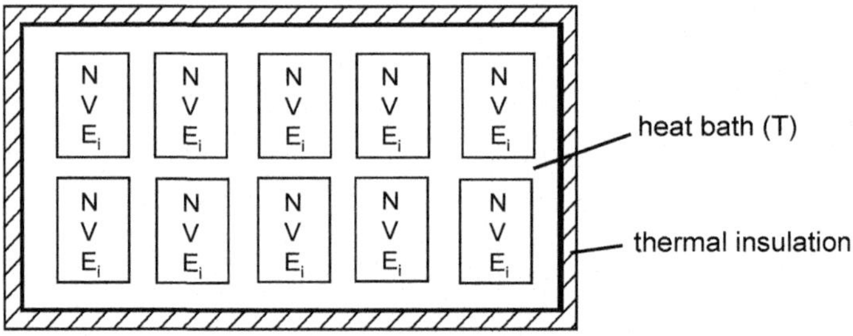

Fig. 2.5 Depiction of the canonical (NVT) ensemble: n copies of a closed system with constant volume V and fixed temperature T, consisting of N equal particles. The internal energy E_i of each copy fluctuates

A particular distribution is characterized by the n_1, n_2, ..., n_i, i.e. number of ensemble members n_i residing in a microstate i with the energy E_i. For the NVT ensemble, the number of ways to obtain a specific distribution is given by

$$W_{NVT} = \frac{n!}{\Pi_{i=0}^{\infty} n_i!}. \tag{2.22}$$

Again, the equilibrium distribution is the most probable distribution realized by the maximum number of different microstates. It is determined by maximizing Eq. 2.22 and employing the Lagrange's method of undetermined multipliers to consider the constraints of the ensemble. When the multiplier for the energy constraint in Eq. 2.20 is $\beta = 1/(k_B T)$, the resulting equation to be solved is

$$\sum_{i=0}^{\infty} \left[\ln\left(\frac{n_i}{n}\right) + \alpha + \frac{E_i}{k_B T} \right] dn_i = 0 \quad \Rightarrow \quad \frac{n_i}{n} = e^{-\alpha} e^{-E_i/k_B T} \tag{2.23}$$

The undetermined multiply α is derived from

$$\sum_{i=0}^{\infty} \frac{n_i}{n} = 1 \quad \Rightarrow \quad e^{-\alpha} = \frac{1}{\sum_{i=0}^{\infty} e^{-E_i/k_B T}} \tag{2.24}$$

Thus, the probability that an ensemble member has the energy E_j is given by

$$\wp_{NVT} = \frac{n_j}{n} = \frac{e^{-E_j/k_B T}}{\sum_{i=0}^{\infty} e^{-E_i/k_B T}} \tag{2.25}$$

with the denominator in Eq. 2.25 being the partition function Q_{NVT} of the canonical ensemble

$$Q_{NVT} = \sum_{i=0}^{\infty} e^{-E_i/k_B T}. \tag{2.26}$$

Combining all energy states belonging to the same system energy level E_j, Q_{NVT} can also be expressed by

$$Q_{NVT} = \sum_{levels\ E_j} Q_{NVE}(E_j, V, N) e^{-E_j/k_B T}, \tag{2.27}$$

wherein the microcanonical partition function $Q_{NVE}(E_j, V, N)$ gives the number of states with the energy E_j. The canonical partition function Q_{NVT} is related to the molecular partition function q by

$$Q_{NVT} = q^N \quad \text{for N distinguishable particles} \tag{2.28}$$

$$Q_{NVT} = \frac{q^N}{N!} \quad \text{for N indistinguishable particles,} \tag{2.29}$$

where for instance particles of an ideal gas are indistinguishable, whereas particles in a solid crystal are distinguishable by their coordinates in the lattice. Equation (2.26) gives the canonical partition function over all quantum states of the system, whereas molecular simulations require the formulation of the Q_{NVT} in the classical phase space Γ. In the classical formulation, the total energy E_i is divided into the kinetic energy E_{kin} and the potential, i.e. configurational energy U_{conf} that arises from interactions between the particles. Thus, the phase space is defined by the positions r and momenta p of the N particles within the system. However, according to the uncertainty principle in quantum mechanics, a state in phase space, defined by $dr\,dp$, can only be located with an uncertainty of h, the Planck's constant [4]. Taking this into account, a state in the phase space $d\Gamma$ in a 3-dimensional system with N undistinguishable particles is given by

$$d\Gamma = \frac{1}{N!}\frac{1}{h^{3N}} dr^{3N} dp^{3N}. \tag{2.30}$$

As the energy levels are quite close together compared to $k_B T$, the sum over all energy states can be approximated by an integral, so that finally the classical formulation of the canonical partition function is given by [1]

$$Q_{NVT} = \frac{1}{N!}\frac{1}{h^{3N}} \int\int e^{-E_i/k_B T} dr^{3N} dp^{3N}. \tag{2.31}$$

2.1.3 The Grand Canonical Ensemble

Similar to the canonical ensemble, the grand canonical ensemble can be visualized by n copies of a system that all have the same volume V, and that are immersed in a large isothermal heat bath so that all members have the same temperature T. But now, the walls of the copies are permeable to allow crossing over of particles from one copy to another. Thus, both the energy E_i and the number of particles N_j in each member of the ensemble fluctuate. Again, the entire collection of the n copies is isolated from the environment so that its total energy U and total number of particles N remain constant. A depiction of the grand canonical ensemble is shown in Fig. 2.6.

When n_{N_j,E_i} is the number of copies that consist of a particular amount of particles N_j and have an energy of E_i, the constraints of the grand canonical ensemble can be expressed by

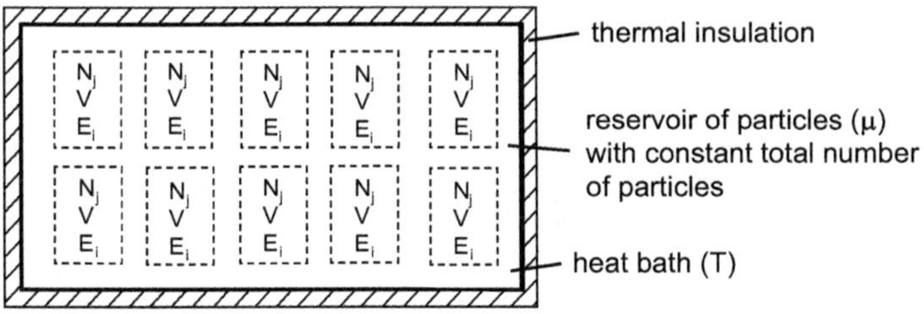

Fig. 2.6 Depiction of the grand canonical (μVT) ensemble: n copies of an open system with constant volume V and fixed temperature T and chemical potential μ. The occupation N_j and the internal energy E_i fluctuate

$$\sum_{N_j=0}^{\infty}\sum_{i=0}^{\infty} n_{N_j,E_i} = n = const. \quad \Rightarrow \quad dn = \sum_{N_j=0}^{\infty}\sum_{i=0}^{\infty} dn_{N_j,E_i} = 0$$

$$\sum_{N_j=0}^{\infty}\sum_{i=0}^{\infty} E_i\, n_{N_j,E_i} = U = const. \quad \Rightarrow \quad dU = \sum_{N_j=0}^{\infty}\sum_{i=0}^{\infty} E_i\, dn_{N_j,E_i} = 0$$

$$\sum_{N_j=0}^{\infty}\sum_{i=0}^{\infty} N_j\, n_{N_j,E_i} = N = const. \quad \Rightarrow \quad dN = \sum_{N_j=0}^{\infty}\sum_{i=0}^{\infty} N_j\, dn_{N_j,E_i} = 0$$

The number of ways to realize a particular distribution n_{N_j,E_i} is given by

$$W_{\mu VT} = \frac{n!}{\Pi_{N_j=0}^{\infty}\Pi_{i=0}^{\infty} n_{N_j,E_i}!}. \tag{2.32}$$

The state of thermodynamic equilibrium, i.e. state with the highest probability is determined from solving $d \ln W_{\mu VT} = 0$ subject to the constraints expressed above. This yields

$$\sum_{N_j=0}^{\infty}\sum_{i=0}^{\infty}\left[\ln\frac{n_{N_j,E_i}}{n} + \alpha + \frac{E_i}{k_B T} + bN_j\right] dn_{N_j,E_i} = 0 \tag{2.33}$$

$$\Rightarrow \quad \frac{n_{N_j,E_i}}{n} = e^{-\alpha} e^{\left(-\frac{E_i}{k_B T} - bN_j\right)} \tag{2.34}$$

Again, the multiplier α is derived from the condition

$$\sum_{N_j=0}^{\infty}\sum_{i=0}^{\infty}\frac{n_{N_j,E_i}}{n} = 1 \tag{2.35}$$

Thus, the probability of finding an ensemble member with the energy E_i and the number of particles N_j is

$$\wp_{\mu VT} = \frac{n_{N_j,E_j}}{n} = \frac{e^{(-E_i/k_BT - bN_j)}}{\sum_{N_j=0}^{\infty}\sum_{i=0}^{\infty} e^{(-E_i/k_BT - bN_j)}} \tag{2.36}$$

where the denominator represents the partition function $Q_{\mu VT}$ of the grand canonical ensemble. The multiplier b can be determined from

$$dS = k_B\, d(\ln W_{\mu VT}) = -k_B \sum_{N_j=0}^{\infty}\sum_{i=0}^{\infty} ln(n_{N_j,E_i}) dn_{N_j,E_i} \tag{2.37}$$

$$= \frac{1}{T}\underbrace{\sum_{N_j=0}^{\infty}\sum_{i=0}^{\infty} E_i\, dn_{N_j,E_i}}_{dU} + k_B b \underbrace{\sum_{N_j=0}^{\infty}\sum_{i=0}^{\infty} N_j\, dn_{N_j,E_i}}_{dN} \tag{2.38}$$

The first double summation in Eq. 2.38 represents the change in the total internal energy dU, whereas the second summation specifies the change in the total number of particles dN. Thus, a comparison with the Gibbs fundamental equation of state for $dV = 0$ yields the relation of the multiplier b and the chemical potential μ

$$-\frac{\mu}{T} = k_B\, b \quad \Rightarrow \quad b = \frac{-\mu}{k_B T}. \tag{2.39}$$

With this, the partition function of the grand canonical ensemble becomes

$$Q_{\mu VT} = \sum_{N_j=0}^{\infty}\sum_{i=0}^{\infty} e^{\frac{-(E_i - \mu N_j)}{k_B T}} \tag{2.40}$$

In the classical formulation, the partition function of the grand canonical ensemble of N indistinguishable particles is given by [1]

$$Q_{\mu VT} = \sum_N \frac{1}{N!}\frac{1}{h^{3N}} e^{\frac{\mu N}{k_B T}} \int\int e^{-E_i/k_BT} dr^{3N} dp^{3N} = \sum_N \left(e^{\frac{\mu N}{k_B T}} Q_{NVT} \right). \tag{2.41}$$

2.1.4 The *NpT*-Ensemble

The *NpT* ensemble represents n copies of a closed system with the same constant number of particles N. The members of the ensemble are arranged in a large isothermal heat bath and pressure chamber to ensure that they all have the same temperature and pressure, whereas their energy E_i and volume V_j fluctuate. However, the entire collection of the n copies is adiabatic and has a constant total volume V, as shown in Fig. 2.7.

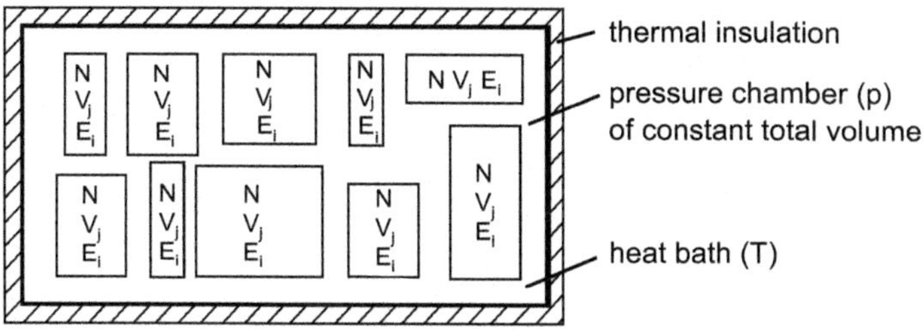

Fig. 2.7 Depiction of the NpT ensemble: n copies of a system with fixed temperature T and pressure p, consisting of N equal particles. The volume V_i and internal energy E_i of each copy fluctuate

For the NpT-Ensemble, n_{V_j,E_i} is the number of copies with a particular volume V_j and energy E_i. The constraints of the ensemble are given by

$$\sum_{V_j=0}^{\infty}\sum_{i=0}^{\infty} n_{V_j,E_i} = n = const. \quad \Rightarrow \quad dn = \sum_{v_j=0}^{\infty}\sum_{i=0}^{\infty} dn_{V_j,E_i} = 0$$

$$\sum_{V_j=0}^{\infty}\sum_{i=0}^{\infty} E_i\, n_{V_j,E_i} = U = const. \quad \Rightarrow \quad dU = \sum_{V_j=0}^{\infty}\sum_{i=0}^{\infty} E_i\, dn_{V_j,E_i} = 0$$

$$\sum_{V_j=0}^{\infty}\sum_{i=0}^{\infty} V_j\, n_{N_j,E_i} = V = const. \quad \Rightarrow \quad dV = \sum_{N_j=0}^{\infty}\sum_{i=0}^{\infty} V_j\, dn_{V_j,E_i} = 0$$

The thermodynamic probability W_{NpT} gives the number of possibilities to realize a specific distribution n_{V_j,E_i}, and can again be derived from combinatorial analysis. This yields

$$W_{NpT} = \frac{n!}{\prod_{V_j=0}^{\infty}\prod_{i=0}^{\infty} n_{V_j,E_i}!}. \tag{2.42}$$

Thus, the most probable distribution subject to the constraints of the NpT ensemble can be determined from solving

$$\sum_{V_j=0}^{\infty}\sum_{i=0}^{\infty}\left[\ln\frac{n_{V_j,E_i}}{n} + \alpha + \frac{E_i}{k_B T} + bV_j\right] dn_{V_j,E_i} = 0 \tag{2.43}$$

$$\Rightarrow \quad \frac{n_{V_j,E_i}}{n} = e^{-\alpha} e^{\left(-\frac{E_i}{k_B T} - bV_j\right)}. \tag{2.44}$$

With the multiplier α derived from the condition

$$\sum_{V_j=0}^{\infty}\sum_{i=0}^{\infty}\frac{n_{V_j,E_i}}{n} = 1, \tag{2.45}$$

the probability $\wp_{NpT}$ that a microstate has a particular combination of energy E_i and volume V_j is

$$\wp_{NpT} = \frac{n_{V_j,E_i}}{n} = \frac{e^{(-E_i/k_B T - bV_j)}}{\sum_{V_j=0}^{\infty} \sum_{i=0}^{\infty} e^{(-E_i/k_B T - bV_j)}}. \tag{2.46}$$

The multiplier b can be determined by comparing the expression for dS of the NpT ensemble

$$dS = k_B\, d(\ln W_{NpT}) = -k_B \sum_{V_j=0}^{\infty} \sum_{i=0}^{\infty} ln(n_{V_j,E_i}) dn_{V_j,E_i} \tag{2.47}$$

$$= \frac{1}{T} \sum_{V_j=0}^{\infty} \sum_{i=0}^{\infty} E_i\, dn_{V_j,E_i} + k_B b \sum_{V_j=0}^{\infty} \sum_{i=0}^{\infty} V_j\, dn_{V_j,E_i} \tag{2.48}$$

with the Gibbs fundamental equation of state for $dN = 0$. This yields

$$k_B b \sum_{V_j=0}^{\infty} \sum_{i=0}^{\infty} V_j\, dn_{V_j,E_i} = k_B b\, dV = \frac{p}{T} dV \qquad \Rightarrow \qquad b = \frac{p}{k_B T} \tag{2.49}$$

Thus, the partition function of the NpT ensemble, which is given by the dominator in Eq. 2.46, is

$$Q_{NpT} = \sum_{V_j=0}^{\infty} \sum_{i=0}^{\infty} e^{\frac{-(E_i + pV_j)}{k_B T}} \tag{2.50}$$

In the classical formulation, the partition function of the NpT ensemble of N indistinguishable particles is [1]

$$Q_{NpT} = \frac{1}{N!} \frac{1}{h^{3N}} \frac{1}{V_0} \int e^{-pV_j/k_B T} dV \int \int e^{-E_i/k_B T} dr^{3N} dp^{3N} \tag{2.51}$$

$$= \frac{1}{V_0} \int Q_{NVT}\, e^{-pV_j/k_B T} dV, \tag{2.52}$$

wherein the basic unit of volume V_0 is included to render Q_{NpT} dimensionless. Another common choice for the volume scale is the factor βp instead of V_0^{-1}.

2.2 Thermodynamic Properties of the Ensemble

As described above, the partition function in statistical thermodynamics is of similar importance as the fundamental equation of state in macroscopic thermodynamics. From the partition function, all thermodynamic properties of the ensemble can be obtained from derivatives or by algebraic operations. The connection of the partition function to thermodynamics is established by the bridge equation [6]

$$\Psi_{ensemble} = -\ln Q_{ensemble}, \tag{2.53}$$

where Ψ is the thermodynamic potential of the ensemble, i.e. the property whose minimum defines the equilibrium condition—in accordance with the thermodynamic potentials in macroscopic thermodynamics. The definitions of thermodynamic potentials of macrostates and microstates are compared in Table 2.1.

Thus, the thermodynamic properties of the **microcanonical ensemble** can be derived from the bridge equation

$$S = k_B \ln Q_{NVE}. \tag{2.54}$$

A comparison with the Gibbs Fundamental equation of state for the entropy S

$$dS = \frac{1}{T}dE + \frac{p}{T}dV - \frac{\mu}{T}dN. \tag{2.55}$$

yields the relation between the molecular partition function Q_{NVE} and the thermodynamic properties T, p and μ

$$\frac{1}{T} = \left(\frac{\partial S}{\partial E}\right)_{V,N} = k_B \left(\frac{\partial \ln Q_{NVE}}{\partial E}\right)_{V,N} \tag{2.56}$$

$$\frac{p}{T} = \left(\frac{\partial S}{\partial V}\right)_{E,N} = k_B \left(\frac{\partial \ln Q_{NVE}}{\partial V}\right)_{E,N} \tag{2.57}$$

Table 2.1 Comparison of thermodynamic potentials in macroscopic and statistical thermodynamics [6]

Macrostate		Microstate	
System	Thermodynamic potential	Ensemble	$\Psi_{ensemble}$
$U, V, n =$ const.	Entropy: $-S \to min.$	Microcanonical	$\Psi_{NVE} = -\frac{S}{k_B}$
$T, V, n =$ const.	Helmholtz: $F \to min.$	Canonical	$\Psi_{NVT} = \frac{F}{k_B T}$
$T, p, n =$ const.	Gibbs: $G \to min.$	NpT	$\Psi_{NpT} = \frac{G}{k_B T}$
$\mu, V, T =$ const.	Hill: $-pV \to min.$	Grand canonical	$\Psi_{\mu VT} = -\frac{pV}{k_B T}$

$$\frac{\mu}{T} = -\left(\frac{\partial S}{\partial N}\right)_{E,V} = -k_B \left(\frac{\partial \ln Q_{NVE}}{\partial N}\right)_{E,V}. \tag{2.58}$$

In a similar way, the correlations between the thermodynamic properties S, p, μ and U and the partition function Q_{NVT} of the **canonical ensemble** can be derived from the bridge equation

$$F = -k_B T \ln Q_{NVT}, \tag{2.59}$$

and the formulation of the Gibbs Fundamental equation of state for the Helmholtz energy F

$$dF = -SdT - pdV + \mu dN. \tag{2.60}$$

This yields

$$S = -\left(\frac{\partial F}{\partial T}\right)_{V,N} = k_B \left(\ln Q_{NVT} + T\left(\frac{\partial \ln Q_{NVT}}{\partial T}\right)_{V,N}\right) \tag{2.61}$$

$$p = -\left(\frac{\partial F}{\partial V}\right)_{T,N} = k_B T \left(\frac{\partial \ln Q_{NVT}}{\partial V}\right)_{T,N} \tag{2.62}$$

$$\mu = \left(\frac{\partial F}{\partial N}\right)_{T,V} = -k_B T \left(\frac{\partial \ln Q_{NVT}}{\partial N}\right)_{T,V} \tag{2.63}$$

$$U = F + TS = k_B T^2 \left(\frac{\partial \ln Q_{NVT}}{\partial T}\right)_{V,N} \tag{2.64}$$

For the **NpT ensemble** the bridge equation becomes

$$G = -k_B T \ln Q_{NpT}. \tag{2.65}$$

By means of the Gibbs Fundamental equation of state for the Gibbs energy G

$$dG = -SdT + Vdp + \mu dN, \tag{2.66}$$

the thermodynamic properties S, V, μ and H can be related to the partition function Q_{NpT} by

$$S = -\left(\frac{\partial G}{\partial T}\right)_{p,N} = k_B \left(\ln Q_{NpT} + T\left(\frac{\partial \ln Q_{NpT}}{\partial T}\right)_{p,N}\right) \tag{2.67}$$

$$V = \left(\frac{\partial G}{\partial p}\right)_{T,N} = -k_B T \left(\frac{\partial \ln Q_{NpT}}{\partial p}\right)_{T,N} \tag{2.68}$$

$$\mu = \left(\frac{\partial G}{\partial N}\right)_{p,T} = -k_B T \left(\frac{\partial \ln Q_{NpT}}{\partial N}\right)_{p,T} \tag{2.69}$$

$$H = G + TS = k_B T^2 \left(\frac{\partial \ln Q_{NpT}}{\partial T} \right)_{p,N} \tag{2.70}$$

Finally, the bridge equation of the **grand canonical ensemble** is given by

$$pV = k_B T \ln Q_{\mu VT}. \tag{2.71}$$

The differential of the Euler equation in consideration of the Gibbs Fundamental equation of state yields

$$d(pV) = SdT + pdV + Nd\mu. \tag{2.72}$$

Thus, the thermodynamic properties S, p and N can be derived from the partition function $Q_{\mu VT}$ by

$$S = \left(\frac{\partial (pV)}{\partial T} \right)_{V,\mu} = k_B \left(\ln Q_{\mu VT} + T \left(\frac{\partial \ln Q_{\mu VT}}{\partial T} \right)_{V,\mu} \right) \tag{2.73}$$

$$p = \left(\frac{\partial (pV)}{\partial V} \right)_{T,\mu} = k_B T \left(\frac{\partial \ln Q_{\mu VT}}{\partial V} \right)_{T,\mu} \tag{2.74}$$

$$N = \left(\frac{\partial (pV)}{\partial \mu} \right)_{V,T} = k_B T \left(\frac{\partial \ln Q_{\mu VT}}{\partial \mu} \right)_{V,T}. \tag{2.75}$$

In general, those properties that are derived from derivatives of the partition functions, i.e. the pressure p, volume V, the energy U and the enthalpy H are referred to as mechanical properties. These properties are easily evaluated in molecular simulations studies. However, the properties Gibbs energy G, the free or Helmholtz energy F, or the entropy S itself are related to the partition function directly and are so called statistical quantities. Thus, special simulation techniques are required (see Sect. 7.4) to obtain acceptable estimates for the entropy related properties.

2.3 Molecular Partition Function of the Ideal Gas

The partition function contains all the thermodynamic information of the system, and its knowledge allows for the calculation of all thermodynamic functions. However, the same problems that exist for setting up a physical based fundamental equation of state for a real compound also arise for formulating its partition function. The physics of real systems are too complex and are still not completely understood so that it is impossible to find a physical and mathematical consistent description of all the energy levels of a compound that is necessary to formulate its partition function. Thus, the exact formulation of a partition function is limited to simple and idealized systems such as the ideal gas, for which the energy levels are known accurately and for which it can be assumed that the system's energy is the sum of energies of its particles [2]. As the total energy of an ideal gas can be divided into independent contributions from translational, rotational, vibrational motions and from the electronic and nuclear energy levels, the total molecular partition functions can be expressed as

$$q = q^{trans} \cdot q^{rot} \cdot q^{vib} \cdot q^{elec} \cdot q^{nucl}. \tag{2.76}$$

Thus, the following subsections will provide an overview over the different contributions to the partition function for an ideal gas.

2.3.1 Partition Function of Translational Motion

The translational partition function can be derived by considering N monoatomic molecules of an ideal gas. Each atom is treated as mass point of mass m and the gas is enclosed in a cubic container of volume V and side length L. The energy level ε_i^{trans} of an atom is its kinetic energy associated with its translational motion in the three directions in space

$$\varepsilon_i^{trans} = \frac{\mathrm{p}_x^2 + \mathrm{p}_y^2 + \mathrm{p}_z^2}{2m}. \tag{2.77}$$

In quantum mechanics, the so called de Broglie relation states that the wavelength λ of an atom is Planck's constant h divided by the atom's momentum

$$\lambda = \frac{h}{\mathrm{p}}. \tag{2.78}$$

The wavelength in turn is related to the container's side length L, as a multiplier of half the wavelength has to fit into the box [2]. Thus,

$$\lambda = \frac{2L}{n}, \tag{2.79}$$

with the quantum number n. Therefore, according to the de Broglie relation, the momentum for each dimensional coordinate can be expressed as

$$\mathrm{p}_x = \frac{hn_x}{2L_x}, \quad \mathrm{p}_y = \frac{hn_y}{2L_y}, \quad \mathrm{p}_z = \frac{hn_z}{2L_z}, \tag{2.80}$$

and hence the kinetic energy level as

$$\varepsilon_i^{trans} = \frac{h^2}{8m}\left(\frac{n_x^2}{L_x^2} + \frac{n_y^2}{L_y^2} + \frac{n_z^2}{L_z^2}\right). \tag{2.81}$$

The molecular partition function of the translational motion then becomes

$$
\begin{aligned}
q^{trans} &= \sum_{i=0}^{\infty} e^{-\frac{\varepsilon_i^{trans}}{k_B T}} \qquad (2.82)\\
&= \left(\sum_{i=1}^{\infty} e^{-\frac{h^2}{8mk_BT}\left(\frac{n_x}{L_x}\right)^2}\right) \cdot \left(\sum_{i=1}^{\infty} e^{-\frac{h^2}{8mk_BT}\left(\frac{n_y}{L_y}\right)^2}\right) \cdot \left(\sum_{i=1}^{\infty} e^{-\frac{h^2}{8mk_BT}\left(\frac{n_z}{L_z}\right)^2}\right)\\
&= q_x^{trans} \cdot q_y^{trans} \cdot q_z^{trans}.
\end{aligned}
$$

As the energy levels ε_i^{trans} are in general very close together, the sums can be approximated by integrals. Thus, the molecular partition function for the x-dimension for instance can be written as

$$
\begin{aligned}
q_x^{trans} &= \sum_{i=1}^{\infty} e^{-\frac{h^2}{8mk_BT}\left(\frac{n_x}{L_x}\right)^2} \approx \int_0^{\infty} e^{-\frac{h^2}{8mk_BT}\left(\frac{n_x}{L_x}\right)^2} dn_x\\
&= \frac{L_x}{2}\sqrt{\frac{8\pi mk_BT}{h^2}} \qquad (2.83)
\end{aligned}
$$

The product of the three side lengths $L_xL_yL_z$ gives the container volume V. Therefore, the total translational partition function as product of the partition functions for the three dimensions becomes

$$
q^{trans} = V\left(\sqrt{\frac{2\pi mk_BT}{h^2}}\right)^3 = \frac{V}{\Lambda^3} \qquad (2.84)
$$

where Λ is the de Broglie wavelength

$$
\Lambda = \sqrt{\frac{h^2}{2\pi mk_BT}}. \qquad (2.85)
$$

According to Eq. 2.28, the canonical partition function of translational motion for a system of N indistinguishable particles then becomes

$$
Q_{NVT}^{trans} = \frac{\left(q^{trans}\right)^N}{N!} = \frac{V^N}{\Lambda^{3N}N!}. \qquad (2.86)
$$

As we will see that the canonical partition function of translational motion is the only volume dependent contribution to the partition function of the ideal gas, we can derive from Eq. 2.86 an expression for the pressure of the ideal gas, by employing the relation given by Eq. 2.62. With the Sterling equation $\ln(x!) = x\ln x - x$, we can express

$$\ln Q_{NVT}^{trans} = N \ln V - 3N \ln \Lambda - (N \ln N - N) \tag{2.87}$$

Thus, Eq. 2.62 yields

$$p^{id} = k_B T \left(\frac{\partial \ln Q_{NVT}^{trans}}{\partial V} \right)_{T,N} = k_B T \frac{N}{V}, \tag{2.88}$$

i.e. the well known ideal gas law. Accordingly, the chemical potential of the monoatomic ideal gas can be derived from Eq. 2.63

$$\mu^{id} = -k_B T \left(\frac{\partial \ln Q_{NVT}^{trans}}{\partial N} \right)_{T,V} = -k_B T \ln \left(\frac{V}{N \Lambda^3} \right) \tag{2.89}$$

2.3.2 Partition Function of Rotational Motion

The energy level ε_i^{rot} of a polyatomic molecule associated with its rotational motion can be subdivided into motions around the three axes x, y, z with the three moments of inertia I_{xx}, I_{yy}, I_{zz}

$$\varepsilon_i^{rot} = \frac{j_x^2}{2I_{xx}} + \frac{j_y^2}{2I_{yy}} + \frac{j_z^2}{2I_{zz}} \tag{2.90}$$

where j_x, j_y, j_z are the three angular momenta.

In case of a linear diatomic molecule for which z is the axis of symmetry, the moment of inertia I_{zz} is too small so that it only has two rotational degrees of freedom. When the diatomic molecule is regarded as rigid and symmetric rotor (dumbball) with $I_{xx} = I_{yy} = I$ and $j_x = j_y = j$, the classical kinetic energy of rotation becomes

$$\varepsilon_i^{rot} = \frac{j^2}{I}. \tag{2.91}$$

In the quantum mechanical formulation that quantized the angular momentum j in units of $\frac{h}{2\pi}$ [2], the rotational kinetic energy levels of the rigid rotor are given by

$$\varepsilon_j^{rot} = \frac{\mathrm{j}(\mathrm{j}+1)h^2}{8\pi^2 I} \tag{2.92}$$

where j is the quantum number j $= 0, 1, 2, ...$ Each energy level has a degeneracy of $2j + 1$ [7], so that the molecular partition function becomes

$$q^{rot} = \sum_{j=0}^{\infty} (2j+1) e^{-\frac{\mathrm{j}(\mathrm{j}+1)h^2}{8\pi^2 I k_B T}} \tag{2.93}$$

By defining a characteristic rotational temperature Θ^{rot} as abbreviation for

$$\Theta^{rot} = \frac{h^2}{8\pi^2 I k_B}, \tag{2.94}$$

the rotation partition function is

$$q^{rot} = \sum_{j=0}^{\infty} (2j+1) e^{-\frac{j(j+1)\Theta^{rot}}{T}}. \tag{2.95}$$

At the high temperature limit with $T >> \Theta^{rot}$, Θ^{rot}/T becomes very small, so that the rotational energy levels are close together, and the sum can be approximated by an integral

$$q^{rot} = \int_{j=0}^{\infty} (2j+1) e^{-\frac{j(j+1)\Theta^{rot}}{T}} dJ \tag{2.96}$$

that gives

$$q^{rot} = \frac{T}{\Theta^{rot}} = \frac{8\pi^2 I k_B T}{h^2}. \tag{2.97}$$

which represents the classical limit of q^{rot} for fully excited rotation [4]. For diatomic molecules of pure compounds, i.e. symmetric molecules such as O_2, N_2 etc., the rotation around 360° only gives distinguishable states for rotations up to 180°. Thus, a symmetry number σ is introduced that gives the number of undistinguishable states for a rotation around 360° [10]

$$q^{rot} = \frac{T}{\Theta^{rot}\sigma}. \tag{2.98}$$

with $\sigma = 2$ for symmetric diatomic molecules. At temperatures below the characteristic temperature for rotation, i.e. $\Theta^{rot} > 0.7\,T$ [7], the rotational partition function can be directly determined from the summation in Eq. 2.95, considering $j = 0, 1, 2, 3$. For intermediate temperatures, approximations for the partition function of the rigid rotor are required, as discussed in the textbook by McQuarrie [7]. However, characteristic temperatures Θ^{rot} are generally quite low, as exemplary shown for some selected molecules in Table 2.2. Therefore, the rotational motion can be regarded as fully excited at ordinary temperatures.

For polyatomic molecules with three relevant moments of inertia I_{ii} with $i = x, y, z$, three characteristic rotational temperatures can be defined by [10]

$$\Theta_i^{rot} = \frac{h^2}{8\pi^2 k_B I_{ii}}. \tag{2.99}$$

Thus, the total rotational partition function becomes [10]

$$q^{rot} = \frac{1}{\sigma}\sqrt{\frac{\pi T^3}{\Theta_x^{rot}\cdot\Theta_y^{rot}\cdot\Theta_z^{rot}}}. \tag{2.100}$$

2.3.3 Partition Function of Vibrational Motion

In the classical approach, the partition function of vibrational motions is derived from the uncoupled contributions of its different vibrational degrees of freedom within the molecules, i.e. the vibrational normal modes. Assuming that the vibrational motion along a bond within a molecule is harmonic, the energy states of this (one-dimensional) harmonic oscillator i are

$$\varepsilon_{i,n}^{vib} = h\nu_i\left(n+\frac{1}{2}\right), \quad n = 0, 1, 2, ... \tag{2.101}$$

where ν_i is the frequency of the vibration that depends on the force constant k_r of the bond. When m_1 and m_2 are the masses of the bonded atoms, a reduced mass can be defined [2]

$$m^* = \frac{m_1\cdot m_2}{m_1+m_2} \tag{2.102}$$

and the frequency of harmonic vibration is given by

$$\nu_i = \frac{1}{2\pi}\sqrt{\frac{k_{r,i}}{m^*}}. \tag{2.103}$$

It should be mentioned that the approximation of an harmonic vibration is only valid for low frequencies ν, i.e. moderate temperatures. For increasing ν, the potential energy curve becomes more anharmonic [2] and is better described by the Morse potential [5] (see Chap. 6). Introducing a characteristic temperature of vibration [10]

$$\Theta_i^{vib} = \frac{h\nu_i}{k_B}, \tag{2.104}$$

the partition function for this vibrational mode can be expressed as

$$q_i^{vib} = \sum_{n=0}^{\infty} e^{-\left(\frac{\Theta_i^{vib}(n+1/2)}{T}\right)} = e^{-\frac{\Theta_i^{vib}}{2T}}\sum_{n=0}^{\infty} e^{-\frac{n\Theta_i^{vib}}{T}}. \tag{2.105}$$

Therein the factor before the sum represents the ground state energy [2] that depends on the chosen ground state along the potential energy well. The sum can be evaluated

Table 2.2 Characteristic temperatures of rotation and vibration for some selected molecules with data from [7–9]. For polyatomic molecules, the range of characteristic temperatures for the different modes is provided

Molecule	Θ^{rot} (K)	Θ^{vib} (K)
H_2	85.2	6140
O_2	2.09	2260
N_2	2.88	3374
CO	2.77	3120
CO_2	0.561	954–3360
SO_2	0.495–2.92	750–1960
CH_4	7.54	1870–4320
H_2O	13.4–40.1	2290-5360

and gives

$$\sum_{n=0}^{\infty} e^{-nx} = \frac{1}{1 - e^{-x}}. \tag{2.106}$$

Thus, the vibrational partition function of an ideal harmonic oscillator becomes

$$q_i^{vib} = \frac{e^{-(\Theta_i^{vib}/2T)}}{1 - e^{-\Theta_i^{vib}/T}}. \tag{2.107}$$

The total vibrational partition function from the contributions of all vibrational normal modes is then

$$q^{vib} = \Pi_i q_i^{vib}. \tag{2.108}$$

As exemplary shown for some molecules in Table 2.2, characteristic temperatures Θ^{vib} are in general very high, so that only the lowest vibrational energy level needs to be considered at ordinary temperatures [8].

2.3.4 *Electronic and Nuclear Partition Function*

Both the electronic and the nuclear partition function can be written as sum over the different electronic and nuclear energy levels ε_i [7]. When a specific energy level has a degeneracy of g_i (see Eq. 2.17), the partition function can be expressed as

$$q^{elec/nucl} = \sum_{1}^{\infty} g_i^{elec/nucl} e^{-\left(\frac{\varepsilon_i^{elec/nucl}}{k_B T}\right)}. \tag{2.109}$$

The first energy level is the ground state, i.e. $\varepsilon_1^{elec/nucl} = 0$.

In general the separations between electronic or nuclear energy levels are large compared to $k_B T$. Thus, for the electronic partition function, only the ground state and possibly the first excited state need to be considered at ordinary temperature. With $\Delta\varepsilon_{12}^{elec}$ being the energy difference between the ground state and the first excited state, the electronic partition function becomes [7]

$$q^{elec} = g_1^{elec} + g_2^{elec} e^{-\left(\frac{\Delta\varepsilon_{12}^{elec}}{k_B T}\right)}. \tag{2.110}$$

For the nuclear partition function no excited state needs to be included at terrestrial temperatures so that it is only given by the degeneracy of the ground state energy [7]

$$q^{nucl} = g_1^{nucl}. \tag{2.111}$$

2.4 Maxwell-Boltzmann Distribution of Velocities

The Maxwell-Boltzmann relation provides a probability distribution of the velocities of particles within a system as a function of the temperature. Equation (2.15) gives the probability that a particle has a specific energy ε_i. When the different energy states have degeneracies of g_j, the probability N_i/N of the energy state ε_i becomes

$$\frac{N_i}{N} = \frac{g_i e^{-\varepsilon_i/k_B T}}{\sum\limits_j g_j e^{-\varepsilon_j/k_B T}}. \tag{2.112}$$

When the energy of a particle is its kinetic energy associated with the translational motion with the velocity v, ε_i is given by

$$\varepsilon_i = \frac{m}{2} v^2. \tag{2.113}$$

(see Sect. 2.3.1). Thus, the probability that any one particle has an energy between ε and $\varepsilon + d\varepsilon$ is

$$\frac{N(\varepsilon)}{N} = C\, g(\varepsilon)\, e^{-\varepsilon/k_B T} = C\, g(v)\, e^{-mv^2/2k_B T} = f(v)\, dv, \tag{2.114}$$

where $f(v)$ is the distribution function of the velocity v, and C is a proportionality factor.

For the velocity distribution, the translational motions in all three dimensions have to be taken into account with

$$v^2 = v_x^2 + v_y^2 + v_z^2. \tag{2.115}$$

Thus, all particles with the same velocity lie on the surface of a sphere with the radius v. Therefore, the number of particles having an energy between ε and $\varepsilon + d\varepsilon$ corresponds to the number of particles whose velocity vector ends in a spherical shell in velocity space with radius v and thickness dv [8]. With this, the degeneracy $g(v)$ is given by the volume of the spherical velocity shell

$$g(v) = 4\pi v^2 dv, \tag{2.116}$$

and the velocity distribution function then becomes

$$f(v)\,dv = 4\pi\, C\, v^2\, e^{-mv^2/2k_BT}\, dv. \tag{2.117}$$

The proportionality factor C can be derived from the postulation that the integration of the velocity distribution function over all possible velocities has to yield $= 1$

$$\int_0^\infty f(v)\,dv = 4\pi\, C \int_0^\infty v^2\, e^{-mv^2/2k_BT}\, dv = 1 \quad \rightarrow \quad C = \left(\sqrt{\frac{m}{2\pi k_B T}}\right)^3. \tag{2.118}$$

This finally gives the Maxwell-Boltzmann velocity distribution

$$f(v)\,dv = 4\pi \left(\sqrt{\frac{m}{2\pi k_B T}}\right)^3 v^2\, e^{-mv^2/2k_BT}\, dv, \tag{2.119}$$

from which three characteristic velocities can be derived. The most probable velocity $\widehat{v}$ at a given temperature T results from the maximum of the distribution function

$$\frac{df(v)}{dv} = 0 \quad \rightarrow \quad \widehat{v} = \sqrt{\frac{2k_B T}{m}}. \tag{2.120}$$

The mean velocity $\langle v \rangle$ is given by the integral over all possible velocities, weighted by the probability distribution $f(v)$

$$\langle v \rangle = \int_0^\infty v f(v)\,dv = \sqrt{\frac{8k_B T}{m\pi}}. \tag{2.121}$$

In the same way, the root mean square velocity $\langle v^2 \rangle$ can be determined

$$\langle v^2 \rangle = \int_0^\infty v^2 f(v)\,dv = \frac{3k_B T}{m}, \tag{2.122}$$

which relates the kinetic energy of a system to the temperature T. For a system of N particles, the total kinetic energy is

$$E_{kin} = \langle \sum^{N} \frac{m}{2} v^2 \rangle = N \frac{m}{2} \langle v^2 \rangle = \frac{3}{2} N k_B T. \tag{2.123}$$

References

1. Allen M, Tildesley DJ (1987) Computer simulation of liquids. Oxford Science Publication
2. Gasser PH, Richards WG (1995) An introduction to statistical thermodynamics. World Scientific
3. Haile JM (1997) Molecular dynamics simulation. Wiley Professional Paperback Series
4. Hill TL (1986) An introduction to statistical thermodynamics. Dover Publications, New York
5. Hinchliffe A (2008) Molecular modelling for beginners, 2nd edn. Wiley
6. Kofke D (1992) Lecture CE530. SUNY Buffalo, Department of Chemical and Biological Engineering
7. McQuarrie DA (2000) Statistical mechanics. University Science Books
8. Sears FW, Salinger GL (1986) Thermodynamics, kinetic theory and statistical thermodynamics. Addison-Wesley
9. Van Carey P (1999) Statistical thermodynamics and microscale thermophysics. Chambridge University Press
10. Weingärtner H (2003) Chemische Thermodynamik. Teubner

Chapter 3
Monte Carlo Simulations

This chapter will provide an introduction into the basic ideas of Monte Carlo simulations, i.e. the importance sampling technique and its application to simulations in various ensembles. Furthermore, biased sampling techniques for complex molecules and reweighting methods are discussed. Section 3.6 then focuses on special simulation techniques for phase equilibria, i.e. the Gibbs Ensemble and Gibbs Duhem method as well as Histogram reweighting. A more detailed introduction to the concepts of Monte Carlo simulation are given in the standard textbooks by Allen and Tildesley [1], or Frenkel and Smit [13]. References [47, 48] provide a more comprehensive discussion on the advanced histogram simulation techniques.

3.1 Introduction to Monte Carlo Simulations

In Monte Carlo simulations, different microstates corresponding to the macroscopic system of interest are generated by stochastic changes to the previous configuration. Therefore, it is also referred to as sampling experiment [1], and the stochastic changes are usually called 'moves'. The moves are generated by employing random numbers, which also coined the naming of the simulation method as it resembles to gambling.

Depending on the simulation ensemble and the types of molecules involved—atoms, rigid small molecules, macromolecules etc.—a multiplicity of different moves are possible and/or required to adequately sample the phase space. Figure 3.1 only provides an overview of basic moves as various variations or combinations of these basic moves are imaginable, such as anisotropic volume, pivot or different kinds of biased moves [23].

A trial transition from a state n to a new state m is generated by employing a random number R_{andom} [0, 1]. A trial particle displacement for instance involves the generation of random numbers for each direction x, y, z, and a new x-coordinate $r_x(m)$ is then derived from the preceding position $r_x(n)$ by

G. Raabe, *Molecular Simulation Studies on Thermophysical Properties*,
Molecular Modeling and Simulation, DOI 10.1007/978-981-10-3545-6_3

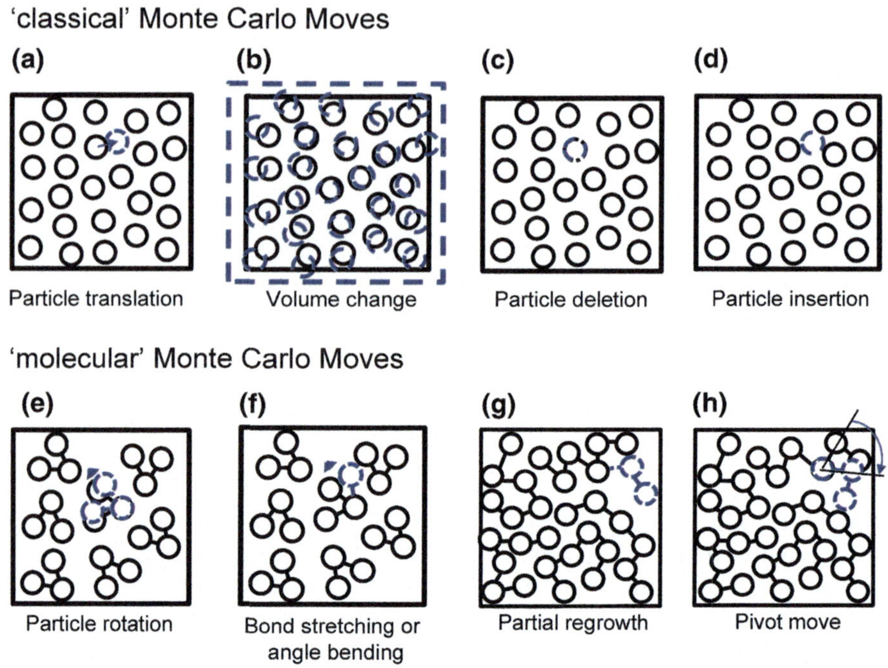

Fig. 3.1 Basic Monte Carlo moves for molecular systems in different ensembles

$$r_x(m) = r_x(n) + (2R_{andom} - 1)\delta r_{max} \tag{3.1}$$

to realize a displacement in the $x-$coordinate in the span of $[-\delta r_{max}; +\delta r_{max}]$. The parameter δr_{max} is the maximum displacement, which is adjusted during the simulation to yield reasonable acceptance rates for the trial moves (see Sect. 3.2). Accordingly, new $y-$ and $z-$coordinates are determined.

Attempted volume changes $V(n) \rightarrow V(m)$ in the $NpT-$ensemble are generated in a similar way by

$$V(m) = V(n) + (2R_{andom} - 1)\delta V_{max}, \tag{3.2}$$

and again, this formulation is used to ensure that both attempts to increase and to decrease the volume are made in equal measure. As before, the maximum volume change δV_{max} is an adjustable parameter during the simulation.

In the μVT ensemble, a particle of a component i to be destroyed in a trial move is randomly selected by

$$R_{andom} * N_i + 1, \tag{3.3}$$

where N_i is the total number of particles of this component in the system. For an attempted insertion of a molecule into a system, its position within the simulation volume is selected randomly. A trial position in the simulation box centered in the origin is generated by

$$r_x(m) = (R_{andom} - 0.5) \cdot L_x. \tag{3.4}$$

Therein L_x is the length of the simulation box in the x-coordinate. Accordingly trial positions in the $y-$ and $z-$coordinates are determined.

Before going into the details of the simulation technique, it is necessary to become aware, what kind of information can be obtained from Monte Carlo simulations. This is exemplarily discussed for the *NVT*-ensemble. In Chap. 2, we derived the classical formulation of the canonical partition function

$$Q_{NVT} = \frac{1}{N!}\frac{1}{h^{3N}} \int\int e^{-E_i/k_BT} dr^{3N} dp^{3N}, \tag{3.5}$$

where the total energy E_i consists of the kinetic energy $E_{kin}(p^{3N})$ as function of the particles momenta, and the configurational energy $U_{conf}(r^{3N})$ that arises from interactions between the particles, and which only depends on their positions. Therefore, the integrals can be separated, and the canonical partition function is written as

$$Q_{NVT} = \frac{1}{N!}\frac{1}{h^{3N}} \int e^{-(E_{kin}(p)/k_BT)} dp^{3N} \int e^{(-U_{conf}(r)/k_BT)} dr^{3N}. \tag{3.6}$$

Therein the kinetic energy from the translational motion of a particle is given by $E_{kin,j} = \frac{p_j^2}{2m}$, so that the integral of momenta can be solved analytically, as outlined in Sect. 2.3.1. It follows

$$\frac{1}{h^{3N}} \int e^{-p^2/(2mk_BT)} dp^{3N} = \frac{1}{h^{3N}} \left(\sqrt{2\pi m k_B T}\right)^{3N} = \frac{1}{\Lambda^{3N}}. \tag{3.7}$$

With this, the canonical partition function becomes

$$Q_{NVT} = \frac{1}{N!}\frac{1}{\Lambda^{3N}} \int e^{-U_{conf}(r)/k_BT} dr^{3N}. \tag{3.8}$$

A comparison with the canonical partition function of translational motion for an ideal gas in Eq. 2.86 yields

$$Q_{NVT} = Q_{NVT}^{id} \cdot \frac{1}{V^N} \int e^{(-U_{conf}(r)/k_BT)} dr^{3N} \tag{3.9}$$

$$= Q_{NVT}^{id} \cdot Q_{NVT}^{res}. \tag{3.10}$$

Thus, the partition function can be divided into a kinetic ideal gas contribution and a residual contribution that describes the deviation from an ideal gas due to interatomic interactions. As a result of the stochastic motions of the particles, Monte Carlo simulations do not offer information on the kinetics of the system. Therefore, Monte Carlo simulations can only provide residual contributions to thermophysical properties, whereas the ideal gas contribution has to be added after the simulation. Furthermore, Monte Carlo methods are in general not suitable to determine transport properties or any information of the dynamics of the system, which is the domain of molecular dynamics simulations.

The integral that covers the residual contribution of interatomic interactions is often referred to as configurational integral Z. For the NVT-ensemble it is [1]

$$Z_{NVT} = \int e^{(-U_{conf}(r)/k_BT)} dr^{3N}. \tag{3.11}$$

This separation of kinetic and configurational properties, i.e. of ideal gas and residual contribution can also easily be accomplished for the NpT-ensemble. According to Eq. 2.51, the classical formulation of the partition function of the NpT-ensemble is given by

$$Q_{NpT} = \frac{1}{N!}\frac{1}{h^{3N}}\frac{1}{V_0}\int e^{-pV_j/k_BT} dV \int\int e^{-E_i/k_BT} dr^{3N} dp^{3N}, \tag{3.12}$$

wherein the total energy E_i of the microstate again consists of $E_{kin}(p^{3N})$ and $U_{conf}(r^{3N})$.

Thus, the configurational integral Z_{NpT} of the NpT-Ensemble is given by [1]

$$Z_{NpT} = \frac{1}{V_0}\int e^{(-pV_j/k_BT)} dV \int e^{(-U_{conf}(r)/k_BT)} dr^{3N}. \tag{3.13}$$

Problems in formulating a configurational integral for the grand canonical ensemble arise from the fact that now the number of particles is also a (discontinuous) variable.

To return to the example of Monte Carlo simulations in the NVT-ensemble, the ensemble average of any residual thermodynamic function $\langle A^{res}\rangle_{NVT}$ is defined as (see Eq. 2.2)

$$\langle A^{res}\rangle_{NVT} = \int A^{res}(r)\, \wp_{NVT}(r)\, dr^{3N} \tag{3.14}$$

$$= \frac{\int A^{res}(r)\, e^{(-U_{conf}(r)/k_BT)}\, dr^{3N}}{\int e^{(-U_{conf}(r)/k_BT)} dr^{3N}} \tag{3.15}$$

The fundamental idea of Monte Carlo simulations is to estimate the integral by sampling the configurational phase space by stochastic trial moves. With this, the ensemble average is derived from

$$\langle A^{res} \rangle_{NVT} = \frac{\sum^{N_{trials}} A^{res}(r)\, e^{(-U_{conf}(r)/k_B T)}}{\sum^{N_{trials}} e^{(-U_{conf}(r)/k_B T)}}. \tag{3.16}$$

However, the phase space of possible combinations of 3N positions is immense so that the accurate estimate of the integrals also requires an enormous number of trials moves. Thus, it is not possible by naive random moves to adequately sample all accessible configurations in phase what is required to achieve an ergodic trajectory. In the course of the sampling though, usually many configurations are generated that have a very small contribution to the ensemble average. These are trial moves that generate overlaps or other configurations with high energies, for which the exponential term becomes small. Thus, in order to obtain accurate estimates of ensemble averages with a practicable number of trial moves, it is essential to use special sampling techniques that concentrate the sampling on states that make substantial contributions to the ensemble averages. These techniques are referred to as importance sampling and are introduced in the following section.

3.2 Importance Sampling

The basic idea of importance sampling is to bias trial moves in such a way that the sample of configurations in phase is representative for the equilibrium state. This is realized by introducing acceptance rules for the trial moves to yield configurations that are distributed according to the probability distribution of the respective ensemble. That means that states with a high probability $\wp$ that contribute significantly to the ensemble average are sampled more frequently than states with a low probability. Metropolis et al. [28] introduced the first importance sampling scheme, which has become widely adopted. Thus, the discussion of importance sampling in this section will concentrate on the Metropolis method. In the Metropolis method, a transition matrix $\pi(n \rightarrow m)$ is introduced that defines the transition probability to move from a configuration n to a configuration m. The choice of the transition probability is somehow arbitrary, provided that the condition of detailed balance or 'microscopic reversibility' is satisfied. This condition of detailed balance postulates that at equilibrium, the net flux between the two states (n) and (m) is zero, i.e. the probability $\wp(n)$ of being in state (n) times the probability of moving to state (m) has to be equal the probability $\wp(m)$ of being in state (m) and moving to state (n)

$$\wp(n) \cdot \pi(n \rightarrow m) = \wp(m) \cdot \pi(m \rightarrow n). \tag{3.17}$$

The key point of the Metropolis scheme is the separation of the transition probability $\pi(n \rightarrow m)$ into the probability of proposing the transition $\alpha(n \rightarrow m)$, and the probability $acc(n \rightarrow m)$ of accepting it

$$\pi(n \to m) = \alpha(n \to m) \cdot acc(n \to m). \tag{3.18}$$

The matrix $\alpha(n \to m)$ is in general chosen to be symmetric

$$\alpha(n \to m) = \alpha(m \to n). \tag{3.19}$$

With this, inserting Eq. 3.18 into the detailed balance yields

$$\wp(n) \cdot acc(n \to m) = \wp(m) \cdot acc(m \to n) \quad \to \frac{acc(n \to m)}{acc(m \to n)} = \frac{\wp(m)}{\wp(n)}. \tag{3.20}$$

There are many acceptance probabilities *acc* thinkable that obey this condition. The original choice of Metropolis et al. for their acceptance criterion is

$$\begin{aligned} acc(n \to m) &= 1 && \text{for } \wp(m) \geq \wp(n) \\ acc(n \to m) &= \frac{\wp(m)}{\wp(n)} && \text{for } \wp(m) < \wp(n), \end{aligned}$$

or summarized in a single expression

$$acc(n \to m) = min\left[1, \frac{\wp(m)}{\wp(n)}\right]. \tag{3.21}$$

Thus, the transition from state (n) to state (m) is accepted with certainty when the new state (m) has a higher probability distribution $\wp(m)$ than the state (n). However, if the new state (m) has a lower probability distribution than the state (n), the move $(n \to m)$ is not generally rejected, but accepted with a probability of $\wp(m)/\wp(n)$. Whether the transition is accepted or not is then decided by comparing the ratio of the probability distribution of the two states (m) and (n) to a uniformly distributed random number $R_{andom}\,[0, 1]$. Thus, the Metropolis can be summarized in

$$\begin{aligned} \pi(n \to m) &= \alpha(n \to m) && \text{for } \wp(m) \geq \wp(n) \Rightarrow \text{accept move} \\ \pi(n \to m) &= \alpha(n \to m)\frac{\wp(m)}{\wp(n)} && \text{for } \wp(m) < \wp(n) \Rightarrow \text{generate } R_{andom}\,[0, 1] \\ & && R_{andom} < \frac{\wp(m)}{\wp(n)} \quad \Rightarrow \text{accept move} \\ & && R_{andom} > \frac{\wp(m)}{\wp(n)} \quad \Rightarrow \text{reject move.} \end{aligned}$$

The Metropolis method creates a so called Markov change of states, as the transition probability $\pi(n \to m)$ only depends on the states (m) and its directly preceding state (n), but not upon any previous transitions.

As the states are already generated according to the correct probability distribution of the respective ensemble, the ensemble averages can now be calculated by counting all sampled microstates equally

$$\langle A^{res} \rangle_{ensemble} = \frac{1}{N_{sample}} \sum^{sample} A^{res}(r^{3N}). \tag{3.22}$$

3.3 Monte Carlo Simulation in Various Ensembles

Due to limited applications for microcanonical Monte Carlo simulations in studies on thermophysical properties of molecular systems, this section focuses on the introduction of basic simulation schemes for the *NVT*, *NpT* and μVT ensembles. It is convenient to use reduced coordinates s, which are scaled by the length L of the simulation box.

$$s = \frac{r}{L} \quad \Rightarrow \quad dr^{3N} = L^{3N} ds^{3N} = V^N ds^{3N}.$$

With this, and by using the reciprocal temperature $\beta = 1/(k_B T)$, the partition functions of the different ensemble can be expressed as

$$Q_{NVT} = \frac{V^N}{N!\Lambda^{3N}} \int e^{-\beta U_{conf}(s \cdot L)} ds^{3N} \tag{3.23}$$

$$Q_{NpT} = \frac{1}{N!\Lambda^{3N}} \frac{1}{V_0} \int V^N e^{-\beta p V} dV \int e^{-\beta U_{conf}(s \cdot L)} ds^{3N} \tag{3.24}$$

$$Q_{\mu VT} = \sum \frac{V^N}{N!\Lambda^{3N}} e^{\beta \mu N} \int e^{-\beta U_{conf}(s \cdot L)} ds^{3N}. \tag{3.25}$$

3.3.1 Canonical Ensemble

In then *NVT*-ensemble the term $V^N/(N!\Lambda^{3N})$ in Eq. 3.23 is constant, so that the probability distribution $\wp_{NVT}$ is only determined by

$$\wp_{NVT} \propto \exp(-\beta U_{conf}). \tag{3.26}$$

The displacement of a single particle j is the simplest Monte Carlo move, for which the basic idea of the Metropolis importance sampling scheme is best illustrated. Thus, first the interaction energy $U^{(n)}_{conf\,j}$ of the particle j in its old position is determined. Then a new position for the particle j is generated according to Eq. 3.1, and its interaction energy $U^{(m)}_{conf\,j}$ is calculated in the new state (m). The ratio of the probability densities of the new and the old state is given by

$$\left(\frac{\wp_{NVT}(m)}{\wp_{NVT}(n)}\right)_{\Delta r_j} = \exp(-\beta(U^{(m)}_{conf,j} - U^{(n)}_{conf,j}))$$
$$= \exp(-\beta\Delta U^{(n\to m)}_{conf,j}). \quad (3.27)$$

Thus, the move to the state (m) is certainly accepted when its configurational energy is lower than in state (n). If the transition to state (m) would result in a higher configurational energy, the move is accepted with a probability given by $\wp(m)/\wp(n) = \exp\left(-\beta\Delta U^{(n\to m)}{}_{conf,j}\right)$ as described in Sect. 3.2. Thus, for a new configuration (m) with a much more unfavorable configurational energy than state (n), its probability $\wp(m) \ll \wp(n)$ and with this $\wp(m)/\wp(n) \ll 1$, so that it is most likely that the ratio is smaller than the generated random number $R_{andom}[0, 1]$, and that the transition will be rejected. An effective simulation depends crucially on a suitable choice of the maximum possible displacement δr_{max} (see Eq. 3.1). A small value of δr_{max} means, that the two states (n) and (m) are located in nearby regions in phase space. Thus, the move involves only small energy changes, and is therefore likely to be accepted. However, to obtain meaningful simulation results, it is required to explore a wide range of configurations, and this is not efficiently achieved by small displacements. Large values for δr_{max} though usually result in low acceptance rates as it becomes more likely to generate a unfavorable configuration with overlapping particles etc. Therefore, the δr_{max} is adjusted during the simulation to yield an acceptance rate of round about 50 %.

3.3.2 *NpT*-Ensemble

The NpT-ensemble is widely used in simulation studies on thermophysical properties, as its constraints match conditions of corresponding experimental studies. The NpT-ensemble is the only ensemble that allows for volume transitions. Thus, moves to sample the configurational space consist of single particle displacements and volume changes. According to Eq. 3.24, with $N!\Lambda^{3N} = const.$, the probability distribution $\wp_{NpT}$ is determined by

$$\wp_{NpT} \propto V^N \exp(-\beta pV)\exp(-\beta U_{conf}) \quad (3.28)$$
$$= \exp\left(N \ln V - \beta pV - \beta U_{conf}\right). \quad (3.29)$$

Single particle moves Δr_j at constant volume ($V(m) = V(n) = V$) are performed as described in the previous section. The ratio of the probability distributions of the states (m) and (n) is now given by

$$\left(\frac{\wp_{NpT}(m)}{\wp_{NpT}(n)}\right)_{\Delta r_j} = \frac{\exp(-N \ln V - \beta pV - \beta U^{(m)}_{conf,j})}{\exp(-N \ln V - \beta pV - \beta U^{(n)}_{conf,j})}$$
$$= \exp(-\beta\Delta U^{(n\to m)}_{conf,j}). \quad (3.30)$$

Though as before in the *NVT*-ensemble, it only depends on the difference of the configurational energies of the particle j in the states (m) and (n).

For a volume transition move, the system volume is changed randomly according to Eq. 3.2. Simultaneously, all center-of-mass positions of the molecules are consistently linearly scaled. Therefore, a trial move from $V(n) = V$ to $V(m) = V + \Delta V$ is associated with a change of the positions of the particles, and with this with a change in the configurational energy of the entire system. Thus, the ratio of the probability distributions of the states (m) and (n), used in the Metropolis acceptance criterion, is determined by

$$\left(\frac{\wp_{NpT}(m)}{\wp_{NpT}(n)}\right)_{\Delta V} = \frac{\exp(-N\ln(V+\Delta V) - \beta p(V+\Delta V) - \beta U_{conf}^{(m)})}{\exp(-N\ln V - \beta pV - \beta U_{conf}^{(n)})}$$
$$= \exp\left[N\ln\left(\frac{V+\Delta V}{V}\right) - \beta\left(p\Delta V + \Delta U_{conf}^{(n\rightarrow m)}\right)\right] \quad (3.31)$$

whereas in single particle moves only the interactions of the selected particle in the two different positions is computed, the volume move requires the full recalculation of all interactions to yield the difference in the configurational energy $\Delta U_{conf}^{n\rightarrow m}$ of the total system of N particles. Thus, volume transitions are computationally more expensive then single particle displacements, and therefore are less frequently performed. This is also justified by the fact that volume translations perturb the system more dramatically than single particle moves. The ratio of attempted single particle moves and volume moves is in the order of magnitude of $N : 1$. As described in the previous section for δr_{max}, a large maximum possible volume change δV_{max} in Eq. 3.2 generally results in low acceptance rates, whereas a small δV_{max} is inefficient. Thus, the value δV_{max} is also adjusted during the simulation to yield an adequate acceptance rate.

3.3.3 Grand Canonical Ensemble

The Grand canonical ensemble is the only ensemble in which the total number of particles N is not imposed. It is therefore the ensemble of choice for simulation studies on adsorption processes to determine the amount of adsorbed particles as function of the chosen external conditions [13]. Attempted moves in the μVT-ensemble are single particle displacement, particle insertion, particle destruction, and in mixtures also an identity swap. It is assumed that the system exchanges particles with an infinite large reservoir ($V_{reservoir} \rightarrow \infty$, $N_{reservoir} \rightarrow \infty$) so that changes in the reservoir do not need to be taken into account in the partition function. According to Eq. 3.25 the probability distribution is then determined by [13]

$$\wp_{\mu VT} \propto \frac{V^N}{N!\Lambda^{3N}} \exp(\beta\mu N)\exp(-\beta U_{conf}) \tag{3.32}$$

$$= \exp\left(N \ln V - \ln N! - 3N \ln \Lambda + \beta\mu N - \beta U_{conf}\right). \tag{3.33}$$

For a single particle move at a constant total number of particles ($N = const.$), the ratio $\wp(m)/\wp(n)$ is again given by the change in the configurational energy of the particle j involved with the transition $n \rightarrow m$ from one position (n) to another (m)

$$\left(\frac{\wp_{\mu VT}(m)}{\wp_{\mu VT}(n)}\right)_{\Delta r_j} = \exp(-\beta\Delta U_{conf,j}^{(n\rightarrow m)}). \tag{3.34}$$

For a particle insertion, a random position within the simulation volume is generated according to Eq. 3.4. Then its interaction energy in the new position is computed to determine the change in the configurational energy of the system $\Delta U_{conf}^{(n\rightarrow m)}$ due to the insertion of the additional particle. The ratio of the probability distributions of the states m ($N(m) = N + 1$) and n ($N(n) = N$) is

$$\begin{aligned}
&\left(\frac{\wp_{\mu VT}(m)}{\wp_{\mu VT}(n)}\right)_{N+1} \\
&= \frac{\exp((N+1)\ln V - \ln(N+1)! - 3(N+1)\ln\Lambda + \beta\mu(N+1) - \beta U_{conf}^{(m)})}{\exp(N\ln V - \ln N! - 3N\ln\Lambda + \beta\mu N - \beta U_{conf}^{(n)})} \\
&= \exp\left(\ln\left(\frac{V}{(N+1)\Lambda^3}\right) + \beta\mu - \beta\Delta U_{conf}^{(n\rightarrow m)}\right) \\
&= \frac{V\exp(\beta\mu)}{(N+1)\Lambda^3}\exp(-\beta\Delta U_{conf}^{(n\rightarrow m)}).
\end{aligned} \tag{3.35}$$

Thus, the probability to accept a trial particle insertion increases with the system volume V, and decreases with the number of particles N.

For a destruction trial, a particle is randomly selected following Eq. 3.3, and its interaction energy is determined to account for the change in the configurational energy of the system $\Delta U_{conf}^{(n\rightarrow m)}$ that comes along with its deletion. With $N(n) = N$ and $N(m) = N - 1$, the destruction of the particle is then accepted with a probability of

$$\begin{aligned}
&\left(\frac{\wp_{\mu VT}(m)}{\wp_{\mu VT}(n)}\right)_{N-1} \\
&= \frac{\exp((N-1)\ln V - \ln(N-1)! - 3(N-1)\ln\Lambda + \beta\mu(N-1) - \beta U_{conf}^{(m)})}{\exp(N\ln V - \ln N! - 3N\ln\Lambda + \beta\mu N - \beta U_{conf}^{(n)})} \\
&= \exp\left(-\ln V + \ln N + 3\ln\Lambda - \beta\mu - \beta\Delta U_{conf}^{(n\rightarrow m)})\right)
\end{aligned}$$

$$= \frac{N\Lambda^3 \exp(-\beta\mu)}{V} \exp(-\beta \Delta U_{conf}^{(n \to m)}). \quad (3.36)$$

Thus, a large number of particles in the system favors particle destruction trials. With increasing size and complexity of the molecules and elevated densities, it becomes more and more difficult to successfully insert or remove a particle, as both trial moves entail large energy changes. In case of particle insertion, this is due to the high risk to generate configurations in which atoms of the inserted molecule overlap with atoms of the other N molecules. For the destruction trial though the removal of the molecule causes a loss of its attractive interaction energy. Thus, special sampling techniques are required for complex molecules and high densities in order to increase the acceptance ratio of the trials, and with this to improve the performance of the simulation. These techniques will be discussed in the following section.

3.4 Biased Sampling

As discussed in the previous section, the rate of successful trial moves significantly decreases with increasing complexity of the molecular structure due to the high risk of generating overlaps with the present configuration. The problem does not only arise for particle insertion moves, but also for simple particle displacements. Thus, so called biased sampling schemes are used that guide the choice of possible positions of the molecule towards those with favorable configurational energies. For instance, the center-of-mass of a molecule is moved to a new position, and then different orientations of the molecule are generated with the aim to find a preferential orientation that fits into the existing structure. However, to ensure that the configurations are generated according to the correct Boltzmann distribution, the biasing sampling scheme has to satisfy the condition of detailed balance (see Eq. 3.17)

$$\wp(n) \cdot \alpha(n \to m) \cdot acc(n \to m) = \wp(m) \cdot \alpha(m \to n) \cdot acc(m \to n). \quad (3.37)$$

In the original Metropolis method, the probability matrix α of proposing a trial position is chosen to be symmetric, i.e. $\alpha(n \to m) = \alpha(m \to n)$. This is no longer true when the trial move is biased towards an acceptable configuration. In this case, the probability $\alpha(n \to m)$ of proposing a transition to state (m) depends on its configurational energy

$$\alpha(n \to m) = f\left(U_{conf}^{(m)}\right) \quad (3.38)$$

and accordingly

$$\alpha(m \to n) = f\left(U_{Conf}^{(n)}\right). \tag{3.39}$$

Inserting this condition into the detailed balance in Eq. 3.37 yields a modified acceptance rule

$$\frac{acc(n \to m)}{acc(m \to n)} = \frac{f\left(U_{conf}^{(n)}\right)}{f\left(U_{conf}^{(m)}\right)} \frac{\wp(m)}{\wp(n)}, \tag{3.40}$$

or more generally

$$\frac{acc(n \to m)}{acc(m \to n)} = \frac{\alpha(m \to n)}{\alpha(n \to m)} \frac{\wp(m)}{\wp(n)}. \tag{3.41}$$

This means, when an arbitrary biasing function $f\left(U_{conf}\right)$ is introduced to guide the trial move towards configurations with enhanced acceptance probabilities, the acceptance rule has to be modified accordingly to remove the bias in order to satisfy the detailed balance.

The basic approach of biasing sampling techniques is best illustrated for the orientational-bias technique. Thus, the following Sect. 3.4.1 will provide a short introduction to this method before we will discuss the widely used configurational-bias method in Sect. 3.4.2.

Apart from these methods, there are also other biasing schemes available such as the force or the torque bias. In this approach, the displacement of the particle is biased in the direction of the force or rather the torque acting on it to mimic the trend of the particle movement in a molecular dynamics simulation. More details on this method are provided in the original work by Pangali et al. [38, 42], or for instance in the work of Van't Hof et al. [57], who applied these biasing methods in their molecular simulation studies on phase equilibria.

3.4.1 Orientational-Bias

In the orientational-bias method, the basic trial move, for one a particle displacement, is applied to the center-of-mass of the molecule, and then different orientations are generated to bias the molecule towards an acceptable orientation. This technique is used to enhance the acceptance rate for trial moves of polar or hydrogen-bonding molecules etc., whose intermolecular interactions strongly depend on their orientation [13]. The configurational energy of the particle is divided into a part that only depends on its (center-of-mass) position ($U_{conf,i}^{pos}$) and into an orientational depending part ($U^{or}_{conf,i}$). Thus, in the new state (m), a number of trial orientations (denoted by $\mathbf{R}_i^{(m)}$) of the molecule are generated, and for each trial position, the interaction energy $U^{or,(m)}_{conf}(\mathbf{R}_i^{(m)})$ is determined. Then a statistical weight $W^{or,(m)}$ is derived as sum over the Boltzmann factors of the configurational energies of all trial orientations.

$$W^{or,(m)} = \sum_{i=1}^{N_{or}} \exp\left(-\beta U^{or,(m)}_{conf}(\mathbf{R}_i^{(m)})\right). \tag{3.42}$$

From these trial orientations, a preferred orientation $\mathbf{R}_m^{(m)}$ with high Boltzmann weight is chosen with a probability of

$$\alpha(n \to m) = \frac{\exp\left(-\beta U^{or,(m)}_{conf}(\mathbf{R}_m^{(m)})\right)}{W^{or,(m)}}. \tag{3.43}$$

To evaluate the acceptance rule, a corresponding statistical weight $W^{or,(n)}$ has to be derived for the old position of the molecule. This requires that also N_{or} different orientations of the molecule are considered in the state (n). In addition to the real existing orientation of the molecule (denoted by $\mathbf{R}_n^{(n)}$), then $(N_{or} - 1)$ additional orientations are generated and their associated interaction energies $U^{or,(n)}_{conf}(\mathbf{R}_i^{(n)})$ are computed. The statistical weight $W^{or,(n)}$ is then given by

$$\begin{aligned} W^{or,(n)} &= \sum_{i=1}^{N_{or}} \exp\left(-\beta U^{or,(n)}_{conf}(\mathbf{R}_i^{(n)})\right) \\ &= \exp\left(-\beta U^{or,(n)}_{conf}(\mathbf{R}_n^{(n)})\right) + \sum_{i=2}^{N_{or}} \exp\left(-\beta U^{or,(n)}_{conf}(\mathbf{R}_i^{(n)})\right), \end{aligned} \tag{3.44}$$

and the probability of the reverse move is defined by

$$\alpha(m \to n) = \frac{\exp\left(-\beta U^{or,(n)}_{conf}(\mathbf{R}_n^{(n)})\right)}{W^{or,(n)}}. \tag{3.45}$$

This yields an acceptance criterion for the particle displacement of

$$\begin{aligned} \frac{acc(n \to m)}{acc(m \to n)} &= \frac{\exp\left(-\beta U^{or,(n)}_{conf}\right)}{\exp\left(-\beta U^{or,(m)}_{conf}\right)} \frac{W^{or,(m)}}{W^{or,(n)}} \\ &\cdot \exp\left(-\beta((U^{pos,(m)}_{conf} + U^{or,(m)}_{conf}) - (U^{pos,(n)}_{conf} + U^{or,(n)}_{conf}))\right) \end{aligned} \tag{3.46}$$

and summarized in a modified acceptance rule

$$acc(n \to m) = min\left[1, \frac{W^{or,(m)}}{W^{or(n)}} \exp\left(-\beta(U_{conf}^{pos,(m)} - U_{conf}^{pos,(n)})\right)\right]. \tag{3.47}$$

3.4.2 Configurational-Bias

The configurational-bias Monte Carlo (CBMC) method is based on a scheme proposed by Rosenbluth and Rosenbluth [43] to sample configurations of a chain molecule on a lattice. In this approach, the molecule is grown segment by segment, and in each step, the segment is biased towards a lower energy position in a similar way as described in the previous Sect. 3.4.1 and illustrated in Fig. 3.2.

With this, the growth of the whole chain is preferentially directed to improve the chance of a successful insertion. The method was extended and applied by several authors [12, 20, 29, 51, 52] for simulations on flexible chain molecules in off-lattice systems. For a chain molecule consisting of l segments, with k trial orientations generated for each segment, the general proceeding of the CBMC method is as follows:

- The first segment is inserted at random position, and its configurational energy $U_{conf,1}^{(m)}$ is determined. The statistical weight of the first segment is computed by

$$w_1^{(m)} = k \exp\left(-\beta U_{conf,1}^{(m)}\right). \tag{3.48}$$

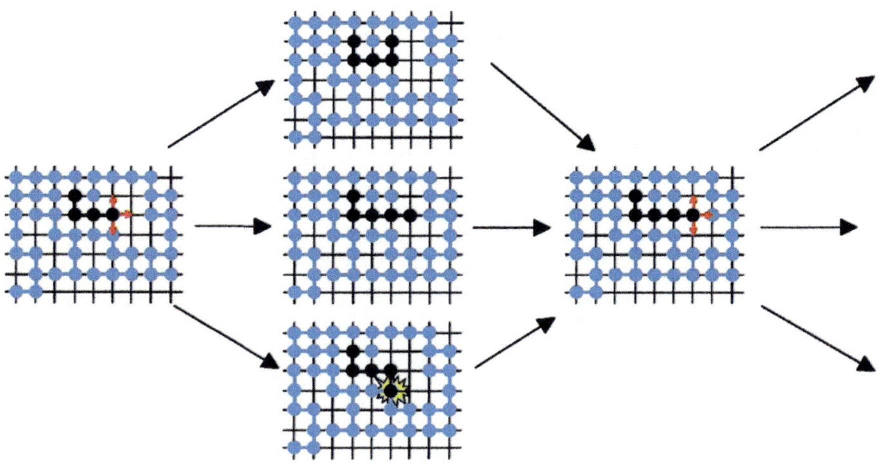

Fig. 3.2 Illustration of the configurational-bias method for a chain molecule on a lattice

- For each appending segment j, k different trial positions are generated, (denoted by $\mathbf{R}_i^{(m)}$) and for each trial position i, the configurational energy $U_{conf,j}^{(m)}(\mathbf{R}_i^{(m)})$ is computed. The statistical weight of each segment j is derived from

$$w_j^{(m)} = \sum_{i=1}^{k} \exp\left(-\beta U_{conf,j}^{(m)}(\mathbf{R}_i^{(m)})\right). \tag{3.49}$$

One of the lower energy positions, say $(\mathbf{R}_m^{(m)})$, is chosen with a probability of

$$\alpha_j^{(m)} = \frac{\exp\left(-\beta U_{conf,j}^{(m)}(\mathbf{R}_m^{(m)})\right)}{\sum_{i=1}^{k} \exp\left(-\beta U_{conf,j}^{(m)}(\mathbf{R}_i^{(m)})\right)} = \frac{\exp\left(-\beta U_{conf,j}^{(m)}(\mathbf{R}_m^{(m)})\right)}{w_j^{(m)}}. \tag{3.50}$$

- When the entire chain is grown, the so called Rosenbluth factor for the state (m) is determined from the statistical weights of the trial positions of the single segments by

$$W^{(m)} = \prod_{j=1}^{l} w_j^{(m)}. \tag{3.51}$$

- The probability $\alpha(n \to m)$ of proposing the transition of the complete chain to the state (m) is given by the product of the probabilities to choose a position for a single segment

$$\begin{aligned}\alpha(n \to m) = \prod_{j=1}^{l} \alpha_j^{(m)} &= \prod_{j=1}^{l} \frac{\exp\left(-\beta U_{conf,j}^{(m)}(\mathbf{R}_m^{(m)})\right)}{w_j^{(m)}} \\ &= \frac{\exp\left(\sum_{j=1}^{l}\left(-\beta U_{conf,j}^{(m)}\right)\right)}{\prod_{j=1}^{l} w_j^{(m)}} \\ &= \frac{\exp\left(-\beta U_{conf,chain}^{(m)}\right)}{W^{(m)}}.\end{aligned} \tag{3.52}$$

When evaluating the acceptance rule, it is always necessary to relate the probability for the transition to state (m) to the probability of the reverse transition to the original state (n). Thus, the same procedure is used to grow the chain molecule at the original position:

- The statistical weight of the first segment at its original position is determined from

$$w_1^{(n)} = k \exp\left(-\beta U_{conf,1}^{(n)}\right). \tag{3.53}$$

- For each other segment j, $k-1$ different trial positions in addition to its real position (denoted by $\mathbf{R}_n^{(m)}$) are generated. The statistical weight of each segment j is then given by

$$w_j^{(n)} = \exp\left(-\beta U_{conf,j}^{(n)}(\mathbf{R}_n^{(m)})\right) + \sum_{i=2}^{k} \exp\left(-\beta U_{conf,j}^{(n)}(\mathbf{R}_i^{(n)})\right). \tag{3.54}$$

  The probability of choosing the position $\mathbf{R}_n^{(m)}$ can be described by

$$\alpha_j^{(n)} = \frac{\exp\left(-\beta U_{conf,j}^{(n)}(\mathbf{R}_n^{(n)})\right)}{w_j^{(n)}}. \tag{3.55}$$

- The Rosenbluth factor of the chain in state (n) is given by

$$W^{(n)} = \prod_{j=1}^{l} w_j^{(n)}. \tag{3.56}$$

- With this, the probability $\alpha(m \to n)$ of proposing the reverse transition to state (n) becomes

$$\alpha(m \to n) = \prod_{j=1}^{l} \alpha_j^{(n)} = \frac{\exp\left(-\beta U_{conf,chain}^{(n)}\right)}{W^{(n)}}. \tag{3.57}$$

For an unbiased displacement of the chain within the same box, the acceptance criterion is

$$acc(n \to m) = min\left[1, \exp\left(-\beta(U_{conf,chain}^{(m)} - U_{conf,chain}^{(n)})\right)\right]. \tag{3.58}$$

In order to remove the effect of the bias by growing the chain in the way described above, the acceptance rule has to be modified according to Eq. 3.41. This yields

$$\begin{aligned} acc(n \to m) &= min\left[1, \frac{\alpha(m \to n)}{\alpha(n \to m)} \exp\left(-\beta(U_{conf,chain}^{(m)} - U_{conf,chain}^{(n)})\right)\right] \\ &= min\left[1, \frac{W^{(m)}}{W^{(n)}}\right]. \end{aligned} \tag{3.59}$$

The acceptance rule for an insertion of the chain in the μVT ensemble (see Eq. 3.36) is modified accordingly by incorporating the ratio of the Rosenbluth weights

$$acc(n \rightarrow m) = min\left[1, \frac{N\Lambda^3 \exp(-\beta\mu)}{V}\frac{W^{(m)}}{W^{(n)}}\right]. \tag{3.60}$$

Though it should be noted that the configurational energy of each segment consists of contributions from both intramolecular, bonded interactions, i.e. bond stretching, angle bending, torsion (see Chap. 6), and intermolecular, non-bonded interactions. Thus, trial positions for the segments are generated according to the intramolecular potential, whereas the probability of selecting a trial position in Eq. 3.59 is only determined by the intermolecular, 'external' interaction energy, denoted by the index 'ext'.

The CBMC method has become extensively used for simulation studies of long chain molecules, and has been modified and refined to improve its efficiency and to deal with complex molecular structures. Vlugt et al. [59] developed the computational cost-saving Dual-cutoff CBMC method in which a reduced cutoff for non-bonded interactions is used during the growth of the molecule. The full cutoff is only used for the computation of the configurational energy of the final grown structure. However, this results in a difference δU^{ext}_{conf} between the configuration energy of the final structures U^{ext}_{conf} and the energy calculated during its growth $\bar{U}^{ext}_{conf}$. It is therefore necessary to account for this energy difference in the acceptance rule. The modified acceptance rule is then given by [59]

$$acc(n \rightarrow m) = min\left[1, \frac{\bar{W}^{ext(m)}}{\bar{W}^{ext(n)}} \exp\left(-\beta\left(\delta U^{ext(m)}_{conf} - \delta U^{ext(n)}_{conf}\right)\right)\right], \tag{3.61}$$

therein the Rosenbluth weights $\bar{W}^{ext}$ were calculated with the reduced cutoff during the growth of the molecules.

Martin and Siepmann [25] and Martin and Thompson [26] refined the CBMC method for simulations of branched molecules. In their "Coupled-Decoupled" formulation, they have decoupled the bond and angle generation from the dihedral and non-bonded terms, which corresponds to a separation between 'hard' and 'soft' degrees of freedom [44]. In their approach (in [26]), bond angles and flexible bonds are generated by a biasing procedure based only on the intramolecular energy terms for the angle bending and bond stretching, respectively. "Fixed Endpoint" methods such as the "Self-Adapting SAFE-CBMC" by Wick and Siepmann [62] were developed for cyclic molecules to bias the growth of the molecule in such a way that it will result in an acceptable ring closure. In [24], Martin and Frischknecht proposed a new sampling scheme for the intramolecular terms by using an arbitrary distribution to generate trial bond length, bond angles and dihedrals. They also modified the Coupled-Decoupled CBMC method by combining the trial generation for the bond length, bond angles and dihedral to a selection step prior to the selection based on the intermolecular, 'external' interactions. The coupling of the selection of the bonded terms in this "Coupled to pre-nonbond" formulation allows to account for the mutual influence of the intramolecular interactions. Reference [24] provides an overview and a more detailed description of the different CBMC formulations. Several of these modifications of the CBMC method are implemented in the MC simulation program Towhee [23].

Another alternative approach to the Coupled-Decoupled formulation that also makes use of the separation of soft and hard degrees of freedom was proposed by Shah and Maginn [44]. They represent complex molecules as a collection of fragments, and first sample hard degrees of freedom such as angle bending in presimulations on the fragments. From these presimulations, they generate a library of fragments with known probability distribution, so that conventional CBMC techniques for the soft degrees of freedom (dihedral) can be employed to reassemble the fragments in order to reconstitute the molecule. Their approach can be employed to molecules with arbitrary complexity such as branched and cyclic compounds, and it is implemented in the Monte Carlo package Cassandra [22] developed in the Maginn group.

3.5 Reweighting Methods

Reweighting techniques represent a framework for analyzing simulation results in order to extract as much information from each simulation as possible. They can be used to combine simulation results from different state points to improve the statistics of the derived thermodynamic properties, or to allow for estimating properties at state points at which no simulations were performed [50]. Thus, reweighting methods can be applied for various purposes and in different ensembles. The application of reweighting methods in phase equilibria studies for instance is discussed in Sect. 3.6.3. These techniques are also often employed in estimating free energy differences as described in Sect. 7.4. Although the reweighting techniques are discussed in this chapter on Monte Carlo simulations for didactical or historical reasons, they can also be employed for analyzing trajectories from molecular dynamics.

The basic idea of reweighting methods is that information on the equilibrium probability distribution obtained from a simulation at one state point is used to determine the equilibrium probability distribution at another (nearby) state point. When a simulation is performed, for instance, in the canonical ensemble at some temperature T_0 (or inverse temperature β_0), the probability distribution is

$$\wp_{NVT}(\beta_0) \propto \exp(-\beta_0 U_{conf}). \tag{3.62}$$

The canonical probability distribution at a different temperature T_1 (β_1) can be related to the distribution at β_0 by

$$\frac{\wp_{NVT}(\beta_1)}{\wp_{NVT}(\beta_0)} \propto \exp\left(-(\beta_1 - \beta_0)U_{conf}\right). \tag{3.63}$$

With this, the probability distribution at β_1 can be estimated from the known distribution $\wp_{NVT}(\beta_0)$ by

$$\wp_{NVT}(\beta_1) = C \exp\left(-(\beta_1 - \beta_0)U_{conf}\right) \cdot \wp_{NVT}(\beta_0), \tag{3.64}$$

provided that both distributions overlap sufficiently.

In the histogram reweighting method that will be introduced in the following sections, estimates for the probability distributions at some state points are derived from histograms that are then reweighted to a different state point. Estimator techniques such as Bennetts acceptance ratio [4] or its variants are aimed at estimating the difference between the state points 0 and 1, and at minimizing the variance in the calculated difference. These techniques will be discussed in Sect. 3.5.3.

3.5.1 Introduction to Histogram Reweighting Techniques

For a basic introduction to the histogram reweighting technique, we first restrict the discussion to the canonical ensemble and to the use of a single histogram. As deduced in Sect. 2.1.2, the probability of a microstate in the canonical ensemble at some inverse temperature β_0 is given by

$$\wp_{NVT}(\beta_0) = \frac{\exp(-\beta_0 U_{conf})}{Q_{NVT}}. \tag{3.65}$$

The probability of the canonical ensemble to have a specified energy U_{conf} at β_0 is described by

$$\wp_{NVT}(U_{conf}, \beta_0) = Q_{NVE}(U_{conf}) \cdot \frac{\exp\left(-\beta_0 U_{conf}\right)}{Q_{NVT}(\beta_0)}, \tag{3.66}$$

with the microcanonical partition function $Q_{NVE}(U_{conf})$ giving the number of microstates that have the energy U_{conf} (see Sect. 2.1.1). The partition function of the canonical ensemble can therefore be derived from (see Eq. 2.27)

$$Q_{NVT} = \int Q_{NVE}(U_{conf}) \cdot \exp(-\beta_0 U_{conf}) dU_{conf}. \tag{3.67}$$

The microcanonical partition function $Q_{NVE}(U_{conf})$ though does not depend on the temperature. Thus, when we relate the canonical probability of some energy U_{conf} (given by Eq. 3.66) at another (nearby) temperature T_1 (i.e. β_1) to that at β_0, the Q_{NVE} term simply cancels, and the relation is given by

$$\frac{\wp_{NVT}(U_{conf}, \beta_1)}{\wp_{NVT}(U_{conf}, \beta_0)} = \frac{\exp(-\beta_1 U_{conf})}{\exp(-\beta_0 U_{conf})} \cdot \frac{Q_{NVT}(\beta_0)}{Q_{NVT}(\beta_1)}. \tag{3.68}$$

With this, the probability distribution at β_1 can be derived from

$$\wp_{NVT}(U_{conf}, \beta_1) = \wp_{NVT}(U_{conf}, \beta_0) \exp\left(-(\beta_1 - \beta_0)U_{conf}\right) \frac{Q_{NVT}(\beta_0)}{Q_{NVT}(\beta_1)}. \tag{3.69}$$

Now we can integrate both sides of Eq. 3.69 over all energies, whereby the ratio of partition functions is independent of U_{conf} and can be factored out of the integrand of the right side. Due to the probability normalization condition, the integration of the left side of Eq. 3.69 yields

$$\int \wp_{NVT}(U_{conf}, \beta_1) dU_{conf} = 1 \tag{3.70}$$

Thus, by integrating Eq. 3.69 over all energies, we can deduce an expression for the ratio of the partition function that is given by

$$\frac{Q_{NVT}(\beta_0)}{Q_{NVT}(\beta_1)} = \frac{1}{\int \wp_{NVT}(U_{conf}, \beta_0) \exp\left(-(\beta_1 - \beta_0)U_{conf}\right) dU_{conf}}. \tag{3.71}$$

With this, Eq. 3.69 becomes

$$\wp_{NVT}(U_{conf}, \beta_1) = \frac{\wp_{NVT}(U_{conf}, \beta_0) \exp\left(-(\beta_1 - \beta_0)U_{conf}\right)}{\int \wp_{NVT}(U_{conf}, \beta_0) \exp\left(-(\beta_1 - \beta_0)U_{conf}\right) dU_{conf}}, \tag{3.72}$$

which means that the probability distribution at β_1 can be estimated from the distribution $\wp_{NVT}(U_{conf}, \beta_0)$ at β_0. Then, the ensemble average of any property A at β_1 can be determined by

$$\begin{aligned} \langle A(\beta_1) \rangle &= \int A \cdot \wp_{NVT}(U_{conf}, \beta_1)\, dU_{conf} \\ &= \frac{\int A \cdot \wp_{NVT}(U_{conf}, \beta_0) \exp\left(-(\beta_1 - \beta_0)U_{conf}\right) dU_{conf}}{\int \wp_{NVT}(U_{conf}, \beta_0) \exp\left(-(\beta_1 - \beta_0)U_{conf}\right) dU_{conf}}. \end{aligned} \tag{3.73}$$

In the histogram methods, the histograms $H\left(U_{conf}\right)$ are used to estimate the probability $\wp_{NVT}(U_{conf}, \beta_0)$. Thereby $H\left(U_{conf}\right)$ counts how many times the configurational energy U_{conf} is observed during the simulation at β_0. With K being the total number of configurations generated during the simulation, the probability distribution can be evaluated from

$$\wp_{NVT}(U_{conf}, \beta_0) = \left.\frac{H\left(U_{conf}\right)}{K}\right|_{\beta_0}, \quad K = \sum_{U_{conf}} H\left(U_{conf}\right). \tag{3.74}$$

With this, an estimate of the ensemble average $\langle A(\beta_1)\rangle$ can be obtained from

$$\langle A(\beta_1)\rangle = \frac{\sum\limits_{U_{conf}} A\,H(U_{conf})\,\exp\left(-(\beta_1-\beta_0)U_{conf}\right)}{\sum\limits_{U_{conf}} H(U_{conf})\exp\left(-(\beta_1-\beta_0)U_{conf}\right)}. \tag{3.75}$$

When histogram reweighting techniques are employed in studies on phase equilibria (see Sect. 3.6.3), the simulations are generally performed in the grand canonical ensemble. An excellent introduction to the histogram reweighting techniques in the μVT ensemble is provided by de Pablo et al. in [7], which is shortly summarized here. For simulations in the grand canonical ensemble, the chemical potential μ and the temperature T (or β) are imposed, whereas the number of particles and the configurational energy of the system fluctuate. Thus, two-dimensional histograms $H(N, U_{conf})$ are employed in which the number of times is stored that a certain number of particles N and a certain energy U_{conf} is observed during the simulation. When the simulation is performed at conditions of T_0 (or β_0) and μ_0, and the total number of configurations generated in the course of the simulation is denoted by $K = \sum_N \sum_{U_{conf}} H(N, U_{conf})$, the probability $\wp(N, U, T_0, \mu_0)$ that the system is observed with a certain combination of N and U_{conf} is

$$\wp(N, U_{conf}, \mu_0, \beta_0) = \frac{H(N, U_{conf})}{K}. \tag{3.76}$$

Then again, the probability distribution for the μVT-ensemble to have a specified energy U_{conf} and a specific number of particles N at the conditions of β_0, μ_0 is in general given by

$$\wp(N, U_{conf}, \mu_0, \beta_0) = \frac{Q_{NVE}(N, U_{conf})\exp\left(-\beta_0 U_{conf} + N\beta_0\mu_0\right)}{Q_{\mu VT}(\mu_0, \beta_0)}. \tag{3.77}$$

Thereby, the grand canonical partition function can be estimated by

$$Q_{\mu VT}(\mu_0, \beta_0) = \sum_N \sum_{U_{conf}} Q_{NVE}(N, U_{conf})\exp\left(-\beta_0 U_{conf} + N\beta_0\mu_0\right). \tag{3.78}$$

Combining the Eqs. 3.76 and 3.77 then yields

$$\begin{aligned} Q_{NVE}(N, U_{conf}) &= \frac{\wp(N, U_{conf}, \mu_0, \beta_0)\, Q_{\mu VT}(\mu_0, \beta_0)}{\exp\left(-\beta_0 U_{conf} + N\beta_0\mu_0\right)} \\ &= \frac{Q_{\mu VT}(\mu_0, \beta_0)}{K} \cdot \frac{H(N, U_{conf})}{\exp\left(-\beta_0 U_{conf} + N\beta_0\mu_0\right)}, \end{aligned} \tag{3.79}$$

which represents an estimate for the microcanonical partition function $Q_{NVE}(N, U_{conf})$. Therein the factor $Q_{\mu VT}(\mu_0, \beta_0)/K$ is a proportionality constant that is often abbreviated by w_i. As $Q_{NVE}(N, U_{conf})$ is not a function of the temperature and the chemical potential, the estimate from Eq. 3.79 can be used to determine the probability $\wp(N, U, \mu, \beta)$ to observe the system with N particles and an energy U_{conf} at another state point of μ and T (i.e. β). Inserting Eq. 3.79 in the general definition of $\wp(N, U_{conf}, \mu, \beta)$ given by

$$\wp(N, U_{conf}, \mu, \beta) = \frac{Q_{NVE}(N, U_{conf}) \exp\left(-\beta U_{conf} + N\beta\mu\right)}{Q_{\mu VT}(\mu, \beta)},$$

then yields

$$\wp(N, U_{conf}, \mu, \beta) = \wp(N, U_{conf}, \mu_0, \beta_0)\frac{Q_{\mu VT}(\mu_0, \beta_0)}{Q_{\mu VT}(\mu, \beta)} \cdot \exp\left(-U_{conf}(\beta - \beta_0) + N(\beta\mu - \beta_0\mu_0)\right). \tag{3.80}$$

This means that $\wp(N, U_{conf}, \mu, \beta)$ can be estimated by simply 'reweighting' the probability $\wp(N, U_{conf}, \mu_0, \beta_0)$ obtained from the simulation at β_0, μ_0. With this estimated probability $\wp(N, U_{conf}, \mu, \beta)$, the ensemble average of any property $A(\mu, \beta)$ can be derived from

$$\langle A \rangle = \sum_{N} \sum_{U_{conf}} A(N, U_{conf}) \wp(N, U_{conf}, \mu, \beta), \tag{3.81}$$

without the need for additional simulations at μ and β—provided that the probability distributions of the two states μ_0, β_0 and μ, β sufficiently overlap, i.e. the states are not too different. Although the grand canonical partition functions $Q_{\mu VT}(\mu_0, T_0)$ and $Q_{\mu VT}(\mu, T)$ in Eq. 3.80 are unknown, they are constant for the specified states, so they can be replaced by constant factors C_i according to

$$\ln \frac{Q_{\mu VT}(\mu_0, T_0)}{Q_{\mu VT}(\mu, T)} = C_0 - C_1 = \Delta C. \tag{3.82}$$

3.5.2 Multiple-Histogram Methods

The basic idea of multiple histogram reweighting methods is to combine data (i.e. histograms) from multiple simulations to derive an optimized estimate for the probability distribution. For the introduction to the multiple histogram technique, we start again with the canonical ensemble where we intend to combine R simulations at different (inverse) temperatures β_i, in which we sample K_i independent configurations, respectively. At each temperature, the probability of a specific energy U_{conf} is estimated by histograms

$$\wp_{NVT}^{(i)}(U_{conf}, \beta_i) = \left. \frac{H_i(U_{conf})}{K_i} \right|_{\beta_i} . \tag{3.83}$$

The true probability distribution is given by

$$\wp_{NVT}^{(i)}(U_{conf}, \beta_i) = \frac{Q_{NVE}^{(i)}(U_{conf}) \cdot \exp\left(-\beta_i U_{conf}\right)}{Q_{NVT}(\beta_i)}, \tag{3.84}$$

wherein the numerator represents the un-normalized probability distribution

$$\mathcal{P}_{NVT}^{(i)}(U_{conf}, \beta_i) = Q_{NVE}^{(i)}(U_{conf}) \cdot \exp\left(-\beta_i U_{conf}\right). \tag{3.85}$$

With the introduction of the dimensionless free energy f_i at β_i, which is directly related to the canonical partition function (see Sect. 2.2)

$$f_i = \beta_i F_i = -\ln Q_{NVT}(\beta_i) \tag{3.86}$$

we can rewrite Eq. 3.84 as

$$\wp_{NVT}^{(i)}(U_{conf}, \beta_i) = Q_{NVE}^{(i)}(U_{conf}) \cdot \exp\left(-\beta_i U_{conf} + f_i\right). \tag{3.87}$$

Thus, from each of the R simulations, an estimate for the microcanonical partition function $Q_{NVT}^{(i)}$ can be derived by storing into histograms $H_i\left(U_{conf}\right)$ how many times the configurational energy U_{conf} is observed during the simulation at β_i

$$Q_{NVE}^{(i)}(U_{conf}) = \wp_{NVT}^{(i)}(U_{conf}, \beta_i) \cdot \exp\left(\beta_i U_{conf} - f_i\right) \tag{3.88}$$

$$= \frac{H_i(U_{conf})}{K_i} \cdot \exp\left(\beta_i U_{conf} - f_i\right). \tag{3.89}$$

Ferrenberg and Swendsen [11] derived a method to improve the statistics in estimating Q_{NVE} by linearly combining the estimates $Q_{NVE}^{(i)}(U_{conf})$ from all R simulations, and introducing weighting factors W_i to the $Q_{NVE}^{(i)}$

$$Q_{NVE}(U_{conf}) = \sum_{i=1}^{R} W_i \cdot Q_{NVE}^{(i)}(U_{conf}), \quad \text{with} \quad \sum_{i=1}^{R} W_i = 1. \tag{3.90}$$

The set of weights W_i that result in the best estimate of Q_{NVE} is then derived by minimizing the statistical error $\delta^2 Q_{NVE}$. The error in determining Q_{NVE} arises from errors in the single $Q_{NVE}^{(i)}$, which in turn are related to the errors in the histograms

$\delta^2 H_i(U_{conf})$, which are due to the fluctuations in the ratio H_i/K_i during the course of the simulation.

The expectation value of the histogram $H_i(U_{conf})$ can be estimated by [18]

$$\overline{H_i(U_{conf})} = K_i \cdot \mathcal{Q}_{NVE}(U_{conf}) \cdot \exp(-\beta_i U_{conf} + f_i). \tag{3.91}$$

Following Ferrenberg and Swendsen [10, 11], the errors in the single histograms are given by

$$\delta^2 H_i(U_{conf}) = g_i \overline{H_i(U_{conf})}, \tag{3.92}$$

wherein $g_i = 1 + 2\tau_i$ accounts for the autocorrelation in the simulations, with τ being the correlation time (see Sect. 7.2.1). When all configurations are independent, the g_i can be set to $g_i = 1$. The optimal estimate of the microcanonical partition function is then derived from [11, 18]

$$\mathcal{Q}_{NVE}(U_{conf}) = \frac{\sum_{i=1}^{R} g_i^{-1} H_i(U_{conf})}{\sum_{j=1}^{R} K_j g_j^{-1} \exp\left(-\beta_j U_{conf} + f_j\right)}. \tag{3.93}$$

Substitution into Eq. 3.85 yields the optimized estimate of the un-normalized probability distribution

$$\mathcal{P}_{NVE}(U_{conf}) = \frac{\sum_{i=1}^{R} g_i^{-1} H_i(U_{conf}) \exp\left(-\beta_j U_{conf}\right)}{\sum_{j=1}^{R} K_j g_j^{-1} \exp\left(-\beta_j U_{conf} + f_j\right)}. \tag{3.94}$$

The dimensionless free energies $\exp(f_i)$ are determined from Eq. 3.86, wherein the canonical partition function can be estimated by

$$\mathcal{Q}_{NVT}(\beta_i) = \sum_{U_{conf}} \mathcal{Q}_{NVE}(U_{conf}) \exp(-\beta_i U_{conf}) = \sum_{U_{conf}} \mathcal{P}_{NVE}(U_{conf}), \tag{3.95}$$

which then gives

$$\exp(-f_i) = \mathcal{Q}_{NVT}(\beta_i) = \sum_{U_{conf}} \mathcal{P}_{NVE}(U_{conf}). \tag{3.96}$$

Starting from an initial guess for the f_i, they are self-consistently derived by iterating the Eqs. 3.94 and 3.96 until convergence. Assuming independent samples, i.e. $g_i = 1$, the free energy at β_i can then be determined by

$$\beta_i F_i = -\ln \sum_{U_{conf}} \frac{\sum_{i=1}^{R} H_i(U_{conf}) \exp\left(-\beta_i U_{conf}\right)}{\sum_{j=1}^{R} K_j \exp\left(-\beta_j U_{conf} + \beta_j F_j\right)}, \tag{3.97}$$

up to an additive constant.

When simulations are performed in the grand canonical ensemble for phase equilibria studies, histograms $H_i(N, U_{conf})$ sampled in $i = 1, ..., R$ simulations at different values for β_i and μ_i are combined to derive optimal estimates of $Q_{NVE}(N, U_{conf})$ and the un-normalized probability $\mathcal{P}_{\mu VT}(N, U_{conf})$. Following the self-consistent method by Ferrenberg and Swendsen [10, 11] and assuming $g_i = 1$, the optimized un-normalized probability is given by [47]

$$\mathcal{P}_{\mu VT}(N, U_{conf}) = \frac{\sum_{i=1}^{R} H_i(N, U_{conf}) \exp\left(-\beta U_{conf} + \beta N \mu\right)}{\sum_{i=1}^{R} K_i \exp\left(-\beta_i U_{conf} + \beta_i N \mu_i - C_i\right)}. \tag{3.98}$$

The constants C_i are the dimensionless Hill potentials and related to the grand canonical partition function (see Sect. 2.2)

$$C_i = \beta_i p_i V = \ln Q_{\mu VT}(\beta_i, \mu_i), \quad Q_{\mu VT} = \sum_N \sum_{U_{conf}} \mathcal{P}_{\mu VT}(N, U_{conf}). \tag{3.99}$$

Thus, the C_i can be derived from

$$\exp(C_i) = \sum_N \sum_{U_{conf}} \mathcal{P}(N, U_{conf}, \mu, T). \tag{3.100}$$

Again the C_i are self-consistently determined by iterating the Eqs. 3.98 and 3.100 until convergence.

When the histogram reweighting method in the grand canonical ensemble is applied to mixtures, the chemical potentials μ_i of all N_C compounds and the temperature T (or β) are imposed, and the number of particles N_i of the different species, and the energy U_{conf} of the system are periodically recorded. Whereas for pure fluids, the values for N and U_{conf} can be directly stored into the two-dimensional histograms $H(N, U_{conf})$, for mixtures only the values N_i, U_{conf} are listed, and the un-normalized probability defined by

$$\mathcal{P}_{\mu_i VT}(N_1, ..., N_{N_C}, U_{conf}) = Q_{NVE}(N_1, ..., N_{N_C}, U_{conf}) \exp\left(-\beta U_{conf} + \sum_{i=1}^{N_C} \beta N_i \mu_i\right) \tag{3.101}$$

is computed from the data when the simulation is completed. The "Weighted Histogram Analysis Method" (*WHAM*) by Kumar et al. [18] represents an extension and generalization of the multiple histogram method by Ferrenberg and Swendsen. WHAM is often used to analyze umbrella sampling simulations. Umbrella sampling is aimed at determining the free energy change along a reaction path, which is described by the reaction coordinate ξ. In order to sample different regions of the reaction path more efficiently, biasing potentials are introduced that restrain the sampling to the small regions (windows) of ξ. For each region, the configurational energy (or more general, the Hamiltonian $\mathcal{H}$) is then replaced by a modified potential $\hat{\mathcal{H}}_{\{\lambda\}}$ [18]

$$\hat{\mathcal{H}}_{\{\lambda\}} = \sum_{l=1}^{L} \lambda_l V_l, \tag{3.102}$$

wherein $\{\lambda\}$ denotes a set of coupling parameters $\lambda_1, \lambda_2, ...; \lambda_L$ [18].

For studies in the *NVT* ensemble, $i = 1, ..., R$ independent simulations are performed at different (inverse) temperatures β_i and sets of the coupling constants $\{\lambda\}_i$. Each simulation samples K_i independent configurations, and the histograms store the occurrence of a certain set of biased potentials $\{V\}$ and a certain reaction coordinate ξ. Adjusting Eq. 3.94 accordingly yields the un-normalized probability

$$\mathcal{P}_{NVE,\{\lambda\},\beta}(\{V\}, \xi) = \frac{\sum_{k=1}^{R} g_k^{-1} H_k(\{V\}, \xi) \exp\left(-\beta \sum_{i=1}^{L} \lambda_i V_i\right)}{\sum_{j=1}^{R} K_j g_j^{-1} \exp\left(-\beta_j \sum_{i=1}^{L} \lambda_{i,j} V_i + f_j\right)}. \tag{3.103}$$

This expression is known as WHAM equation. The use of WHAM in umbrella sampling simulations or to derive potential of mean force (PMF) profiles is, for instance, described by Kumar et al. [18, 19] or Souaille and Roux [54].

3.5.3 Estimator Methods BAR and MBAR

The Bennett acceptance ratio method (*BAR*, [4]) was originally derived as Monte Carlo method to estimate free energy differences. As the free energy difference between two states 0 and 1 is related to the ratio of partition functions $Q_{NVT}^{(0)}$, $Q_{NVT}^{(1)}$, Bennett [4] introduced his method by firstly describing how the ratio of partition functions can be estimated from standard MC simulation employing the Metropolis acceptance criterion. This introduction to the BAR method is summarized here.

In Sect. 3.2, we deduced the Metropolis acceptance criterion for the transition from state 0 to state 1 in the canonical ensemble

$$acc(0 \rightarrow 1) = min\left[1, \exp\left(-\beta \Delta U_{conf}^{(0\rightarrow 1)}\right)\right]. \tag{3.104}$$

The acceptance criterion can be expressed as 'Metropolis function'

$$M(x) = min\left[1, \exp(-x)\right], \tag{3.105}$$

which has the property

$$\frac{M(x)}{M(-x)} = \exp(-x). \tag{3.106}$$

Now we consider a trial move in which the configuration remains unchanged, but the potential function is switched from $U_{conf}^{(0)}$ to $U_{conf}^{(1)}$, i.e.

$$x = -\beta\left(U_{conf}^{(1)} - U_{conf}^{(0)}\right). \tag{3.107}$$

Inserting in Eq. 3.106 yields

$$\frac{M\left(\beta(U_{conf}^{(1)} - U_{conf}^{(0)})\right)}{M\left(\beta(U_{conf}^{(0)} - U_{conf}^{(1)})\right)} = \exp\left(-\beta(U_{conf}^{(1)} - U_{conf}^{(0)})\right) = \frac{\exp(-\beta U_{conf}^{(1)})}{\exp(-\beta U_{conf}^{(0)})}, \tag{3.108}$$

or transformed

$$M\left(\beta(U_{conf}^{(1)} - U_{conf}^{(0)})\right) \cdot \exp(-\beta U_{conf}^{(0)}) = M\left(\beta(U_{conf}^{(0)} - U_{conf}^{(1)})\right) \cdot \exp(-\beta U_{conf}^{(1)}).$$

When the left side is multiplied with $Q_{NVT}^{(0)}/Q_{NVT}^{(0)}$ and the right side with $Q_{NVT}^{(1)}/Q_{NVT}^{(1)}$, and both sides are integrated over all configurations, this yields

$$\frac{Q_{NVT}^{(0)} \int M\left(\beta(U_{conf}^{(1)} - U_{conf}^{(0)})\right) \cdot \exp(-\beta U_{conf}^{(0)}) dr^{3N}}{Q_{NVT}^{(0)}}$$
$$= \frac{Q_{NVT}^{(1)} \int M\left(\beta(U_{conf}^{(0)} - U_{conf}^{(1)})\right) \cdot \exp(-\beta U_{conf}^{(1)}) dr^{3N}}{Q_{NVT}^{(1)}}$$
$$\Rightarrow Q_{NVT}^{(0)} \left\langle M\left(\beta(U_{conf}^{(1)} - U_{conf}^{(0)})\right)\right\rangle_0 = Q_{NVT}^{(1)} \left\langle M\left(\beta(U_{conf}^{(0)} - U_{conf}^{(1)})\right)\right\rangle_1.$$

Thus, the ratio of partition functions can be determined by

$$\frac{Q_{NVT}^{(0)}}{Q_{NVT}^{(1)}} = \frac{\left\langle M\left(\beta(U_{conf}^{(0)} - U_{conf}^{(1)})\right)\right\rangle_1}{\left\langle M\left(\beta(U_{conf}^{(1)} - U_{conf}^{(0)})\right)\right\rangle_0}. \tag{3.109}$$

When both the numerator and denominator of the ratio of partition functions is multiplied by an integral over configurations involving an arbitrary weighting function W

$$\frac{Q_{NVT}^{(0)}}{Q_{NVT}^{(1)}} = \frac{Q_{NVT}^{(0)}}{Q_{NVT}^{(1)}} \cdot \frac{\int W \exp(-\beta U_{conf}^{(0)}) \exp(-\beta U_{conf}^{(1)}) dr^{3N}}{\int W \exp(-\beta U_{conf}^{(0)}) \exp(-\beta U_{conf}^{(1)}) dr^{3N}},$$

the ratio of partition functions can also be determined by

$$\frac{Q_{NVT}^{(0)}}{Q_{NVT}^{(1)}} = \frac{\left\langle W \exp(-\beta U_{conf}^{(0)})\right\rangle_1}{\left\langle W \exp(-\beta U_{conf}^{(1)})\right\rangle_0}. \tag{3.110}$$

In the *NVT* ensemble, the ratio $Q_{NVT}^{(0)}/Q_{NVT}^{(1)}$ determines the free energy difference between the two states (see Eq. 7.85), i.e.

$$\beta\left(F^{(1)} - F^{(0)}\right) = -\ln\left(\frac{Q_{NVT}^{(1)}}{Q_{NVT}^{(0)}}\right) \tag{3.111}$$

$$= \ln\left\langle W \exp(-\beta U_{conf}^{(0)})\right\rangle_1 - \ln\left\langle W \exp(-\beta U_{conf}^{(1)})\right\rangle_0. \tag{3.112}$$

Bennett then studied which choice of the weighting function W results in a minimum error in the free energy estimate. He found the optimal choice is given by

$$W = \frac{const.}{\frac{Q_{NVT}^{(0)}}{K_0} \exp(-\beta U_{conf}^{(1)}) + \frac{Q_{NVT}^{(1)}}{K_1} \exp(-\beta U_{conf}^{(0)})}, \tag{3.113}$$

with K_0 and K_1 being the total number of statistically independent configurations sampled in the simulations for $U_{conf}^{(0)}$ and $U_{conf}^{(1)}$, respectively. Inserting the expression for the weighting function into Eq. 3.110 yields

$$\frac{Q_{NVT}^{(0)}}{Q_{NVT}^{(1)}} = \frac{\left\langle f\left(U_{conf}^{(0)} - U_{conf}^{(1)} + C\right)\right\rangle_1}{\left\langle f\left(U_{conf}^{(1)} - U_{conf}^{(0)} - C\right)\right\rangle_0} \cdot \exp(\beta C). \tag{3.114}$$

This expression resembles Eq. 3.109, only that the Metropolis function is replaced by the Fermi function $f(x) = 1/(1 + \exp(x))$, and that a constant C is introduced to optimize the overlap between the distributions at 0 and 1. Bennett also determined the best choice for the constant C to be

$$\beta C = \ln\left(\frac{Q_{NVT}^{(0)} \cdot K_1}{Q_{NVT}^{(1)} \cdot K_0}\right). \tag{3.115}$$

As the constant C also depends on the ratio of partition functions, Eq. 3.114 must be solved in a self-consistent way. The ensemble averages of the Fermi functions can be estimated by

$$\left\langle f\left(U_{conf}^{(0)} - U_{conf}^{(1)} + C\right)\right\rangle_1 = \frac{1}{K_1}\sum^{(1)} f\left(U_{conf}^{(0)} - U_{conf}^{(1)} + C\right) \tag{3.116}$$

$$\left\langle f\left(U_{conf}^{(1)} - U_{conf}^{(0)} - C\right)\right\rangle_0 = \frac{1}{K_0}\sum^{(0)} f\left(U_{conf}^{(1)} - U_{conf}^{(0)} - C\right). \tag{3.117}$$

Inserting these expressions into Eq. 3.114 yields an estimate for the free energy difference that is given by

$$\begin{aligned}\beta\Delta F &= \ln\left(\frac{Q_{NVT}^{(0)}}{Q_{NVT}^{(1)}}\right)\\ &= \ln\frac{\sum^{(1)} f\left(U_{conf}^{(0)} - U_{conf}^{(1)} + C\right)}{\sum^{(0)} f\left(U_{conf}^{(1)} - U_{conf}^{(0)} - C\right)} + \ln\frac{K_0}{K_1} + \beta C.\end{aligned} \tag{3.118}$$

On the other hand, when deriving the free energy difference from the ratio of partition functions described by Eq. 3.115, this yields

$$\beta\Delta F = \ln\frac{K_0}{K_1} + \beta C. \tag{3.119}$$

Thus, both estimates for the free energy difference only agree, if

$$\sum^{(1)} f\left(U_{conf}^{(0)} - U_{conf}^{(1)} + C\right) = \sum^{(0)} f\left(U_{conf}^{(1)} - U_{conf}^{(0)} - C\right). \tag{3.120}$$

This represents the self-consistency criterion for selecting C, which is treated as adjustable parameter and is varied until Eq. 3.120 is satisfied. The BAR method only considers two states 0 and 1 to estimate free energy differences. Thus, Shirts and Chodera [50] extended the method to the "multistate Bennett acceptance ratio estimator" (*MBAR*) to allow for the optimal use of samples from multiple states. Again, the method aims to derive a global estimator for some ratio of partition functions $Q^{(i)}$ and $Q^{(j)}$. Following Eq. 3.110, for simulations in the *NVT* ensemble, the ratio can be estimated by

$$\frac{Q_{NVT}^{(i)}}{Q_{NVT}^{(j)}} = \frac{\left\langle \alpha_{ij} \exp(-\beta U_{conf}^{(i)}) \right\rangle_j}{\left\langle \alpha_{ij} \exp(-\beta U_{conf}^{(j)}) \right\rangle_i}, \tag{3.121}$$

or

$$Q_{NVT}^{(i)} \left\langle \alpha_{ij} \exp(-\beta U_{conf}^{(j)}) \right\rangle_i = Q_{NVT}^{(j)} \left\langle \alpha_{ij} \exp(-\beta U_{conf}^{(i)}) \right\rangle_j \tag{3.122}$$

with the weights α_{ij}. When R is again the number of simulations at different thermodynamic states, and K_i and K_j represent the number of uncorrelated configurations sampled in state i and j, Eq. 3.122 can be estimated by

$$\sum_{j=1}^{R} \frac{Q_{NVT}^{(i)}}{K_i} \sum_{k=1}^{K_i} \alpha_{ij} \exp\left(-\beta U_{conf}^{(j)}(r_{i,k}^{3N})\right) = \sum_{j=1}^{R} \frac{Q_{NVT}^{(j)}}{K_j} \sum_{k=1}^{K_j} \alpha_{ij} \exp\left(-\beta U_{conf}^{(i)}(r_{j,k}^{3N})\right). \tag{3.123}$$

This represents a set of R estimating equations that are parameterized by the choice of the weights α_{ij}. In statistics, these are known as 'extended bridge sampling estimators' [56]. Thus, Chodera and Shirts employed results from the field of statistical inference to derive a set of α_{ij} that minimizes the variances of free energy differences between all states simultaneously. The best choice for the estimator is then given by

$$\alpha_{ij} = \frac{K_j / Q_{NVT}^{(j)}}{\sum_{r=1}^{R} K_r / Q_{NVT}^{(r)} \exp(-\beta U_{conf}^{(r)})}. \tag{3.124}$$

When inserting Eq. 3.124 into Eq. 3.123, the term $Q_{NVT}^{(i)}/K_i$ on the left side can be taken out of the summation, and the order of summation can be switched [32]. The summations on the left side then simply reduce to K_i, so that an estimate of the $Q_{NVT}^{(i)}$ can be deduced

$$Q_{NVT}^{(i)} = \sum_{j=1}^{R} \sum_{k=1}^{K_j} \frac{\exp(-\beta U_{conf}^{(i)})}{\sum_{r=1}^{R} K_r / Q_{NVT}^{(r)} \exp(-\beta U_{conf}^{(r)})}. \tag{3.125}$$

The partition function in the denominators can be replaced by

$$\frac{1}{Q_{NVT}^{(r)}} = \exp(\beta_r F_r). \tag{3.126}$$

With this, an expression for the free energy can be derived

$$\beta F_i = -\ln \sum_{j=1}^{R} \sum_{k=1}^{K_j} \frac{\exp(-\beta U_{conf}^{(i)})}{\sum_{r=1}^{R} K_r \exp(-\beta U_{conf}^{(r)} + \beta_r F_r)} \tag{3.127}$$

that resembles the expression from the multiple-histogram analysis in Eq. 3.97. This illustrates that MBAR and WHAM are equivalent in the limiting case that the bin width in WHAM is set to zero. The MBAR equation also reduces to the BAR equation when only two states are considered. Again, Eq. 3.127 must be solved self-consistently to yield estimates for the set of free energies F_i, which though can only be determined up to an additive constant. The application of MBAR method is not limited to simulations in the *NVT* ensemble, but can be extended to any potential functions. For simulations in the *NpT* ensemble, the potential term is represented by

$$\exp\left(-\beta(U_{conf} + pV)\right), \tag{3.128}$$

and the MBAR method then allows for estimating differences in the Gibbs free energy G_i from

$$\beta G_i = -\ln \sum_{j=1}^{R} \sum_{k=1}^{K_j} \frac{\exp\left(-\beta U_{conf}^{(i)} + pV_{jk}\right)}{\sum_{r=1}^{R} K_r \exp\left(-\beta(U_{conf}^{(r)} + pV_{rk}) + \beta_r G_r\right)}. \tag{3.129}$$

In [33], Paliwal and Shirts employed MBAR for predicting properties at poorly or completely unsampled states. In [32], they present an extension of the method for free energy calculations involving changes in geometry. A Python implementation of MBAR is available on [49].

3.6 MC Simulation of Phase Equilibria

The thermodynamic criteria for phase coexistence between two phases *I* and *II* are given by

thermal equilibrium	$T^I = T^{II}$
mechanical equilibrium	$p^I = p^{II}$
material equilibrium	$\mu_i^I = \mu_i^{II}, \quad i = 1, ..., N_C$

The number of degrees of freedom N_F, i.e. intensive variables that can be independently specified, is given by the Gibbs phase rule

$$N_F = N_C - N_P + 2 \tag{3.130}$$

with N_C being the number of components, and N_P being the number of phases in equilibrium.

In a simple set-up for experimental studies on fluid phase equilibria, a known amount of the substances is charged into an equilibrium cell that is equipped with sapphire windows. The system is set to the desired temperature and thermoregulated by thermostats. Equilibration can be observed by phase separation and the formation of an interface between the coexistence phases. Often a stirrer is employed to accelerate the equilibration, still it might take several hours to reach equilibration. With this, phase transitions occur at time scales that are not directly accessible by MD simulations [36]. Monte Carlo methods are therefore more efficient to study phase equilibria as they do not follow the time evolution of the system. However, the direct simulation of phase transition following the experimental procedure described above is also challenging for MC simulations, as the interface represents a high free energy barrier (surface tension) between the phases, and material equilibration through the interface is slow. Furthermore, the properties of a significant fraction of the particles are influenced by the interface, and therefore do not correspond to the saturation properties of the bulk phases. Thus, very large simulation systems are required to minimize the effect of the interfacial region on the determination of the saturation properties. Additional difficulties in the direct simulation of phase transition arise from hysteresis effects depending on the history of the system. This means that for the example of a vapor-liquid equilibrium, the liquid phase will be overheated and the gas phase will be supercooled before the phase transition occurs, so that the temperature at which the transition is observed is not the equilibrium temperature. Simulation studies of phase equilibria were greatly enhanced, when Panagiotopoulos [35] proposed his Gibbs Ensemble Monte Carlo (GEMC) method that avoids the simulation of the interphase by using two simulation regions, each corresponding to one of the two equilibrium phases. The GEMC technique represents a direct and robust method for simulating fluid phase equilibria with moderate system sizes. Though the GEMC method has become widely used for simulations on vapor-liquid equilibria, it is not suited for studies on phase transition involving solid phases, and it becomes unstable near critical points [36]. We therefore also discuss the Gibbs-Duhem integration and the Histogram reweighting technique as alternative simulation methods for phase equilibria.

3.6.1 Gibbs Ensemble Monte Carlo (GEMC)

The GEMC simulation methodology avoids difficulties associated with the presence of the interface between two coexistent phases by using two simulation boxes,

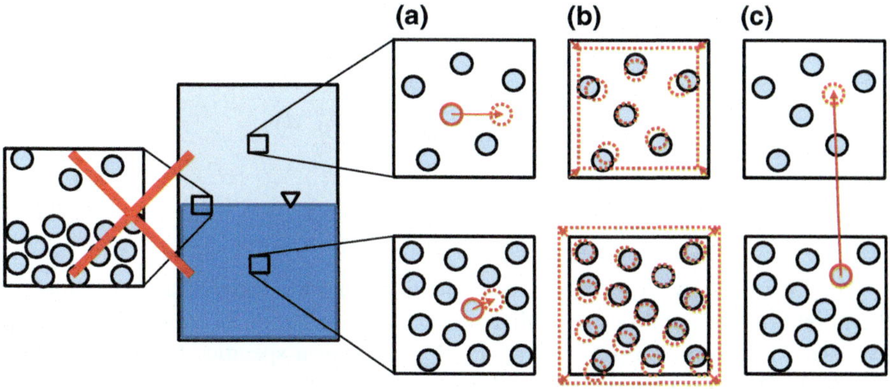

Fig. 3.3 Concept of the Gibbs Ensemble method: avoiding simulations in presence of the interface by using separate simulation boxes for each phase. Moves in the GEMC method: **a** single particle displacement within a box, **b** volume change, **c** particle transfer between the boxes

each representing a homogenous region of the phases that are in thermodynamic equilibrium. As the simulation boxes are in no physical contact, distinct moves are performed to assure that the two systems satisfy the conditions for phase equilibria. As illustrated by Fig. 3.3, these moves comprise single particle displacements within each box to attain internal equilibrium for the imposed temperature, volume fluctuations to obtain mechanical equilibrium, and finally particle transfer between the boxes to reach material equilibrium. Through the volume change and particle exchange, the two simulation systems are indirectly coupled. The different moves are either performed according to a predetermined or random sequence. Each GEMC simulation provides one point on the phase envelope. In the following sections, we provide the partition functions for GEMC simulation studies on pure compounds and mixtures, and deduce the acceptance rules for the moves from the resulting probability densities.

In general, there is also a molecular dynamics implementation for the Gibbs ensemble available [34], but this will not be further discussed here.

3.6.1.1 GEMC for Pure Compounds

For the equilibrium between two phases of a pure compound, the Gibbs phase rule yields $N_F = 1$, i.e. only one intensive variable, the temperature, can be imposed, whereas the vapor pressure results as function of the temperature. Thus, GEMC simulations for pure compounds are performed in the NVT-ensemble. The total number of molecules $N = N_I + N_{II}$ and the total volume $V = V_I + V_{II}$ is conserved, but the values for both boxes will vary due to the particle transfer and volume change. Thus, the partition function of the total system is the product of the partition functions of both simulation boxes subject to the restrictions that N_I may vary from $N_I = 0 - N$, and $V_I = 0 - V$, but then N_{II} is given by $N_{II} = N - N_I$ and accordingly $V_{II} = V - V_I$.

When using scaled coordinates as described above, the partition function is [36]

$$Q_{NVT}^{GEMC} = \frac{1}{\Lambda^3 N!} \sum_{N_I=0}^{N} \frac{N!}{N_I! N_{II}!} \int_{V_I=0}^{V} V_I^{N_I} V_{II}^{N_{II}} dV_I \cdot \int e^{(-\beta U_{conf,I}(N_I))} ds^{3N_I} \cdot \int e^{(-\beta U_{conf,II}(N_{II}))} ds^{3N_{II}}. \tag{3.131}$$

The factor $N!/(N_I!N_{II}!)$ results from the combinatoric analysis and accounts for the number of possibilities to realize a specific distribution of molecules over the two phases. The probability density $\wp$ of observing a specific state described by (N_I, N, V_I, V, T) is then proportional to [36]

$$\wp_{NVT}^{GEMC} \propto \frac{N!}{N_I! N_{II}!} e^{[N_I \ln V_I + N_{II} \ln V_{II} - \beta U_{conf,I}(N_I) - \beta U_{conf,II}(N_{II})]}. \tag{3.132}$$

To realize internal equilibrium for the given temperature, single particle displacements are performed within the boxes while the number of particles and the volumes are unchanged. Thus, the ratio of the probability densities $\wp(m)/\wp(n)$ again only depends on the difference in the configuration energy of the displaced particle j in both positions, and the acceptance rule is given by

$$acc(n \rightarrow m)_{\Delta r_j}^{GEMC} = min\left[1, \exp\left(-\beta \Delta U_{conf,j}^{(n\rightarrow m)}\right)\right]. \tag{3.133}$$

In the volume change move to control the equilibrium pressure, the volume $V_I(n) = V_I$ of one phase is changed randomly accordingly to Eq. 3.2 to yield a trial volume in state (m) of $V_I(m) = V_I + \Delta V$. As the total volume is conserved, the volume of the second phase correspondingly decreases to $V_{II}(m) = V_{II} - \Delta V$. With a conserved number of particles N_I, N_{II} in both boxes, the ratio $\wp(m)/\wp(n)$ is

$$\frac{\wp(m)}{\wp(n)} = \frac{\left[N_I \ln(V_I + \Delta V) + N_{II} \ln(V_{II} - \Delta V) - \beta U_{conf,I}^{(m)}(N_I) - \beta U_{conf,II}^{(m)}(N_{II})\right]}{\left[N_I \ln V_I + N_{II} \ln V_{II} - \beta U_{conf,I}^{(n)}(N_I) - \beta U_{conf,II}^{(n)}(N_{II})\right]}.$$

Thus, the resulting acceptance criterium for a coupled trial volume change is

$$acc(n \rightarrow m)_{\Delta V} = min\left[1, \exp\left(N_I \ln \frac{V_I + \Delta V}{V_I} + N_{II} \ln \frac{V_{II} - \Delta V}{V_{II}} - \beta \Delta U_{conf,I}^{(n\rightarrow m)}(N_I) - \beta \Delta U_{conf,II}^{(n\rightarrow m)}(N_{II})\right)\right]. \tag{3.134}$$

In the particle transfer move, a randomly selected molecule is removed from one phase, say box II, and inserted at a random position within the simulation volume of the other box I. The volumes of both boxes are conserved during the particle exchange. Thus, with $N_I(n) = N_I$, $N_I(m) = N_I + 1$ and accordingly $N_{II}(n) = N_{II}$, $N_{II}(m) = N_{II} - 1$, the ratio of the probability densities $\wp(m)/\wp(n)$ is

$$\frac{\wp(m)}{\wp(n)} = \frac{N_I!N_{II}!}{(N_I+1)!(N_{II}-1)!} \cdot \frac{\exp\left[(N_I+1)\ln V_I + (N_{II}-1)\ln V_{II} - \beta U^{(m)}_{conf,I}(N_I+1) - \beta U^{(m)}_{conf,II}(N_{II}-1)\right]}{\exp\left[N_I \ln V_I + N_{II}\ln V_{II} - \beta U^{(n)}_{conf,I}(N_I) - \beta U^{(n)}_{conf,II}(N_{II})\right]}.$$

This yields an acceptance criterion for the particle transfer from II to I

$$acc(n \to m)_{\Delta N} = min\left[1, \frac{N_{II}V_I}{(N_I+1)V_{II}} \exp\left(-\beta\Delta U^{(n\to m)}_{conf,I} - \beta\Delta U^{(n\to m)}_{conf,II}\right)\right]. \quad (3.135)$$

Though the Gibbs ensemble does not require the calculation of the chemical potentials and pressures in both phases to evaluate the criteria of phase equilibria, it is recommended to derive these properties to independently verify that equilibrium has been attained. The pressure is determined by the virial equation given by Eq. 7.9 in Chap. 7, whereas the chemical potentials can be obtained by the Widom method (see Sect. 7.4.4) within the particle transfer step.

3.6.1.2 GEMC for Mixtures

According to the Gibbs phase rule, the number of degrees of freedom increase with the number of components. Thus, in GEMC studies on phase equilibria in mixtures, the pressure is now also specified, and the simulations are performed in the NpT−ensemble. From this it follows that the volumes of the two simulation boxes, i.e. phases, are allowed to change independently. Still the total number of molecules is conserved so that N_I may vary from $N_I = 0$ to N, but N_{II} is given by $N_{II} = N - N_I$. With this, the partition function of the total system when using scaled coordinates is given by

$$Q^{GEMC}_{NpT} = \frac{\beta p}{\Lambda^3 N!}\sum_{N_I=0}^{N}\frac{N!}{N_I!N_{II}!}\int_0^\infty V_I^{N_I}e^{-\beta p V_I}dV_I \cdot \int_0^\infty V_{II}^{N_{II}}e^{-\beta p V_{II}}dV_{II} \cdot \int e^{(-\beta U_{conf,I}(N_I))}ds^{3N_I} \cdot \int e^{(-\beta U_{conf,II}(N_{II}))}ds^{3N_{II}}, \quad (3.136)$$

and the probability density $\wp$ to observe a specific state described by $(N_I, N, V_I, V_{II}, p, T)$ becomes proportional to [36]

$$\wp^{GEMC}_{NpT} \propto \frac{N!}{N_I!N_{II}!}e^{[N_I\ln V_I + N_{II}\ln V_{II} - \beta U_{conf,I}(N_I) - \beta U_{conf,II}(N_{II}) - \beta p(V_I+V_{II})]}. \quad (3.137)$$

As described before for the pure compounds, the internal equilibrium in each phase is achieved by single particle displacements within the boxes, and also the acceptance rule for this trial move remains unchanged. The main difference compared to the pure component simulation is the volume change move, as volume change attempts are now independently performed for both phases. When for instance the volume of phase I, $V_I(n) = V_I$, is changed randomly to $V_I(m) = V_I + \Delta V$, accompanied by a change in the configurational energy $\Delta U^{(n\rightarrow m)}_{conf,N_I}(N_I)$, the acceptance criterium becomes [36]

$$acc(n \rightarrow m)_{\Delta V_I} = \min\left[1, \exp\left(N_I \ln \frac{V_I + \Delta V}{V_I} - \beta \Delta U^{(n\rightarrow m)}_{conf,I}(N_I) - \beta p \Delta V\right)\right]. \tag{3.138}$$

The particle transfer move now has to ensure that the chemical potentials of each component are equalized in both phases. The acceptance rule for a particle transfer move of a molecule of the component i from one phase, say II, to the phase I is similar to the criterion derived for the pure component in Eq. 3.135. The only difference is that the total number of molecules in both boxes N_I, N_{II} has to be replaced by the number of molecules of the species i that is transfers, i.e. $N_{i,I}$, $N_{i,II}$ [36]. The resulting acceptance criterion for the particle transfer of a molecule of species i from phase II to I is

$$acc(n \rightarrow m)_{\Delta N_i} = \min\left[1, \frac{N_{i,II} V_I}{(N_{i,I} + 1) V_{II}} \exp\left(-\beta \Delta U^{(n\rightarrow m)}_{conf,I} - \beta \Delta U^{(n\rightarrow m)}_{conf,II}\right)\right]. \tag{3.139}$$

For both, studies on pure compounds and on mixtures, the Gibbs Ensemble relies on the particle transfer to ensure that the condition of material equilibrium is satisfied. As discussed before, the chance of successful particle insertions in a dense liquid phase significantly decreases with increasing complexity of the molecules. Thus, CBMC methods (see Sect. 3.4.2) are employed to allow for fluid phase equilibria studies on multiatom molecules.

3.6.1.3 GEMC for Multiphase Equilibria

Although the GEMC method is most commonly used for studies on vapor-liquid equilibria, Canongia Lopes and Tildesley [6] have shown that the method can be easily extended to multiphase equilibria by increasing the number of boxes accordingly. As example, they have performed studies on the vapor-liquid-liquid equilibrium (VLLE) of binary mixtures, and on the VLLLE of ternary mixtures of Lennard-Jones atoms. The volume change and particle transfer moves were only performed between pairs of boxes, which were selected randomly. By only considering pairs of boxes for the exchange moves, the procedure and acceptance criteria for the moves can directly be adopted from the GEMC simulation methods for 2-phase equilibria (VLE) discussed in the previous section. Canongia Lopes and Tildesley noted that the equilibrium between the liquid phases is much easier achieved in presence of the vapor phase. They justified this with the much higher acceptance rate for the particle transfer between the vapor and the liquid phase compared to the transfer between two liquids, so that the particle exchange between the liquid phases effectively occurs through the vapor phase. Based on these observations, it was concluded that GEMC studies on liquid-liquid equilibria (LLE) also require the incorporation of a third box representing a vapor phase as transfer medium. This approach for GEMC simulations on LLE was for instance also taken by Bai and Siepmann [3] in their studies on the mixture of dipropylene glycol dimethyl ether and water. In this work, Bai and Siepmann state that the vapor phase box has to be populated with ghost particles to ensure that it maintains a suitable volume. The number of ghost particles then has to be considered in the acceptance rule for the volume move (see Eq. 3.138). Apart from this, the ghost particles are treated as ideal, this means that they do not interact with the molecules of the studies mixture, and that they do not move, i.e. they are confined to the vapor box. It should be noted that the number of required ghost particles varies with the temperature.

3.6.2 Gibbs-Duhem Integration (GDI)

This method proposed by Kofke [16] makes use of a well known thermodynamic relation, the Gibbs-Duhem-equation

$$S_m dT - V_m dp + \sum_i^{N_c} x_i d\mu_i = 0, \tag{3.140}$$

with S_m and V_m being the molar entropy and volume of the system. The Gibbs-Duhem equation states that the intensive variables, p, T, μ_i cannot change independently. In phase coexistence, this relation applies to each equilibrium phase. Thus, in phase equilibrium between two phases I and II for a pure compound with $N_c = 1, x_1 = 1$, it says

$$\begin{aligned} S_{m,I}\, dT_I \;-\; V_{m,I}\, dp_I + d\mu_{1,I} &= 0 \\ S_{m,II}\, dT_{II} \;-\; V_{m,II}\, dp_{II} + d\mu_{1,II} &= 0. \end{aligned}$$

According to the thermodynamic criteria of phase coexistence, $dT_I = dT_{II} = dT$, $dp_I = dp_{II} = dp$ and $d\mu_{1,I} = d\mu_{1,II}$. With this, the substraction of the relations for phase II and I yields

$$\left(\frac{\partial p}{\partial T}\right)_s = \frac{S_{m,II} - S_{m,I}}{V_{m,II} - V_{m,I}} = \frac{H_{m,II} - H_{m,I}}{T\left(V_{m,II} - V_{m,I}\right)}. \tag{3.141}$$

This is the well-known Clapeyron equation that describes the gradient of the saturation line in the $p - T$-phase diagram, i.e. for instance the vapor pressure curve in vapor-liquid equilibrium. When using the reciprocal temperature $\beta = 1/(k_B T)$, this equation can be expressed as

$$\left(\frac{\partial p}{\partial \beta}\right)_s = \frac{\partial p}{\partial T} \cdot \frac{\partial T}{\partial \beta} = -\frac{H_{m,II} - H_{m,I}}{\beta \left(V_{m,II} - V_{m,I}\right)}. \tag{3.142}$$

Thus, Kofke [16] suggested to directly evaluate phase coexistence by integrating this differential equation numerically. Thereby, the right-hand side of the equation is evaluated by molecular simulations. The thermodynamic integration along the saturation line ensures that the pressure changes with temperature by maintaining coexistence so that (at least for pure fluids) no particle exchange is required to satisfy the conditions for phase equilibrium. With this, the Gibbs-Duhem integration does not feature the limitation of the GEMC methods that is involved with its reliance on the particle exchange, and it can therefore also be applied to simulation studies on solid-liquid-equilibria. In contrast to the GEMC approach that only provides one point of the phase envelope per simulation, the Gibbs-Duhem integration allows for the direct determination of the saturation line through a series of simulations. However, the Gibbs-Duhem method requires the prior knowledge of one coexistence point as starting point for the numerical integration, and this initial point has to be provided by another simulation method. In the following section we will present details on the integration and simulation procedure of the Gibbs-Duhem method using the example of the vapor-liquid equilibrium of a pure compound. Section 3.6.2.2 will then describe its application to study phase coexistence in mixtures.

In general, the GDI method can also be performed using molecular dynamics simulation. References [21, 39] represent examples for studies on solid-liquid equilibria by MD simulations using Gibbs-Duhem integration. However, GDI-MD simulations will not further be discussed here.

3.6.2.1 Gibbs-Duhem Integration for the VLE of Pure Compounds

In the Gibbs-Duhem integration for the VLE of pure compounds, the differential Eq. 3.142 is usually expressed in terms of $\ln p$

$$\left(\frac{\partial \ln p}{\partial \beta}\right)_s = -\frac{H_{m,II} - H_{m,I}}{\beta p \left(V_{m,II} - V_{m,I}\right)} = F(p, \beta). \tag{3.143}$$

wherein the function $F(p, \beta)$ is introduced as abbreviation for the preceding equation to be evaluated. A predictor-corrector scheme is then applied to numerically integrate this equation. Thus, for a new temperature, the pressure is estimated in a predictor step, and NpT-simulations are performed in two simulation boxes that represent homogenous regions of the liquid and vapor phase. In general, a simulation cycle consists of attempted displacements of each particle and an attempted volume change for each box, with acceptance rules according to Eqs. 3.133 and 3.138, respectively. In the course of the simulation, running averages for the energies and volumes in both boxes are determined in certain intervals ($\approx$ every 10 cycles) to evaluate the function F in Eq. 3.143. With this instantaneous value of F, the estimate for the pressure is refined in the corrector step, and the simulation continues with the refined pressure value. After an equilibration period, the running averages are re-zeroed, and then instantaneous values for the pressure, the densities and energies of both phases are collected to determine their ensemble averages and to estimate their statistical errors [16]. Then the Gibbs-Duhem integration proceeds to the next temperature, i.e. β-value.

The accuracy and stability of the procedure is determined by the step size in the reciprocal temperature $\Delta\beta$ between the current and preceding simulation, and the order of the predictor-corrector formula as integrator (see for instance [16] for a detailed discussion). Higher-order predictor-corrector schemes require data for the function F from several prior simulations at different β-values to approximate the pressure by a polynomial. Thus, to initialize the Gibbs-Duhem integration, a low-order trapezoid formula is used in the first simulation as it only requires one data point $[p_0(\beta_0), F_0(\beta_0)]$ to estimate the pressure $p_1(\beta_1 = \beta_0 + \Delta\beta)$. However, this initial point $[p_0(\beta_0), F_0(\beta_0)]$ has to be provided by another simulation method as discussed above. In the course of the integration, when more data $[p_n(\beta_n), F_n(\beta_n)]$ are available, the order of the predictor-corrector formula is updated to the midpoint method and finally to the Adams predictor-corrector. The resulting expressions to estimate the pressure in the predictor (P) and corrector (C) step, depending on the progress of the integration procedure, are summarized in Table 3.1.

Figure 3.4 shows simulation results for the vapor pressure of mercury from [41] (see also Sect. 6.4.1). The vapor pressures in the temperature range $T = (945, 27 - 1575, 45)\,K$, i.e. $T^{-1} = (0.1006 - 6.3474E - 4)\,K^{-1}$ were derived by the GEMC method described in Sect. 3.6.1.1. One of these simulation data derived from GEMC

Table 3.1 Predictor-corrector formulas to estimate the vapor pressure according to Eq. 3.143 in the Gibbs-Duhem integration. The employed order of the predictor-corrector scheme depends on the progress of the integration; F_0, p_0 are initial conditions derived by another simulation method

Integration Progress	Name	Type	Predictor-corrector formula
1. Simulation at β_1	Trapez	P	$p_1 = p_0 \exp(\Delta\beta F_0)$
	Trapez	C	$p_1 = p_0 \exp\left[\frac{\Delta\beta}{2}(F_1 + F_0)\right]$
2. Simulation at β_2	Midpoint	P	$p_2 = p_0 \exp(2\Delta\beta F_1)$
	Midpoint	C	$p_2 = p_0 \exp\left[\frac{\Delta\beta}{3}(F_2 + F_1 + F_0)\right]$
3. Simulation at β_3	Midpoint	P	$p_3 = p_1 \exp(2\Delta\beta F_2)$
	Adams	C	$p_3 = p_2 \exp\left[\frac{\Delta\beta}{24}(9F_3 + 19F_2 - 5F_1 + F_0)\right]$
n. Simulation at β_n	Adams	P	$p_n = p_{n-1} \exp\left[\frac{\Delta\beta}{24}(55F_{n-1} - 59F_{n-2} + 37F_{n-3} - 9F_{n-4})\right]$
$n \geq 4$	Adams	C	$p_n = p_{n-1} \exp\left[\frac{\Delta\beta}{24}(9F_n + 19F_{n-1} - 5F_{n-2} + F_{n-3})\right]$

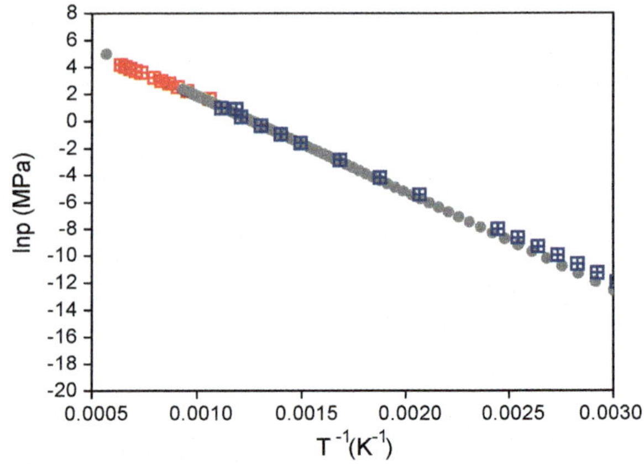

Fig. 3.4 Simulation results for the vapor pressure of mercury [41] from the GEMC method (*red crossed squares*) and Gibbs-Duhem integration (*blue crossed squares*), in comparison with experimental data [15, 55, 58] (colour online)

were then used as initial point for the Gibbs-Duhem integration to derive the vapor pressures at temperatures $T \leq 898\,\mathrm{K}$ ($T^{-1} \geq 0.00111\,\mathrm{K}^{-1}$).

When approaching the critical point, the Gibbs-Duhem integration suffers from similar problems as the GEMC method. Due to the decreasing free energy difference between the two phases, is becomes difficult to maintain the identity of each phase [36].

3.6.2.2 Gibbs-Duhem Integration Method for Mixtures

In [27], Metha and Kofke introduced the extension of the Gibbs-Duhem integration to study phase equilibria in mixtures. We here describe this method by taking the VLE of a binary mixture as an example. The starting point is the Gibbs-Duhem relation given in Eq. 3.140, applied for each equilibrium phase. The thermodynamic criteria of phase coexistence states that $dT_I = dT_{II} = dT$, $dp_I = dp_{II} = dp$ and $d\mu_{i,I} = d\mu_{i,II}$, so that the substraction of the relations for phase II and I yields

$$(S_{m,II} - S_{m,I})\,dT - (V_{m,II} - V_{m,I})\,dp + (x_{1,II} - x_{1,I})\,d\mu_1 + (x_{2,II} - x_{2,I})\,d\mu_2 = 0$$

Again the relation can be expressed in terms of $\ln p$ by replacing $dp = p\,d\ln p$. The change in the chemical potential $d\mu_i$ can be described by the change in the fugacity f_i of the component by

$$d\mu_i = RTd\ln f_i. \tag{3.144}$$

In mixtures, we only study isothermal phase equilibria, so that $dT = 0$. By introducing the compressibility factor

$$Z = \frac{pV_m}{RT} \tag{3.145}$$

Equation 3.144 then becomes

$$-(Z_{II} - Z_I) d\ln p + (x_{1,II} - x_{1,I})\, d\ln f_1 + (x_{2,II} - x_{2,I})\, d\ln f_2 = 0. \quad (3.146)$$

Now a fugacity fraction ξ_2 is introduced

$$\xi_2 = \frac{f_2}{f_1 + f_2}, \quad (3.147)$$

so that the change in the fugacity $d\ln f_1$ of component 1 can be replaced by

$$d\ln f_1 = d\ln f_2 - \frac{d\xi_2}{\xi_2(1 - \xi_2)}. \quad (3.148)$$

Taking into account the constraint $\sum x_i = 1$, i.e.

$$x_{1,II} = 1 - x_{2,II}, \qquad x_{1,I} = 1 - x_{2,I}, \quad (3.149)$$

Equation 3.146 can finally be summarized to

$$(Z_{II} - Z_I) d\ln p + (x_{2,I} - x_{2,II}) \frac{d\xi_2}{\xi_2(1 - \xi_2)} = 0. \quad (3.150)$$

This gives the working equation for the Gibbs-Duhem integration in the so-called semigrand ensemble formulation [27]

$$\left(\frac{\partial \ln p}{\partial \xi_2}\right)_T = \frac{x_{2,II} - x_{2,I}}{\xi_2(1 - \xi_2)(Z_{II} - Z_I)} = F(p, \xi_2). \quad (3.151)$$

Again the Gibbs-Duhem method requires the knowledge of one coexistence point to initialize the integration, and the predictor-corrector technique is chosen as integrator. Hence, the fugacity fraction ξ_2 as independent variable is incremented, and the pressure is estimated in the predictor step by using the formulations given in Table 3.1 with $\Delta\beta$ replaced by $\Delta\xi_2$. MC simulations are performed in two boxes representing the liquid and vapor phase of the mixture. In the progress of the simulation, running averages for the equilibrium properties in both boxes are determined to evaluate the function F and to correct the pressure according to Table 3.1. The continuous refinement of the pressure in the corrector step with current F values is repeated until convergence. The simulation continues with the production run for the given ξ_2 value, in which instantaneous values for the pressure, the densities and composition of both phases are sampled to derive their ensemble averages and statistical errors. Then the Gibbs-Duhem integration proceeds to the next ξ_2 value.

In studies on mixtures, the fugacity or fugacity fraction is used as independent variable, hence the simulations cannot be conducted in the standard NpT-ensemble. To obtain a suitable ensemble, a Legendre transformation is performed in which the number of particles of the species 2, N_2, as independent variable is replaced by its conjugated coordinate, the chemical potential μ_2 or rather the fugacity f_2. Additionally, the total number of particles N is fixed. Details on the derivation of the partition function of the resulting semigrand isothermal-isobaric ensemble are for instance given by Kofke in [17], or in the textbook by Frenkel and Smit [13]. The expression for the partition function for a binary mixture, expressed in terms of the fugacity fractions ξ_i can be found [2, 30]. When using scaled coordinated s, the partition function for the binary mixtures becomes

$$Q_{NpT\xi_2} = \sum_{N_2=0}^{N} \frac{(\xi_2/\xi_1)^{N_2}}{N_2!N_1!} \frac{1}{\Lambda^{3N}} \int V^N e^{-\beta pV} dV \int e^{-\beta U_{conf}} d^{3N} s. \tag{3.152}$$

In order to sample different compositions for a fixed total number of particles, trial moves consist of attempted identity changes of the two species. Thus, the partition function is often rewritten in such a way that the sum of number of particles of species 2 is replaced by a sum over the two possible identities of all N particles [13]. To correct for double counting, the partition function needs to be divided by $N!/(N_2!N_1!)$ [13], which results in

$$Q_{NpT\xi_2} = \sum_{identities} \frac{(\xi_2/\xi_1)^{N_2}}{N!} \frac{1}{\Lambda^{3N}} \int V^N e^{-\beta pV} dV \int e^{-\beta U_{conf}} d^{3N} s. \tag{3.153}$$

In a trial move in which a particle of species 1 changes its identity to become a particle of species 2, so that $N_2^{(m)} = N_2^{(n)} + 1$, the ratio of probability densities is given by [27]

$$\frac{\wp^{(m)}}{\wp^{(n)}} = \exp\left(-\beta\left(U_{conf}^{(m)} - U_{conf}^{(n)}\right) + \ln\frac{\xi_2}{1-\xi_2}\right). \tag{3.154}$$

Note that for $\xi_2 = const$, also $\xi_1 = 1 - \xi_2 = const$. Accordingly, for a trial move with an identity change of one particle from species 2 to 1, i.e. $N_2^{(m)} = N_2^{(n)} - 1$, the acceptance criterion becomes

$$acc(n \to m)_{N_2^-} = min\left[1, \exp\left(-\beta\left(U_{conf}^{(m)} - U_{conf}^{(n)}\right) - \ln\frac{\xi_2}{1-\xi_2}\right)\right]. \tag{3.155}$$

In addition to the identity change moves, again attempted displacements of the particles and volume changes of both boxes are performed. More details on the procedure of Gibbs-Duhem integration in mixtures are given in [27]. Therein, Metha and Kofke also provide an alternative expression for the differential equation to be integrated, the so-called osmotic form

$$\left(\frac{\partial \ln p}{\partial f_2}\right)_T = \frac{x_{1,II}/\varphi_{2,I} - x_{1,I}/\varphi_{2,II}}{x_{1,II}Z_I - x_{1,I}Z_{II}} = F(p, f_2), \tag{3.156}$$

in terms of the fugacity coefficient φ_2 of the component 2. The procedure of Gibbs-Duhem integration is similar to that described above for the semigrand ensemble, but now the attempted moves in the MC simulation consist of particle insertion attempts instead of identity swaps. More details are provided in the original work by Metha and Kofke [27].

Again, the GDI method can also be applied to studies on the solid-liquid-equilibria of mixtures. An example for this is the work by Nam et al. [31] where the Gibbs-Duhem integration in the semigrand ensemble has been used to study metallic alloys.

3.6.3 Histogram Reweighting GCMC Studies on the VLE

When simulations on phase equilibria are performed in the grand canonical ensemble, the phase coexistence criteria of equal temperature and chemical potential are satisfied per definition as these are the imposed variables of the simulation. However, in the two-phase region of a VLE, the probability distribution $\wp(N, U_{conf})$ determined at given values of T and μ should in principle exhibit two peaks, corresponding to the coexistent liquid and vapor phase. According to Eq. 3.99, the integral under the probability distribution is proportional to the pressure. Therefore, the condition of pressure equality as third thermodynamic criterion for phase coexistence is satisfied when the areas of the liquid and vapor peaks are equal. Thus, histogram reweighting is applied to determine the value of the chemical potential $\mu(T)$ that results in equal peak areas for both phases.

However, at temperatures below the critical point, a high free energy barrier separates the liquid and vapor phase that cannot be crossed by standard μVT-simulations. Thus, both the liquid and the vapor peak cannot be sampled within the same simulation, and therefore, separate simulations are performed in the two phases. The combination of data from multiple histograms though requires a reasonable overlap between neighboring histograms, so that liquid and vapor regions are bridged by simulations near the critical point, as illustrated by Fig. 3.5. Thus, histogram reweighting studies on the VLE (see for instance [37, 40]) in general are started by performing a series of μVT-simulations in the vicinity of the expected critical point to derive estimates for T_c and μ_c. For this purpose, a series of short simulations with different μ values are performed near the critical temperature to find the value for μ for which the simulation samples both phases. Once this estimate for μ_c is found, a longer simulation is performed at these conditions to collect the bridging histogram. Then additional simulations are conducted at lower temperatures in both the liquid and the vapor phase to obtain additional histograms on both sides of the coexistence curve.

As described in the previous section, the histograms from the different simulations are then combined to derive an optimized estimate for the $Q_{NVE}(N, U_{conf})$ for the range of state points of the phase diagram. This allows for the reweighting of the probability distribution to derive the phase equilibria properties at different temperatures. Thereby for each temperature, the chemical potential is adjusted until the areas

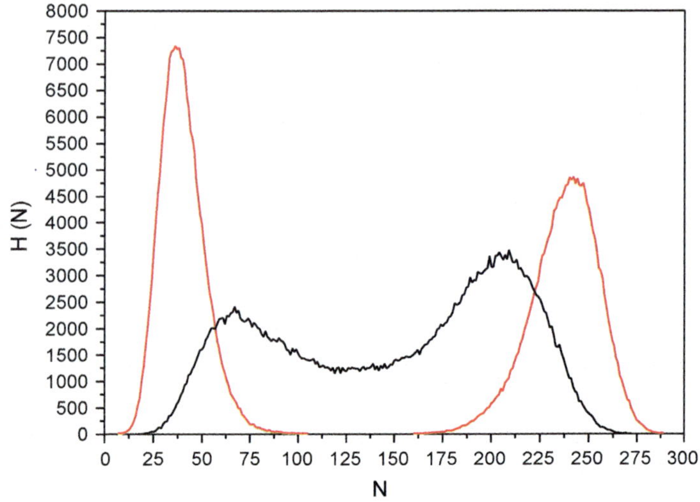

Fig. 3.5 Histograms derived from μVT simulations for R-32. The *black line* represent the histogram $H(N)$ near the critical point at $T = 343$ K and $\mu = 6196.5$ K. The *red lines* give the $H^V(N)$ and $H^L(N)$ at $T = 340$ K determined in separate simulations in the liquid and vapor phase at $\mu^V = 6110$ K and $\mu^L = 6180$ K. The single phase simulations consisted of 50.000.000 moves, the simulation near the critical point of 100.000.000 moves. Histograms were evaluate every 250 moves. The volume was kept constant to $V = 27\,\text{nm}^3$ for all simulations

Fig. 3.6 Saturated densities of the refrigerant R-32 derived by Histogram Reweighting GCMC Studies (*red circles*) and GEMC simulations (*blue circles*). The critical point was estimated by fitting the GEMC simulation results to the scaling law (see Sect. 7.6.1.1)

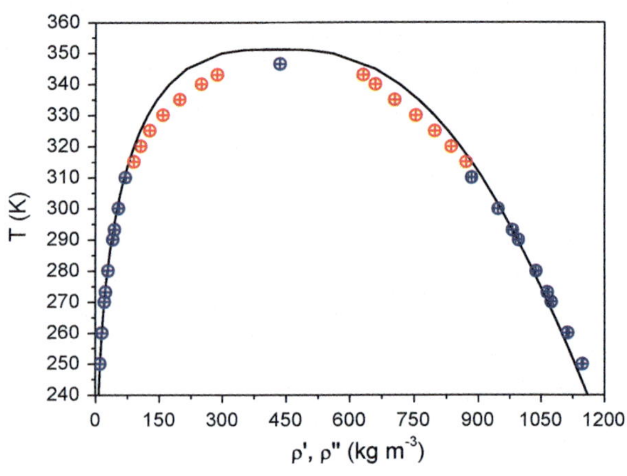

under the peaks for the probability distribution for the liquid and vapor phase become the same. The coexistence densities are then derived from the weighted average of particles under the peaks and the imposed volume. Figure 3.6 shows exemplarily saturated densities (as red circles) of the refrigerant R-32 derived from Histogram Reweighting studies close to its critical temperature. For mixtures, the mole fraction is obtained when additionally the weighted average of particles of each species is calculated. The equilibrium pressure is derived from Eq. 3.101. More details on analyzing simulation results to determine equilibrium properties are given in Sect. 7.6.1.

3.6.4 Flat Histogram Reweighting

As discussed in the previous section, phase coexistence is characterized by a bimodal density distribution $\wp\left(\varrho = \frac{N}{V}\right)$ with each peak corresponding to one of the equilibrium phases I and II. The peaks are separated by a trough, which is well illustrated in Fig. 3.5 for the histograms in the liquid and vapor phase of R-32 at $T = 340\,\mathrm{K}$ (red lines). The trough between the peaks means that configurations located between the peaks, i.e. the intermediate states, have a low probability to be visited. Therefore, transitions between the equilibrium phases are impeded by this probability barrier, which is associated with a high free energy barrier. Thus, biased sampling methods have been developed to enable the crossing of the free energy barrier and with this to sample both equilibrium phases within one simulation. These biased sampling techniques are referred to as flat histogram methods, as they weight the intermediate state with higher probability to enhance their occurrence. Thus, the trough is lifted so that over the entire range of densities enclosed by ϱ_I and ϱ_{II}, all densities have approximately the same probability, and the histogram becomes 'flat'.

For simulations in the grand canonical ensemble, the probability distribution $\wp_{\mu VT}$ (see. Eq. 3.32) is modified by introducing a weighting factor $\eta(N)$ to yield a weighted $\widetilde{\wp}_{\mu VT}$ from [48, 63]

$$\begin{aligned}\widetilde{\wp}_{\mu VT}(N) &\propto \frac{V^N}{N!\Lambda^{3N}} \exp\left(-\beta U_{conf} + \beta\mu N - \eta(N)\right) \\ &= \exp(-\eta(N))\wp_{\mu VT}(N).\end{aligned} \tag{3.157}$$

Simulations in the grand canonical ensemble comprise as usual attempted particle insertion and deletion steps, however now with a modified acceptance criterion, i.e.

$$acc(n \to m) = min\left[1, \frac{\widetilde{\wp}_{\mu VT}(m)}{\widetilde{\wp}_{\mu VT}(n)}\right]. \tag{3.158}$$

According to Eq. 3.35, the ratio of modified probability distributions for a particle insertion becomes

$$\left(\frac{\widetilde{\wp}_{\mu VT}(m)}{\widetilde{\wp}_{\mu VT}(n)}\right)_{N+1} = \frac{V}{(N+1)\Lambda^3} \exp\left(-\beta\Delta U_{conf}^{(n\to m)} + \beta\mu - \Delta\eta^{(n\to m)}\right). \tag{3.159}$$

For an attempted particle deletion, the modified acceptance rule is given by [48]

$$acc(n \to m)_{N-1} = min\left[1, \frac{N\Lambda^3}{V} \exp\left(-\beta\Delta U_{conf}^{(n\to m)} - \beta\mu - \Delta\eta^{(n\to m)}\right)\right]. \tag{3.160}$$

In the course of the simulation, the weighted probability distribution is measured by histograms, which means that the simulation does not sample the correct Boltzmann distribution. It is therefore necessary to convert the weighted $\tilde{\wp}$ back to the unweighted distribution $\wp$ to allow for the computation of physically significant ensemble averages of the grand canonical ensemble. Thus, in the post processing, the Boltzmann distribution is recovered from [63]

$$\wp_{\mu VT}(N) = \exp\left(\eta(N)\right) \tilde{\wp}_{\mu VT}(N). \quad (3.161)$$

Once an accurate estimate for the probability density has been determined, histogram reweighting techniques are employed to find the conditions of phase equilibria as described in the previous section. Thus, the major task in flat histogram techniques is to find an optimal weighting factor $\eta(N)$ such that the histogram becomes approximately flat.

When the probability distribution of the μVT-ensemble given by Eq. 3.77 is introduced in Eq. 3.157, the modified distribution $\tilde{\wp}_{\mu VT}$ is expressed as

$$\tilde{\wp}_{\mu VT}(N) = \exp(-\eta(N)) \frac{Q_{NVE} \exp\left(-\beta U_{conf} + N\beta\mu\right)}{Q_{\mu VT}}, \quad (3.162)$$

For the imposed conditions of the simulation in the grand canonical ensemble, its partition function $Q_{\mu VT}$ is constant. The suitable choice of $\eta(N)$ shall result in a flat histogram, which means that $\tilde{\wp}_{\mu VT}(N) = const.$ for all N (or densities). From Eq. 3.162 then follows

$$\begin{aligned} \exp\left(\eta(N)\right) &\propto Q_{NVE} \exp\left(-\beta U_{conf} + \beta\mu N\right) \\ \eta(N) &\propto \ln Q_{NVE} - \beta U_{conf} + \beta\mu N. \end{aligned} \quad (3.163)$$

The bridge equation Eq. 2.54 relates the microcanonical partition function to the entropy

$$S = k_B \ln Q_{NVE} \quad \rightarrow \quad \ln Q_{NVE} = \beta TS. \quad (3.164)$$

With this, Eq. 3.163 becomes

$$\begin{aligned} \eta(N) &\propto \beta\left(ST - U_{conf}\right) + \beta\mu N \\ &= -\beta F + \beta\mu N. \end{aligned} \quad (3.165)$$

This means that the exact determination of the weighting function would require the knowledge of the free energy function $F(N)$. In turn, once a suitable weighting function is determined, it includes information on the free energy. Several approaches are available to iteratively optimize the weighting functions in the course of the simulation so that a flat histogram is obtained. A histogram is thereby defined as sufficiently flat when the values $H(N)$ are not less than a specified percentage (usually > 80%) of the average value $\langle H(N) \rangle$. The following sections provide an introduction to the widely used flat histogram techniques, the multicanonical sampling, the Wang-Landau-method and the transition matrix method.

3.6.4.1 Multicanonical Method

In the multicanonical sampling method by Berg and Neuhaus [5], a series of simulations is performed to iteratively optimize the weighting function $\eta(N)$. From the measured histogram $H_j(N, \eta_j)$ in the jth simulation performed with the weighting factor $\eta_j(N)$, the weighting factor for the next $j+1$ simulation is obtained from [48]

$$\eta_{j+1}(N) = \ln\left(H(N, \eta_j)\right) + \eta_j + const. \tag{3.166}$$

This equation though becomes ill-defined for $H(N) = 0$, i.e., when there is no entry in the histogram for a specific number of particles. Thus, Smith and Bruce [53] suggested a slightly modified expression to avoid this problem

$$\eta_{j+1}(N) = \ln\left[H(N, \eta_j) + 1\right] + \eta_j(N) + const. \tag{3.167}$$

The iteration is repeated until the histogram is sufficiently flat.

3.6.4.2 Wang-Landau Method

The Wang-Landau method [60, 61] was originally developed for lattice systems to generate the density of state of the canonical distribution in a self-consistent way. The density of state $g(U_{conf})$ is the number of possible configurations for an energy level U_{conf}, so it corresponds to the microcanonical partition function $Q_{NVE}(U_{conf})$ and is related to the systems entropy. With $g(U_{conf})$, the canonical partition function given in Eq. 3.23 can be rewritten as

$$\begin{aligned} Q_{NVT} &= \frac{V^N}{N!\Lambda^{3N}} \int\limits_{configurations} e^{-\beta U_{conf}(s\cdot L)} ds^{3N} \\ &= \frac{V^N}{N!\Lambda^{3N}} \int\limits_{\text{levels of } U_{conf}} g(U_{conf}) e^{-\beta U_{conf}(s\cdot L)} dU_{conf}. \end{aligned} \tag{3.168}$$

Thus, once the density of state is generated accurately enough, the canonical partition function can be obtained. By applying the bridge equation of the *NVT*-ensemble, $F = -k_B T \ln Q_{NVT}$, also the free energy can be derived up to an additive constant [14]. In the Wang-Landau method, the density of state is successively optimized to achieve a flat histogram. The basic idea of Wang and Landau is that a flat histogram in energy is generated when the trial moves are accepted with a probability proportional to the reciprocal of the density of state

$$acc(n \to m) = min\left[1, \frac{g\left(U_{conf}^{(n)}\right)}{g\left(U_{conf}^{(m)}\right)}\right]. \tag{3.169}$$

At the beginning of the simulation, the unknown density of state is set uniformly to $g(U_{conf}) = 1$ [61]. After each move, the estimate of $g(U_{conf})$ is updated by a modification factor f

$$g\left(U_{conf}\right) = g\left(U_{conf}\right) \cdot f, \tag{3.170}$$

and additionally, the histogram $H(U_{conf})$ is updated. When the histogram is sufficiently flat, all histogram entries are reset to zero, and the simulation is continued in its next stage with a reduced modification factor, for instance $f_{j+1} = \sqrt{f_j}$ with $f_0 = e^1$ [61]. When again a flat histogram is obtained, the modification factor is once more reduced, the histogram reset, and the simulation continued. This procedure is repeated until the modification factor goes below a predefined value, and the density of state is converged. It should be noted that the Wang-Landau method does not exactly satisfy the condition of detailed balance as the probability distribution is continuously modified in the course of the simulation. With each stage of the simulation, the modification factor decreases, so that the probability distribution converges to its true value and the detailed balance is asymptotically satisfied.

The Wang-Landau method has been adopted by several authors (for instance [14, 45, 64]) for flat histogram techniques to simulate phase equilibria. For simulations in the μVT-ensemble, a two-dimensional density of state $g\left(U_{conf}, N\right)$ needs to be determined, and also a two-dimensional histogram $H\left(U_{conf}, N\right)$ is constructed in the simulation. Again, trial moves consist of attempted particle insertions and destructions, with modified acceptance rules given by [64]

$$acc(n \to m)_{N+1} = min\left[1, \frac{V}{(N+1)\Lambda^3} \frac{g\left(U_{conf}^{(n)}, N\right)}{g\left(U_{conf}^{(m)}, N+1\right)}\right] \tag{3.171}$$

$$acc(n \to m)_{N-1} = min\left[1, \frac{N\Lambda^3}{V} \frac{g\left(U_{conf}^{(n)}, N\right)}{g\left(U_{conf}^{(m)}, N-1\right)}\right] \tag{3.172}$$

in order to achieve a flat histogram. The simulations could also be performed in the NpT−ensemble, in which the histogram $H\left(U_{conf}, V\right)$ is then constructed, and the density of state $g\left(U_{conf}, V\right)$ is successively obtained (see for instance [14]).

3.6.4.3 Transition Matrix Method

In [9] Errington and Shen present a transition matrix Monte Carlo simulation for phase equilibria in multicomponent systems. It employs a macrostate transition probability $\Pi(I \rightarrow J)$ for the transition of the system between two different macrostates, in accordance to the microstate transition probability $\pi(n \rightarrow m)$ introduced in Sect. 3.2. The total probability $\wp(I)$ of finding the system in a macrostate I is then given by the sum over all microstates i that were assigned to this macrostate

$$\wp(I) = \sum_{i \in I} \wp(i). \tag{3.173}$$

When the detailed balance formulated for the microstate transition (see Sect. 3.2) is summed up over all possible microstate

$$\sum_{i \in I} \sum_{j \in J} \wp(i)\pi(i \rightarrow j) = \sum_{i \in I} \sum_{j \in J} \wp(j)\pi(j \rightarrow i), \tag{3.174}$$

it can be shown [48] that the detailed balance is also satisfied for the macrostate transition

$$\wp(I)\Pi(I \rightarrow J) = \wp(J)\Pi(J \rightarrow I). \tag{3.175}$$

Thus, once the macrostate transition probabilities $\Pi(I \rightarrow J)$ are determined, Eq. 3.175 yields the macrostate probabilities $\wp(I)$, which are related to the partition function of the system.

The transition probabilities are estimated from the so-called collection matrix $C(I \rightarrow J)$ that counts the number of transitions from state I to any state K.

$$\Pi(I \rightarrow J) = \frac{C(I \rightarrow J)}{\sum C(I \rightarrow K)}. \tag{3.176}$$

The collection matrix is continuously updated in the simulation by

$$C(I \rightarrow J) = C(I \rightarrow J) + acc(i \rightarrow j), \tag{3.177}$$

irrespectively whether the moves $(i \rightarrow j)$ are accepted or not [9].

In the transition matrix method to determine phase equilibria, the algorithm is employed to derive the probability $\wp(N)$ (or $\wp(\varrho)$) in the entire range of densities enclosed by ϱ_I and ϱ_{II}. Thus, the macrostate transition probabilities Π describe the transition between neighboring pairs of particle numbers, $\Pi(N \rightarrow N+1)$ and $\Pi(N+1 \rightarrow N)$, and the collection matrix C charts the statistics of particle insertion and deletion moves. In order to obtain a flat histogram, the weighting function $\eta(N)$ is set proportionally to the reciprocal macrostate probability $\wp(N)$ [9]

$$\eta(N) = -\ln(\wp(N)). \tag{3.178}$$

Inserting Eq. 3.178 into Eq. 3.175 then yields [47]

$$\eta(N+1) = \eta(N) - \ln \frac{\Pi(N \to N+1)}{\Pi(N+1 \to N)}. \tag{3.179}$$

Again, estimates for the transition probabilities Π are derived from the collection matrix $C(N, N+1)$, which are then used to periodically update the weighting factors [47]. Applications and modification of the transition matrix method for the evaluation of phase equilibria are for instance given by Errington et al. [8] or Scott et al. [46].

References

1. Allen M, Tildesley DJ (1987) Computer simulation of liquids. Oxford Science Publication
2. Apte PA, Kusaka I (2005) Direct calculation of solid-liquid coexistence points of a binary mixture by thermodynamic integration. J Chem Phys 123:194503
3. Bai P, Siepmann JI (2011) Gibbs Ensemble Monte Carlo simulations for the liquid-liquid phase equilibria of dipropylene glycol dimethyl ether and water: 1. Preliminary report. Fluid Phase Equilib 310:11–18
4. Bennett CH (1976) Efficient estimation for free energy differences from Monte Carlo data. J Comput Phys 22:245–268
5. Berg BA, Neuhaus T (1992) Multicanonical ensemble: a new approach to simulate first-order phase transitions. Phys Rev Lett 68:9–12
6. Canongia Lopes JN, Tildesley DJ (1997) Multiphase equilibria using the Gibbs ensemble Monte Carlo method. Mol Phys 92:187–195
7. dePablo JJ, Yan Q, Escobedo FA (1999) Simulation of phase transitions in fluids. Annu Rev Phys Chem 50:377–411
8. Errington JE, Shen VK (2005) Direct evaluation of multicomponent phase equilibria using flat-histogram methods. J Chem Phys 123:164103
9. Errington JR (2003) Direct calculation of liquid-vapor phase equilibria from transition matrix Monte Carlo simulation. J Chem Phys 118:9915–9925
10. Ferrenberg AM, Swenden RH (1988) New Monte Carlo technique for studying phase transition. Phys Rev Lett 61(23):2635–2638
11. Ferrenberg AM, Swenden RH (1989) Optimized Monte Carlo data analysis. Phys Rev Lett 63(12):1195–1198
12. Frenkel D, Mooij GCAM, Smit B (1992) Novel scheme to study structural and thermal properties of continuously deformable molecules. J Phys Condens Matter 4:3053–3076
13. Frenkel D, Smit B (1996) Understanding molecular simulation. Academic Press
14. Ganzenmüller G, Camp PJ (2007) Applications of Wang-Landau sampling to determine phase equilibria in complex fluids. J Chem Phys 127:154504
15. Götzlaff W (1988) Zustandsgleichung und elektrischer Transport am kritschen Punkt des fluiden Quecksilbers. PhD Thesis Marburg, Germany
16. Kofke DA (1993) Direct evaluation of phase coexistence by molecular simulation via integration along the saturation line. J Chem Phys 98(5):4149–4162
17. Kofke DA (1999) Semigrand canonical Monte Carlo simulation: integration along coexistence lines. In: Ferguson DM, Siepmann JI, Truhlar DG(eds) Advances in chemical physics, Monte Carlo methods in chemical physics, vol 105. Wiley

18. Kumar S, Bouzida D, Swendsen RH, Kollmann PA, Rosenberg JM (1992) The weighted histogram analysis method for free-energy calculations on biomolecules I, The method. J Comput Chem 13:1011–1021
19. Kumar S, Rosenberg JM, Bouzida D, Swendsen RH, Kollmann PA (1995) Multidimensional free-energy calculations using the weighted histogram analysis method. J Comput Chem 16(11):1339–1350
20. Laso M, dePablo JJ, Suter UW (1992) Simulation of phase equilibria of chain molecules. J Chem Phys 97:2817–2819
21. Lisal M, Vacek V (1997) Direct evaluation of the solid-liquid equilibria by molecular dynamics using Gibbs-Duhem-Integration. Mol Sim 19:43–61
22. Maginn E, Shah J, Rimoldi EM, Rai N, Paluch A, Rosch T, Khan S SI2-SSE: development of Cassandra, a general, efficient and parallel Monte Carlo multiscale modeling software platform for materials research
23. Martin MG (2013) MCCCS Towhee: a tool for Monte Carlo molecular simulation. Mol Simul 39:1212–1222
24. Martin MG, Frischknecht AL (2006) Using arbitrary trial distributions to improve intramolecular sampling in configurational-bias Monte Carlo. Mol Phys 104:2439–2456
25. Martin MG, Siepmann JI (1999) Novel configurational-bias Monte Carlo method for branched molecules. Transferable potentials for phase equilibria. 2. United-atom description of branched alkanes. J Phys Chem B 103:4508–45171
26. Martin MG, Thompson AP (2004) Industrial property prediction using Towhee and LAMMPS. Fluid Phase Equilib 217:105–110
27. Metha M, Kofke DA (1994) Coexistence diagrams of mixtures by molecular simulation. Chem Eng Sci 49(16):2633–2645
28. Metropolis N, Rosenbluth A, Rosenbluth MN, Teller AH, Teller E (1953) Equation of state calculations by fast computing machines. J Chem Phys 21:1087–1092
29. Mooij GCAM, Frenkel D, Smit B (1992) Direct simulation of phase equilibria of chain molecules. J Phys Condens Matter 4:L255–L259
30. Morrow TI, Maginn EJ (2005) Isomolar semigrand ensemble molecular dynamics. Development and application to liquid-liquid equilibria. J Chem Phys 122:054504
31. Nam HS, Menelev MI, Srdovitz DJ (2007) Solid-liquid phase diagrams for binary metallic alloys. Adjustable interatomic potential. Phys Rev 75:014204
32. Paliwal H, Shirts MR (2013) Multistate reweighting and configuration mapping together accelerate the efficiency of thermodynamic calculations and a function of molecular geometry by orders of magnitude. J Chem Phys 138:152108
33. Paliwal H, Shirts MR (2013) Using multistate reweighting to rapidly and efficiently explore molecular simulation parameter space for non-bonded interactions. J Chem Theory Comput 9:4700–4712
34. Palmer BJ, Lo C (1994) Molecular dynamics implementation of the Gibbs-Ensemble calculation. J Chem Phys 101(12):10899–10907
35. Panagiotopoulos AZ (1987) Direct determination of phase coexistence properties of fluids by Monte Carlo simulation in a new ensemble. Mol Phys 61(4):813–836
36. Panagiotopoulos AZ (2000) Monte Carlo methods for phase equilibria of fluids. J Phys: Condens Matter 12(3):R25
37. Panagiotopoulos AZ, Wong Y, Floriano MA (1998) Phase equilibria of lattice polymers from histrogam reweighting Monte Carlo simulations. Macromolecules 31:912–918
38. Pangali C, Rao M, Berne BJ (1978) On a novel Monte Carlo scheme for simulating water and aqueous solutions. Chem Phys Lett 55:413–417
39. Polson JM, Frenkel D (1999) Numerical prediction of the melting curve of n-octane. J Chem Phys 111:1501–1510
40. Potoff JJ, Panagiotopoulos AZ (1998) Critical point and phase behavior of the pure fluids and a Lennard-Jones mixture. J Chem Phys 109:10914–10920
41. Raabe G, Sadus RJ (2003) Monte Carlo simulation of the vapor-liquid coexistence of mercury. J Chem Phys 119(13):6691–6697

42. Rao M, Berne BJ (1979) On the force bias Monte Carlo simulation of simple liquids. J Chem Phys 71:129–132
43. Rosenbluth MN, Rosenbluth AW (1950) Monte Carlo simulations of the average extension of molecular chains. J Chem Phys 23:356–359
44. Shah JK, Maginn EJ (2011) A general and efficient Monte Carlo method for sampling intramolecular degrees of freedom of branched and cyclic molecules. J Chem Phys 135:134121
45. Shell MS, Debenedetti PG, Panagiotopoulos AZ (2002) Generalization of the Wang-Landau method for off-lattice simulations. Phys Rev E 90:056703
46. Shell MS, Debenedetti PG, Panagiotopoulos AZ (2003) An improved Monte Carlo method for direct calculation of the density of states. J Chem Phys 119(18):9406–9411
47. Shell MS, Panagiotopoulos AZ (2007) Methods for examining phase equilibria. In: Chipot C, Pohorille A (eds) Free energy calculations. Springer, pp 353–388
48. Shell MS, Panagiotopoulos AZ, Pohorille A (2007) Methods based on probability distributions and histograms. In: Chipot C, Pohorille A (eds) Free energy calculations. Springer, pp 77–188
49. Shirts MR, Chodera JD A Phyton implementation of the Multistate Bennett Acceptance Ratio (MBAR)
50. Shirts MR, Chodera JD (2008) Statistically optimal analysis of samples from multiple equilibrium states. J Chem Phys 129:129105
51. Siepmann JI, Frenkel D (1992) Configurational bias Monte Carlo: a new sampling scheme for flexible chains. Mol Phys 75:59–70
52. Siepmann JI, McDonald IR (1992) Monte Carlo simulations of mixed monolayer. Mol Phys 75:255–259
53. Smith GR, Bruce AD (1995) A study of the multicanonical Monte Carlo method. J Phys A 28:6623–6643
54. Souaille M, Roux B (2001) Extension to the weighted histogram analysis method: combining umbrella sampling with free energy calculations. Comput Phys Commun 135:40–57
55. Sugawara S, Sato T, Minamiyama T (1962) Thermodynamic and electrical properties of fluid Hg up to the liquid-vapor critical point. Bull Jap Soc Mech 5:711
56. Tan Z (2004) On a likelihood approach for Monte Carlo integration. J Am Stat Assoc 99:1027–1036
57. Van't Hof A, De Leeuw SW, Hall CK, Peters CJ (2004) Molecular simulation of binary vapour-liquid equilibria with components differing largely in volatility. Mol Phys 102:301–317
58. Vargaftik NB, Vinogradov YK, Yargin VS (1996) Handbook of physical properties of liquid and gases, 3rd edn. Begell House, New York
59. Vlug TJH, Martin MG, Smit B, Siepmann JI, Krishna R (1998) Improving the efficiency of the configurational-bias Monte Carlo algorithm. Mol Phys 94:727–733
60. Wang F, Landau DP (2001) Determining the density of states for classical statistical models: a ramdom walk algorithm to produce a flat histogram. Phys Rev E 64:056101
61. Wang F, Landau DP (2001) Efficient, multiple-range random walk algorithm to calculate the density of states. Phys Rev Lett 86(10):2050–2053
62. Wick CD, Siepmann JI (2000) Self-Adapting fixed-end-point configurational-bias Monte Carlo method for the regrowth of interior segments of chain molecules with strong intramolecular interactions. Macromolecules 33:7207–7218
63. Wilding NB (2001) Computer simulation of fluid phase transitions. Am J Phys 69(11):1147–1155
64. Yan Q, Faller R, dePablo JJ (2002) Density-of-state Monte Carlo method for simulation of fluids. J Chem Phys 116(20):8745–8749

Chapter 4
Molecular Dynamics Simulations

This chapter provides an introduction to the basics of molecular dynamics simulations. Using the simple example of the translational motion of a spherical particle, the fundamental equations that describe the dynamics of the system, and various finite-difference methods for the numerical integration of the equations of motion, will be discussed. Then different approaches to allow for simulations at constant temperatures and pressures, i.e. so-called thermostats and barostats are presented. Molecules, however, do not only perform translational motions, but may also rotate. Furthermore, forces acting on the atoms of a molecule do not only result from intermolecular interactions but also from bonding forces within the molecules. Thus, in Sect. 4.5, we discuss how molecules are 'handled' in MD simulations. A more comprehensive introduction to MD simulations is provided in the textbooks by Haile [14] and Rapaport [25], or the standard textbooks on molecular simulation techniques by Allen and Tildesley [1], or Frenkel and Smit [10].

4.1 Fundamentals

Molecular dynamics (MD) simulations are aimed to generate the system's trajectory in phase space, i.e. the time evaluation of the positions and momenta of the particles. Thereby, the thermophysical properties of the system are determined at regular time intervals, so that their time averages can be estimated according to Eq. 2.3. As Newtons's laws naturally conserve the total energy E of the system, standard equilibrium MD simulations are performed in the *NVE*-ensemble, i.e. in an isolated system with a constant number of particles N in a fixed volume V. The total energy E of the system is given by the sum of the kinetic energies of the particles and the total potential energy. The trajectory of the system is obtained by integrating the *Newton's equation of motion* for all particles. This equation, also known as Newton's second law, states

G. Raabe, *Molecular Simulation Studies on Thermophysical Properties*,
Molecular Modeling and Simulation, DOI 10.1007/978-981-10-3545-6_4

that the force $\mathbf{f}_i$ acting on a spherical particle i results in its acceleration $\mathbf{a}_i$, which is proportional to the force

$$\mathbf{f}_i(t) = m_i \cdot \mathbf{a}_i(t), \quad \text{with} \quad \mathbf{a}_i(t) = \ddot{\mathbf{r}}_i(t). \tag{4.1}$$

The acceleration is the second time derivative of the particle's position $\mathbf{r}_i$, and m_i is the mass of the particle i. The bold letters indicate vectors, the double dots the second derivative. The force acting on the particle is caused by its interaction with the $N-1$ other particles, and is derived from the gradient of the potential (configuration) energy of the system

$$\mathbf{f}_i(t) = -\frac{\partial U_{conf}\left(\mathbf{r}^N(t)\right)}{\partial \mathbf{r}_i}. \tag{4.2}$$

The force $\mathbf{f}_i(t)$ equates to the time derivative of the particle's momentum given by $\mathbf{p}_i$

$$\mathbf{f}_i(t) = \dot{\mathbf{p}}_i(t) = m_i\dot{\mathbf{v}}_i \quad \Rightarrow \quad \mathbf{f}_i(t) = 0 \rightarrow \mathbf{v}_i(t) = const. \tag{4.3}$$

Thus, when there is no (net) force experienced by a particle, it continues to move with its velocity $\mathbf{v}_i(t)$ according to Newton's first law, also known as law of inertia.

Newton's third law states that the force exerted by particle j on particle i must be equal to the force exerted by particle i on particle j ("Actio aequat reactionem")

$$\mathbf{f}_{ij} = -\mathbf{f}_{ji} \tag{4.4}$$

The force $\mathbf{f}_{ij}$ is aligned to the distance vector $\mathbf{r}_{ij}$ that joins the centers of the particles i and j, as illustrated by Fig. 4.1.

The Newtonian dynamics described above require the solution of $3N$ second order differential equations to determine the position vectors $\mathbf{r}_i$ of the N particles. An alternative description of the system's dynamic is given by the so called *Hamiltonian Dynamics*. The Hamilton $\mathcal{H}$ is the total energy of the isolated system as sum of the kinetic and potential energy. It is therefore given as function of the momenta and positions of the particles by [14]

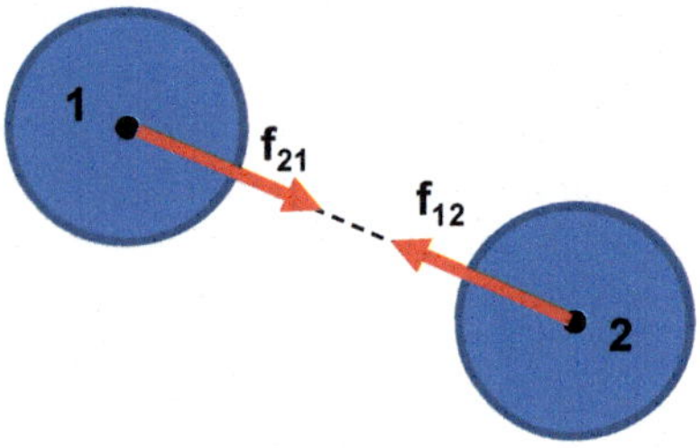

Fig. 4.1 Illustration of the Newton's third law "Actio aequat reactionem"

$$\mathcal{H}\left(\mathbf{r}^N, \mathbf{p}^N\right) = \sum_{i=1}^{N} \frac{\mathbf{p}_i^2}{2m_i} + U_{conf}\left(\mathbf{r}^N\right). \tag{4.5}$$

As the total energy of the isolated system is constant, the Hamilton has no time dependence, and the total time derivative of $\mathcal{H} = f(\mathbf{p}^N, \mathbf{r}^N)$ is given by

$$\frac{d\mathcal{H}}{dt} = \sum_{i=1}^{N} \frac{\partial \mathcal{H}}{\partial \mathbf{p}_i}\dot{\mathbf{p}}_i + \sum_{i=1}^{N} \frac{\partial \mathcal{H}}{\partial \mathbf{r}_i}\dot{\mathbf{r}}_i = 0. \tag{4.6}$$

The time derivative of Eq. 4.5 on the other hand yields

$$\frac{d\mathcal{H}}{dt} = \sum_{i=1}^{N} \frac{\mathbf{p}_i \cdot \dot{\mathbf{p}}_i}{m_i} + \sum_{i=1}^{N} \frac{dU_{conf}}{d\mathbf{r}_i}\dot{\mathbf{r}}_i = 0 \tag{4.7}$$

The comparison of the Eqs. 4.6 and 4.7 then provides the Hamilton's equations of motion

$$\frac{\partial \mathcal{H}}{\partial \mathbf{p}_i} = \frac{\mathbf{p}_i}{m_i} = \dot{\mathbf{r}}_i \tag{4.8}$$

$$\frac{\partial \mathcal{H}}{\partial \mathbf{r}_i} = \frac{dU_{conf}}{d\mathbf{r}_i} = -\mathbf{f}_i = -\dot{\mathbf{p}}_i, \tag{4.9}$$

which represent $6N$ first order differential equations to determine the position vectors $\mathbf{r}_i$ and momenta $\mathbf{p}_i$ of the N particles.

A more general and fundamental description of the system dynamics is given by the *Lagrangian mechanics*, as it allows for easier transformation it into different coordinate systems. The so-called Lagrangian $\mathcal{L}$ is a function of the generalized coordinates $\mathbf{q}_i$, their derivatives $\dot{\mathbf{q}}_i$, and time, i.e. $\mathcal{L}(\mathbf{q}_i, \dot{\mathbf{q}}_i, t)$. The time integral of the Lagrangian $\mathcal{L}(\mathbf{q}_i, \dot{\mathbf{q}}_i, t)$ from a starting point t_a to an end-point t_e defines the 'action' of the mechanical system. According to Hamilton's variational principle, the system will follow a path (trajectory) between t_a and t_e that minimizes the action. Thus, it can be shown that the Lagrangian $\mathcal{L}(\mathbf{q}_i, \dot{\mathbf{q}}_i, t)$ satisfies the equation

$$\frac{d}{dt}\left(\frac{\partial \mathcal{L}}{\partial \dot{\mathbf{q}}_i}\right) - \frac{\partial \mathcal{L}}{\partial \mathbf{q}_i} = 0, \tag{4.10}$$

which represent the Lagrangian equation of motion (see [25]). For a classical system that obeys the Newton's law, the Lagrangian is given by

$$\mathcal{L} = E_{kin} - U_{conf} = \sum_{i=1}^{N} \frac{m_i}{2}\dot{\mathbf{q}}_i^2 - U_{conf}\left(\mathbf{q}_i\right). \tag{4.11}$$

This yields

$$\frac{\partial \mathcal{L}}{\partial \mathbf{q}_i} = -\frac{\partial U_{conf}}{\partial \mathbf{q}_i} = \mathbf{f}_i \tag{4.12}$$

$$\frac{\partial \mathcal{L}}{\partial \dot{\mathbf{q}}_i} = m_i \dot{\mathbf{q}}_i \Rightarrow \frac{d}{dt}\left(\frac{\partial \mathcal{L}}{\partial \dot{\mathbf{q}}_i}\right) = m_i \ddot{\mathbf{q}}_i = \mathbf{f}_i, \tag{4.13}$$

which illustrates that Eq. 4.10 is satisfied.

The advantage of Lagrangian mechanics is that it allows to include restrictions (constraints) into the equation of motion. With $\mathbf{g}_i$ being the constraint force acting on the particle i, its Lagrangian equation of motion becomes

$$\frac{d}{dt}\left(\frac{\partial \mathcal{L}}{\partial \dot{\mathbf{q}}_i}\right) - \frac{\partial \mathcal{L}}{\partial \mathbf{q}_i} = \mathbf{g}_i. \tag{4.14}$$

When the Lagrangian of the system is given, the Hamiltonian can be derived from

$$\mathcal{H} = \sum \dot{\mathbf{q}}_i \mathbf{p}_{q,i} - \mathcal{L}, \tag{4.15}$$

wherein $\mathbf{p}_{q,i}$ is the momentum conjugated with the generalized coordinate $\mathbf{q}_i$.

4.2 Finite-Difference Methods

In simulation studies with 'realistic' molecular models as described in Chap. 6, the potential energy and with this the force between two particles varies continuously with their distance. However, as each particle interacts simultaneously with all others in the system, the motions of all particles are coupled. Thus, the analytical integration of the $3N$ second order differential equations of this many-body problem is impossible. The standard approach for the numerical integration of the Newton's equations of motion is then the use of finite-difference methods, in which the integration over the time t is broken into a series of short time steps δt. At each time step, the forces $\mathbf{f}_i(t)$ experienced by the particles in their current positions $\mathbf{r}_i(t)$ are computed. It is then assumed that the $\mathbf{f}_i(t)$ remain constant during the small time step δt so that they can be combined with known dynamic information (position, velocities etc.) at t (or preceding timesteps) to predict new positions etc. at the next time step $t + \delta t$. For this purpose, the finite-difference methods generally make use of truncated Taylor expansions for the position $\mathbf{r}(t)$ and its derivatives, i.e. the velocity $\mathbf{v}(t) = \dot{\mathbf{r}}(t)$ as first derivative, the acceleration $\mathbf{a}(t) = \ddot{\mathbf{r}}(t)$ as second derivative, $\mathbf{b}(t) = \dddot{\mathbf{r}}(t)$ terming the third derivative etc.

$$\mathbf{r}_i(t+\delta t) = \mathbf{r}_i(t) + \mathbf{v}_i(t)\delta t + \frac{1}{2}\mathbf{a}_i(t)\delta t^2 + \frac{1}{6}\mathbf{b}_i(t)\delta t^3 + \mathcal{O}(\delta t^4) \quad (4.16)$$

$$\mathbf{v}_i(t+\delta t) = \mathbf{v}_i(t) + \mathbf{a}_i(t)\delta t + \frac{1}{2}\mathbf{b}_i(t)\delta t^2 + \mathcal{O}(\delta t^3) \quad (4.17)$$

$$\mathbf{a}_i(t+\delta t) = \mathbf{a}_i(t) + \mathbf{b}_i(t)\delta t + \mathcal{O}(\delta t^2) \quad (4.18)$$

$$\mathbf{b}_i(t+\delta t) = \mathbf{b}_i(t) + \mathcal{O}(\delta t). \quad (4.19)$$

Therein, $\mathcal{O}(\delta t^N)$ defines the order of the truncation error of the Taylor expansion, and it is measured by the first term that has been omitted. Many different finite-difference methods are available for MD studies, and there are several aspects that need to be considered for choosing a suitable one. These are

- *Computational expense*: The evaluation of the forces is in general the most time-consuming part of a MD simulation. Thus, the algorithm should not require the calculation of the forces more than once per timestep.
- *Accuracy*: The finite-difference methods should reproduce the systems trajectory as accurately as possible. The accuracy of the algorithm is related to the truncation error of the Taylor expansion, but also depends on the time step δt used.
- *Stability*: The algorithmic stability is related to the time step δt. Thus, a desirable quantity of a finite-difference method is that it enables the use of preferably long timesteps.
- *Energy conservation*: As described before, standard equilibrium MD simulations are performed in the *NVE*-ensemble, and the algorithm should satisfy the conservation law for energy. When there is no external force acting on the system, its total momentum also has to be conserved according to the law of inertia (see Eq. 4.3).

Other issues are time reversibility, simplicity, memory requirement etc. The following sections provide an introduction to the most widely used finite-difference methods in MD simulations.

4.2.1 Verlet

The Verlet algorithm is based on a Taylor series expansion for the position $r(t)$ with $\mathcal{O}(\delta t^4)$. It thereby combines the expansion for forwarding the current position $r(t)$ to $(t + \delta t)$ with the expansion for the backward step to $(t - \delta t)$

$$\mathbf{r}_i(t+\delta t) = \mathbf{r}_i(t) + \mathbf{v}_i(t)\delta t + \frac{1}{2}\mathbf{a}_i(t)\delta t^2 + \frac{1}{6}\mathbf{b}_i(t)\delta t^3 + \mathcal{O}(\delta t^4) \quad (4.20)$$

$$\mathbf{r}_i(t-\delta t) = \mathbf{r}_i(t) - \mathbf{v}_i(t)\delta t + \frac{1}{2}\mathbf{a}_i(t)\delta t^2 - \frac{1}{6}\mathbf{b}_i(t)\delta t^3 + \mathcal{O}(\delta t^4). \quad (4.21)$$

The addition of the two expansions yields

$$\mathbf{r}_i(t+\delta t) = 2\mathbf{r}_i(t) - \mathbf{r}_i(t-\delta t) + \mathbf{a}_i(t)\delta t^2 + \mathcal{O}(\delta t^4), \quad (4.22)$$

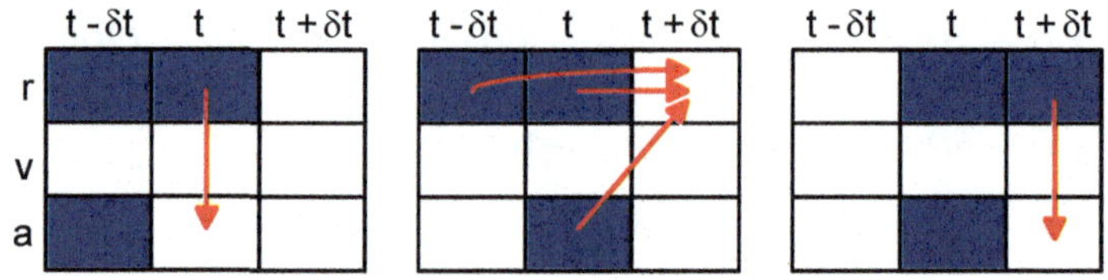

Fig. 4.2 Illustration of the Verlet method (after Allen and Tildesley [1]): accelerations $\mathbf{a}_i(t)$ are derived in the current positions $\mathbf{r}_i(t)$; the current positions $\mathbf{r}_i(t)$ and accelerations $\mathbf{a}_i(t)$, as well as the previous positions $\mathbf{r}_i(t-\delta t)$ are used to determine the new positions $\mathbf{r}_i(t+\delta t)$

which represents a quite simple estimate of the new positions, which is, however, accurate to order $\mathcal{O}(\delta t^4)$. The accelerations $\mathbf{a}_i(t)$ are obtained from the forces experienced by the particles in their current positions $\mathbf{r}_i(t)$ according to Newton's second law. Then the current acceleration $\mathbf{a}_i(t)$, the current position $\mathbf{r}_i(t)$ and the previous position $\mathbf{r}_i(t-\delta t)$ are used to estimate the new position $\mathbf{r}_i(t+\delta t)$, as illustrated in Fig. 4.2.

The use of information from the current and previous time step makes the Verlet algorithm a two-step method, which also means that it is not a self-starting algorithm, as it requires an estimated position $\mathbf{r}_i(-\delta t)$ at $t=0$. The main shortcoming of the Verlet method is that it does not include any velocity equation. As the velocities of the particles are required to determine the kinetic energy of the system (see Eq. 2.123), they can be estimated from [1]

$$\mathbf{v}_i(t) = \frac{\mathbf{r}_i(t+\delta t) - \mathbf{r}_i(t-\delta t)}{2\delta t}. \tag{4.23}$$

This, however, means that the kinetic energy of the system $E_{kin}(t)$ can only be calculated after the positions are forwarded to $\mathbf{r}_i(t+\delta t)$, i.e. delayed by δt to the determination of the configuration energy $U_{conf}\left(\mathbf{r}^N(t)\right)$. A numerical problem of the Verlet method may arise from the fact that the new positions are derived by first calculating the difference between the two large quantities $\mathbf{r}_i(t)$ and $\mathbf{r}_i(t-\Delta t)$ and then adding the smaller quantity $\mathbf{a}_i(t)$. Thus, modifications of the Verlet method have been proposed to overcome these shortcomings, i.e. the Leap Frog or the Velocity Verlet that will be introduced in the next sections.

4.2.2 *Leap Frog*

A simple algorithm with explicit velocity function is provided in the Leap Frog method [16]. It only employs Taylor expansions accurately to $\mathcal{O}(\delta t^2)$ for both the positions and velocities. However, it determines the velocities at a mid-step to increase the accuracy and stability of the algorithm

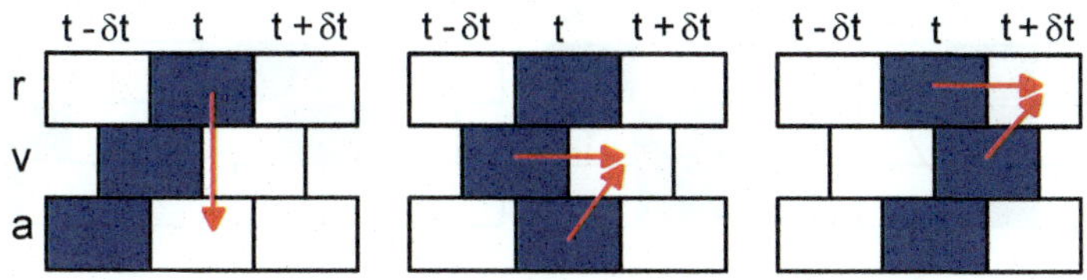

Fig. 4.3 Illustration of the Leap Frog algorithm (after [1]): accelerations $\mathbf{a}_i(t)$ are derived in the current positions $\mathbf{r}_i(t)$; those then yield together with the $\mathbf{v}_i\left(t - \frac{1}{2}\delta t\right)$ the new velocities $\mathbf{v}_i\left(t + \frac{1}{2}\delta t\right)$. The new velocities $\mathbf{v}_i\left(t + \frac{1}{2}\delta t\right)$ and the current positions $\mathbf{r}_i(t)$ are used to determine the new positions $\mathbf{r}_i(t + \delta t)$

$$\mathbf{r}_i(t + \delta t) = \mathbf{r}_i(t) + \mathbf{v}_i\left(t + \frac{1}{2}\delta t\right) + \mathcal{O}(\delta t^2) \tag{4.24}$$

$$\mathbf{v}_i\left(t + \frac{1}{2}\delta t\right) = \mathbf{v}_i\left(t - \frac{1}{2}\delta t\right) + \mathbf{a}_i(t)\delta t + \mathcal{O}(\delta t^2). \tag{4.25}$$

The current positions $\mathbf{r}_i(t)$ are used to calculate the forces acting on the particles to yield their accelerations $\mathbf{a}_i(t)$. These accelerations are then used together with the velocities $\mathbf{v}_i\left(t - \frac{1}{2}\delta t\right)$ at the previous mid-step to derive estimates for the velocities $\mathbf{v}_i\left(t + \frac{1}{2}\delta t\right)$ at the next mid-step. The velocities thereby 'leap' over the positions $\mathbf{r}_i(t)$, which gives the algorithm its name. Then the positions $\mathbf{r}_i(t + \Delta t)$ are obtained from the previous positions $\mathbf{r}_i(t)$ and the velocities $\mathbf{v}_i\left(t + \frac{1}{2}\delta t\right)$, i.e. the positions now leap over the velocities. This scheme is illustrated in Fig. 4.3.

Though the algorithm involves a velocity function, the velocity calculation has an offset of $\frac{1}{2}\delta t$ to the determination of the positions, which also means that the kinetic energy and configurational energy of the system cannot be derived at the same time. To obtain the velocity at the time step at which the positions are known, $\mathbf{v}(t)$ can be estimated from

$$\mathbf{v}_i(t) = \frac{\mathbf{v}_i\left(t + \frac{1}{2}\delta t\right) - \mathbf{v}_i(t - \frac{1}{2}\Delta t)}{2}. \tag{4.26}$$

4.2.3 Velocity Verlet

The Velocity Verlet algorithm by Swope et al. [28] is probably the most widely used finite-difference method today, as it offers simplicity, numerical stability and an adequate handling of the velocities [1]. The positions are derived from a Taylor expansion accurate to $\mathcal{O}(\delta t^3)$, whereas the velocity function only involves a Taylor expansion with $\mathcal{O}(\delta t^2)$. Though the velocities are forwarded in two stages that again involve a mid-step $\left(t + \frac{1}{2}\delta t\right)$

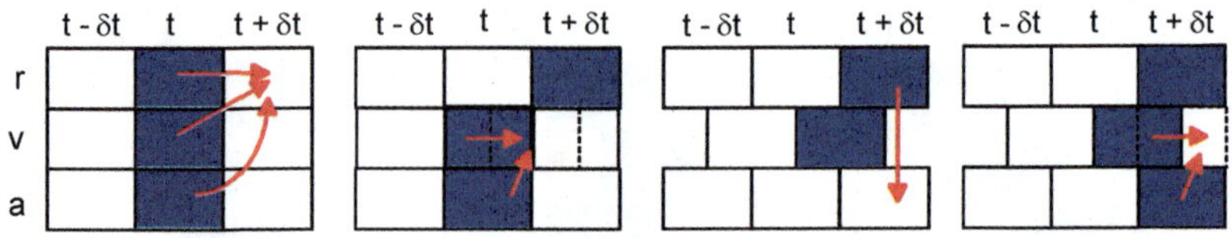

Fig. 4.4 Illustration of the Velocity Verlet algorithm (after [1]): current positions $\mathbf{r}_i(t)$, velocities $\mathbf{v}_i(t)$ and accelerations $\mathbf{a}_i(t)$ are used to estimate the new positions $\mathbf{r}_i(t+\delta t)$ the $\mathbf{v}_i(t)$ and $\mathbf{a}_i(t)$ values are then also used to forward the velocities to $\mathbf{v}\left(t+\frac{1}{2}\delta t\right)$. Accelerations $\mathbf{a}_i(t+\delta t)$ are derived in the new positions $\mathbf{r}_i(t+\delta t)$, which then together with the $\mathbf{v}\left(t+\frac{1}{2}\delta t\right)$ give the new velocities $\mathbf{v}\,(t+\delta t)$

$$\mathbf{r}_i(t+\delta t) = \mathbf{r}_i(t) + \mathbf{v}_i(t)\delta t + \frac{1}{2}\mathbf{a}_i(t)\delta t^2 + \mathcal{O}(\delta t^3) \tag{4.27}$$

$$\mathbf{v}_i\left(t+\frac{1}{2}\delta t\right) = \mathbf{v}_i(t) + \frac{1}{2}\mathbf{a}_i(t)\delta t + \mathcal{O}(\delta t^2) \tag{4.28}$$

$$\mathbf{v}_i\,(t+\delta t) = \mathbf{v}_i\left(t+\frac{1}{2}\delta t\right) + \frac{1}{2}\mathbf{a}_i(t+\delta t)\delta t + \mathcal{O}(\delta t^2) \tag{4.29}$$

First the positions $\mathbf{r}_i(t+\delta t)$ are estimated from the current positions $\mathbf{r}_i(t)$, velocities $\mathbf{v}_i(t)$ and accelerations $\mathbf{a}_i(t)$. The current values $\mathbf{v}_i(t)$ and $\mathbf{a}_i(t)$ then also yield the velocities at the mid-step $\mathbf{v}\left(t+\frac{1}{2}\delta t\right)$, as illustrated in Fig. 4.4. For the new positions $\mathbf{r}_i(t+\delta t)$, the forces, i.e. accelerations $\mathbf{a}_i(t+\delta t)$ are evaluated, which are then used together with the velocity at the mid-step $\mathbf{v}\left(t+\frac{1}{2}\delta t\right)$ to calculate the velocity $\mathbf{v}_i(t+\delta t)$. Then, both positions and velocities are available at $(t+\delta t)$, so that also the configurational energy and kinetic energy can be derived at the same timestep.

4.2.4 Gear Predictor-Corrector

The predictor-corrector method by Gear [11, 12] represents a general family of numerical integration algorithms for ordinary differential equations of various orders. Applied to the integration of the Newton's equation of motion, the three basics steps of the algorithm are given by

1. Predict the positions $\mathbf{r}_i$, and its derivatives $\mathbf{v}_i$, $\mathbf{a}_i$, $\mathbf{b}_i$ etc. at the time step $(t+\delta t)$ from the Taylor expansion given in Eq. 4.16
2. Evaluate the forces $\mathbf{f}_i$ for the predicted positions $\mathbf{r}_i^p(t+\delta t)$ to derive the accelerations $\mathbf{a}_i(t+\delta t)$
3. Correct the predictions from step 1 according to the deviation Δa between the predicted accelerations $\mathbf{a}_i^p(t+\delta t)$ from step 1 and the evaluated $\mathbf{a}_i(t+\delta t)$ from step 2

When we rewrite the terms of the Taylor expansion by $\mathbf{r}_{0,i} = \mathbf{r}_i$, $\mathbf{r}_{1,i} = \mathbf{v}_i(t)\delta t$, $\mathbf{r}_{2,i} = \frac{1}{2}\mathbf{a}_i(t)\delta t^2$ and $\mathbf{r}_{3,i} = \frac{1}{6}\mathbf{b}_i(t)\delta t^3$, Eq. 4.16 can be expressed in matrix form by [1]

$$\begin{pmatrix} \mathbf{r}_i(t+\delta t) \\ \mathbf{r}_{1,i}(t+\delta t) \\ \mathbf{r}_{2,i}(t+\delta t) \\ \mathbf{r}_{3,i}(t+\delta t) \end{pmatrix}^p = \begin{pmatrix} 1 & 1 & 1 & 1 \\ 0 & 1 & 2 & 3 \\ 0 & 0 & 1 & 3 \\ 0 & 0 & 0 & 1 \end{pmatrix} \cdot \begin{pmatrix} \mathbf{r}_i(t) \\ \mathbf{r}_{1,i}(t) \\ \mathbf{r}_{2,i}(t) \\ \mathbf{r}_{3,i}(t) \end{pmatrix}. \tag{4.30}$$

The inclusion of the third derivative $\mathbf{r}_{3,i}$ defines the algorithm as 4th-order predictor. Then the forces acting on the particles in the predicted positions $\mathbf{r}^p_{0,i}(t+\delta t)$ are calculated to derive their accelerations $\mathbf{r}_{2,i}(t+\delta t)$. The deviations between the calculated and predicted accelerations yield an estimate of the error of the predictor step

$$\Delta a_i = \mathbf{r}_{2,i}(t+\delta t) - \mathbf{r}^p_{2,i}(t+\delta t), \tag{4.31}$$

which is then used to derive corrected positions, velocities etc. according to

$$\begin{pmatrix} \mathbf{r}_i(t+\delta t) \\ \mathbf{r}_{1,i}(t+\delta t) \\ \mathbf{r}_{2,i}(t+\delta t) \\ \mathbf{r}_{3,i}(t+\delta t) \end{pmatrix}^c = \begin{pmatrix} \mathbf{r}_i(t+\delta t) \\ \mathbf{r}_{1,i}(t+\delta t) \\ \mathbf{r}_{2,i}(t+\delta t) \\ \mathbf{r}_{3,i}(t+\delta t) \end{pmatrix}^p + \begin{pmatrix} C_0 \\ C_1 \\ C_2 \\ C_3 \end{pmatrix} \cdot \Delta a_i. \tag{4.32}$$

Gear suggested best values for the corrector coefficients C_i that result in optimal accuracy and numerical stability of the numerical integration. In general, the optimal coefficients depend on the order of the differential equation and the order of the Taylor expansion, i.e. of the predictor. For the second order differential equations of the Newton's dynamics and the 4th-order predictor scheme discussed above, the optimal values are $C_0 = \frac{19}{120}$, $C_1 = \frac{3}{4}$, $C_2 = 1$, $C_3 = \frac{1}{2}$ [1, 14].

Theoretically, the predictor-corrector scheme should be iterated for each time step until the deviation Δa goes below a predefined value. However, this would require the recalculation of the forces each time the prediction of the positions is refined. Due to the computational expense of the force evaluation, there is usually only one corrector step applied per timestep [1].

4.3 Thermostats—Simulations in *NVT*-Ensemble

As described in the introduction of this chapter, solving the Newton's equation of motion results in simulations in the *NVE*-ensemble. However, this does not correspond to the conditions at which experimental studies are generally conducted. To derive thermophysical properties from MD simulations that can be compared with experimental data, it is therefore necessary to perform simulations at constant temperature and/or constant pressure, i.e. experimentally relevant conditions. In this section, approaches to perform simulations at constant temperature, so-called 'thermostats', are discussed. An overview and discussion on different thermostats in MD simulations is, for instance, given by Hünenberger [15], whereas Basconi and Shirt

[4] provide additional studies on the damping effect of different thermostats on the systems dynamics and their impact on the determination of transport properties.

In Sect. 2.4 we have introduced the Maxwell-Boltzmann relation, which provides the probability distribution of the velocities of particles within a system as a function of the temperature. Equation 2.123 directly relates the kinetic energy of a system to its temperature. Solving the equation for the instantaneous temperature $\mathcal{T}$ gives

$$\mathcal{T}(t) = \frac{\sum_{i=1}^{N} \frac{m_i}{2} \mathbf{v}_i^2(t)}{N_f k_B}, \tag{4.33}$$

wherein N_f represents the degrees of freedom of the system. This comprises the translational motion of the N particles in the three directions. Thus, for a system with zero total momentum, i.e. $\mathbf{p}_{total} = 0$, this results in $N_f = 3N - 3$. Thus, the obvious way to control the temperature is to scale the velocities by a factor λ in such a way that the resulting kinetic energy of the system corresponds to the imposed temperature T, i.e.

$$T = \frac{\sum_{i=1}^{N} \frac{m_i}{2} \left(\lambda(t)\mathbf{v}_i(t)\right)^2}{N_f k_B}. \tag{4.34}$$

Hence, the required scaling factor is given by

$$\lambda(t) = \sqrt{\frac{T}{\mathcal{T}(t)}}. \tag{4.35}$$

Although the velocity scaling is the simplest way to adjust the system's temperature, it does not generate a canonical distribution. Thus, it should at the most be used in the equilibrium stage of a simulation.

The *Berendsen thermostat* [5] does not involve a 'full' velocity scaling at each time step, but employs a time scale τ_T, so that the scaling factor λ for updating the velocity is given by

$$\lambda^2 = 1 + \frac{\delta t}{\tau_T}\left(\frac{T}{\mathcal{T}(t)} - 1\right). \tag{4.36}$$

However, the Berendsen thermostat also does not generate a canonical distribution of configurations [15].

In order to realize MD simulations in a real *NVT*-ensemble, one has to recall how the canonical ensemble was introduced in Sect. 2.1.2: as a large number of microstates that are all placed in a large isothermal heat bath at the imposed temperature T. Thus, more sophisticated approaches to control the temperature involve interactions of the system's particles with a heat bath: either in a stochastic way such as in the Andersen thermostat, or in a deterministic and dynamic approach as in the case of the Nosé-Hoover or the Gauss thermostat that will be discussed in the following sections.

4.3.1 Andersen Thermostat

In the Andersen thermostat [2], the coupling of the system with a heat reservoir is realized by stochastic collisions of its particles with imaginary heat bath particles. In practice, this means that at certain intervals a particle is selected randomly, and it is assigned a new velocity according to the Maxwell-Boltzmann distribution of velocities (see Sect. 2.4) for the imposed temperature. As a result of the collision with the heat bath particle, the system is advanced from one energy level to the next, at which it proceeds with its Hamiltonian motion [1] in the *NVE* ensemble until the next collision occurs. The strength of the coupling to the heat reservoir is therefore determined by the collision rate.

The stochastic energy changes allow the system to visit all accessible energy levels according to their Boltzmann weight [10], so that the Andersen thermostat generates effectively a canonical probability distribution. Though it should be noted that the system dynamics induced by the Andersen thermostat is unphysical [10]. This method should therefore not be used in MD studies aimed to determine transport coefficients, time correlation functions, etc.(see Sect. 7.2).

4.3.2 Extended Ensemble—Nosé-Hoover Thermostat

Nosé [22] proposed a deterministic approach to generate a canonical distribution by introducing an extra degree of freedom $\mathtt{s}$, which represents a thermal heat reservoir in contact with the system. This variable $\mathtt{s}$ has a conjugated momentum $p_{\mathtt{s}}$ and an effective mass Q_T. The Lagrangian of the N-particle system from Eq. 4.11 is then extended by a potential energy term $U_{\mathtt{s}}$ and a kinetic energy term $E_{kin,\mathtt{s}}$ associated with the dynamic variable $\mathtt{s}$

$$\mathcal{L} = E_{kin} + E_{kin,\mathtt{s}} - U_{conf}\left(\mathbf{r}^N\right) - U_{\mathtt{s}}. \tag{4.37}$$

Therefore, this approach represents an example of an 'extended ensemble' method of thermostatting, as the heat bath becomes an integral part of the system. The parameter $\mathtt{s}$ scales the momentum of the particles to adjust the temperature of the system to the required value T. The scaling of the momentum can be interpreted as rescaling of the timestep δt. The scaled momentum $\mathbf{p}_i = \mathbf{p}_i^*/\mathtt{s}$ is referred to as real momentum, whereas $\mathbf{p}^*$ is interpreted as virtual momentum [10]. The real variables

$\mathbf{r}_i$, $\mathbf{p}_i$, δt are related to the virtual variables $\mathbf{r}_i^*$, $\mathbf{p}_i^*$, δt^* of the extended Lagrangian formulation by [10]

$$\mathbf{r}_i = \mathbf{r}_i^* \tag{4.38}$$

$$\mathbf{p}_i = \frac{\mathbf{p}_i^*}{\mathrm{s}} \tag{4.39}$$

$$\delta t = \frac{\delta t^*}{\mathrm{s}}. \tag{4.40}$$

The Hamiltonian of the extended ensemble in virtual variables is given by

$$\mathcal{H}^* = \sum_{i=1}^{N} \frac{\mathbf{p}_i^{*2}}{2m_i \mathrm{s}^2} + U_{conf}\left(\mathbf{r}^{*N}\right) + \frac{p_{\mathrm{s}}^{*2}}{2Q_T} + k_B T N_f \ln \mathrm{s}. \tag{4.41}$$

From the Hamiltonian, the equations of motion of the virtual variables can be derived according to Eqs. 4.8 and 4.9

$$\frac{d\mathbf{r}_i^*}{dt^*} = \frac{\partial \mathcal{H}^*}{\partial \mathbf{p}_i^*} = \frac{\mathbf{p}_i^*}{m_i \mathrm{s}^2} \tag{4.42}$$

$$\frac{d\mathbf{p}_i^*}{dt^*} = -\frac{\partial \mathcal{H}^*}{\partial \mathbf{r}_i^*} = -\frac{\partial U_{conf}\left(\mathbf{r}^{*N}\right)}{\partial \mathbf{r}_i^*} \tag{4.43}$$

$$\frac{d\mathrm{s}}{dt^*} = \frac{\partial \mathcal{H}^*}{\partial p_{\mathrm{s}}^*} = \frac{p_{\mathrm{s}}^*}{Q} \tag{4.44}$$

$$\frac{dp_{\mathrm{s}}^*}{dt^*} = -\frac{\partial \mathcal{H}^*}{\partial \mathrm{s}} = \sum_{i=1}^{N} \frac{\mathbf{p}_i^{*2}}{m_i \mathrm{s}^3} - \frac{N_f k_B T}{\mathrm{s}}. \tag{4.45}$$

With s being a dynamic variable, the timestep δt in real time fluctuates according to Eq. 4.40. This is inconvenient and results in difficulties when calculating dynamic and transport properties, such as autocorrelation function or diffusion coefficients (see Sect. 7.2). Hoover [17] further developed the method by Nosé by simplifying the transformation from virtual variables back to real variables. Therefore, he introduced a friction parameter ζ, defined by

$$\zeta = \frac{1}{\mathrm{s}} \frac{d\mathrm{s}}{dt} = \frac{d\mathrm{s}}{dt^*} = \frac{p_{\mathrm{s}}^*}{Q_T}. \tag{4.46}$$

With this, the equations of motion of the real variables $\mathbf{r}_i$ and $\mathbf{p}_i$ in real time are given by

$$\frac{d\mathbf{r}_i}{dt} = \frac{d\mathbf{r}_i^*}{dt^*} \cdot \frac{dt^*}{dt} \cdot \frac{d\mathbf{r}_i}{d\mathbf{r}_i^*} = \frac{\mathbf{p}_i^*}{m_i \mathrm{s}^2} \cdot \mathrm{s} \cdot 1 = \frac{\mathbf{p}_i^*}{m_i \mathrm{s}} = \frac{\mathbf{p}_i}{m_i} \tag{4.47}$$

$$\frac{d\mathbf{p}_i}{dt} = \frac{d(\mathbf{p}_i^*/\mathrm{s})}{dt^*} \cdot \frac{dt^*}{dt} = \left[\frac{1}{\mathrm{s}}\frac{d\mathbf{p}_i^*}{dt^*} - \frac{\mathbf{p}_i^*}{\mathrm{s}^2}\frac{d\mathrm{s}}{dt^*}\right] \cdot \mathrm{s} \tag{4.48}$$

$$= -\frac{\partial U_{conf}\left(\mathbf{r}^{*N}\right)}{\partial \mathbf{r}_i^*} - \mathbf{p}_i \frac{d\mathrm{s}}{dt^*} \tag{4.49}$$

$$= -\frac{\partial U_{conf}\left(\mathbf{r}^{N}\right)}{\partial \mathbf{r}_i} - \zeta \mathbf{p}_i. \tag{4.50}$$

The friction parameter ζ evolves in time according to

$$\frac{d\zeta}{dt} = \frac{1}{Q_T}\frac{dp_\mathrm{s}^*}{dt} = \frac{1}{Q_T}\frac{dp_\mathrm{s}^*}{dt^*}\frac{dt_i^*}{dt} = \frac{\mathrm{s}}{Q_T}\left[\sum_{i=1}^{N} \frac{\mathbf{p}_i^{*2}}{2m_i \mathrm{s}^3} - \frac{N_f k_B T}{\mathrm{s}}\right] \tag{4.51}$$

$$= \frac{1}{Q_T}\left[\sum_{i=1}^{N} \frac{\mathbf{p}_i^2}{m_i} - N_f k_B T\right]. \tag{4.52}$$

The first term in the bracket is related to the instantaneous kinetic temperature of the system $\mathcal{T}$

$$\sum_{i=1}^{N} \frac{\mathbf{p}_i^2}{m_i} = N_f k_B \mathcal{T}, \tag{4.53}$$

so that the time evolution of the friction parameter is described by the difference between the actual kinetic temperature of the system $\mathcal{T}(t)$ and the imposed temperature T

$$\frac{d\zeta}{dt} = \frac{N_f k_B}{Q_T}[\mathcal{T}(t) - T]. \tag{4.54}$$

In Hoover's formulation, $N_f = 3N$ according to the translational degrees of freedom of the N particles of the system. The coupling of the system to the heat bath is determined by the effective mass Q_T, which is therefore an adjustable parameter in the simulation. A large value of Q_T thereby corresponds to a slow energy exchange between the system and the heat reservoir.

The Nosé-Hoover thermostat can be implemented in standard Finite-Difference methods with explicit velocity functions. Thereby, the function to evolve the friction parameter ζ is formulated according to the velocity function of the algorithm. The implementation of the Nosé-Hoover thermostat into the Velocity-Verlet or Leap-Frog algorithm is, for instance, described in the manual of the *DL_POLY_2* (classical) MD simulation package [27].

4.3.3 Constraint Methods—Gauss Thermostat

The Gauss thermostat [7, 19] is also referred to as Gauss Constraint [27] as it represents a constraint method, meaning that the system is forced to satisfy an imposed constraint by solving the equation of motion. In this case, a constraint friction force is introduced into the Newton's equation of motion to generate constant kinetic temperature dynamics [1]. The resulting equation of motion is given by

$$\underbrace{m_i \cdot \mathbf{a}_i(t) = \mathbf{f}_i(t)}_{\text{Newtonian}} \; \underbrace{-\zeta m_i \cdot \mathbf{v}_i(t)}_{\text{friction force}} . \tag{4.55}$$

In terms of the momentum, Eq. 4.55 is expressed as

$$\dot{\mathbf{p}}_i(t) = \mathbf{f}_i(t) - \zeta \mathbf{p}_i(t). \tag{4.56}$$

The unknown constraint force is introduced by a Lagrange multiplier ζ that can be interpreted as friction coefficient. It is determined to constrain the system's temperature to a constant value, i.e. to results in $\dot{\mathcal{T}} = 0$. From Eq. 4.33 then follows

$$\dot{\mathcal{T}} = 0 \quad \rightarrow \quad \sum_{i=1}^{N} m_i \mathbf{v}_i \mathbf{a}_i = 0. \tag{4.57}$$

When inserting the equation of motion from Eq. 4.55 into Eq. 4.57, a conditional equation for the friction coefficient can be deduced.

$$\zeta = \frac{\sum_{i=1}^{N} \mathbf{f}_i(t) \cdot \mathbf{v}_i(t)}{\sum_{i=1}^{N} m_i \mathbf{v}_i^2(t)}. \tag{4.58}$$

According to the Gauss principle of least constraint, the constraint force is determined to perturb the Newtonian trajectory as little as possible.

Again the Gauss thermostat can be implemented in all standard Finite-Difference methods with explicit velocity functions, Brown and Clarke [6] for instance proposed its implementation in the Leap Frog algorithm. The Gauss thermostat is often used with the Gear Predictor-Corrector method introduced in Sect. 4.2.4. Then, the velocities of the predictor step and the forces determined for the predicted positions are used to determine the friction parameter ζ. The friction parameter is then employed to correct the estimate of the error of the predictor step Δa_i^*, i.e. in Eq. 4.31 the first term is replaced by Eq. 4.55 to yield

$$\Delta a_i^* = \frac{\mathbf{f}_i(t+\delta t) - \zeta \mathbf{p}_i^p(t)}{m_i} - \mathbf{a}_i^p(t+\delta t). \tag{4.59}$$

4.4 Barostats—Simulations in NpT-Ensemble

Experimental studies are often performed under condition of constant pressure, for instance at atmospheric pressure. Thus, MD studies at conditions relevant for experiments require simulations not only to be performed at an imposed temperature, but often also at constant pressure. A macroscopic system reacts on an imposed pressure p by changing its volume. Accordingly, pressure control in simulation studies involves volume fluctuations of the simulation box. Therefore, the simplest approach to control the pressure is to scale the simulation volume. The counterpart to the Berendsen thermostat (see Sect. 4.3) is the Berendsen barostat, which employs a scaling factor χ for the volume given by

$$\chi(t) = 1 - \kappa_T \frac{\delta t}{\tau_P}(p - \wp), \tag{4.60}$$

τ_P is the time scale of the volume scaling, $\wp$ is the instantaneous pressure of the system, and κ_T is the isothermal compressibility. The volume of the simulation box is then scaled by χ, whereas the center-of-mass coordinates of the particles and the cell factors are scaled by $\chi^{1/3}$.

For a dynamic coupling of the system to a barostat, analogous methods were developed as for the thermostatting, i.e. extended ensemble and Gauss constraint methods. The following section first provides an introduction to the general idea of both barostats before combining the barostat with the thermostat to allow for simulations in the NpT-ensemble.

4.4.1 Extended Ensemble Barostat

The extended ensemble barostat approach is attributed to Andersen [2], who introduced an equation of motion for the volume V to allow for a dynamic volume exchange of the system with the surrounding. The coupling of the system to the external volume is realized by a piston that acts on the system to effect an isotropic (uniform) expansion or compression. The Lagrangian of the system is extended by a potential energy term U_V and a kinetic energy term $E_{kin,V}$ associated with this piston

$$U_V = pV, \quad E_{kin,V} = \frac{Q_P}{2}\dot{V}^2. \tag{4.61}$$

Therein Q_p is the mass of the piston. The coordinates $\mathbf{r}_i$ are replaced by scaled coordinates $\mathbf{s}_i$

$$\mathbf{s}_i = \frac{\mathbf{r}_i}{V^{1/3}}. \tag{4.62}$$

With this, the kinetic energy of the system can be expressed in terms of the scaled coordinates [1] by

$$E_{kin} = \sum_{i=1}^{N} \frac{\mathbf{p}_i^2}{2m_i} = \frac{1}{2}\sum_{i=1}^{N} m_i \mathbf{v}_i^2 = \frac{1}{2} V^{2/3} \sum_{i=1}^{N} m_i \dot{\mathbf{s}}_i^2 \tag{4.63}$$

The Lagrangian of the extended system is then given by [2]

$$\mathcal{L}\left(\mathbf{s}_i, \dot{\mathbf{s}}_i, V, \dot{V}\right) = \frac{1}{2} V^{2/3} \sum_{i=1}^{N} m_i \dot{\mathbf{s}}_i^2 - U_{conf}\left(V^{1/3}\mathbf{s}^N\right) + \frac{1}{2} Q_P \dot{V}^2 - pV. \tag{4.64}$$

From the Lagrangian, the momenta conjugated with the scaled coordinates and the volume, $\mathbf{p}_{s,i}$ and p_V can be derived

$$\mathbf{p}_{s,i} = \frac{\partial \mathcal{L}}{\partial \dot{\mathbf{s}}_i} = m_i V^{2/3} \dot{\mathbf{s}}_i \tag{4.65}$$

$$\mathrm{p}_V = \frac{\partial \mathcal{L}}{\partial \dot{V}} = Q_P \dot{V}. \tag{4.66}$$

Thus, the Lagrangian can also be expressed as

$$\mathcal{L}\left(\mathbf{s}_i, \mathbf{p}_{s,i}, V, \mathrm{p}_V\right) = \frac{1}{2V^{2/3}} \sum_{i=1}^{N} \frac{\mathbf{p}_{s,i}^2}{m_i} - U_{conf}\left(V^{1/3}\mathbf{s}^N\right) + \frac{1}{2Q_P}\mathrm{p}_V^2 - pV. \tag{4.67}$$

According to Eq. 4.15, the Hamiltonian can be derived from the Lagrangian by

$$\mathcal{H} = \sum_{i=1}^{N} \dot{\mathbf{s}}_i \mathbf{p}_{s,i} + \dot{V}\mathrm{p}_V - \mathcal{L} \tag{4.68}$$

$$= \frac{1}{2V^{2/3}} \sum_{i=1}^{N} \frac{\mathbf{p}_{s,i}^2}{m_i} + U_{conf}\left(V^{1/3}\mathbf{s}^N\right) + \frac{1}{2Q_P}\mathrm{p}_V^2 + pV. \tag{4.69}$$

The conserved quantity, which is represented by the Hamiltonian, is the enthalpy H of the system. From the Hamiltonian, now the equations of motion can be obtained

$$\dot{\mathbf{s}}_i = \frac{\partial \mathbf{s}_i}{\partial t} = \frac{\partial \mathcal{H}}{\partial \mathbf{p}_{s,i}} = \frac{\mathbf{p}_{s,i}}{m_i V^{2/3}} \tag{4.70}$$

$$\dot{\mathbf{p}}_{s,i} = \frac{\partial \mathbf{p}_{s,i}}{\partial t} = -\frac{\partial \mathcal{H}}{\partial \mathbf{s}_i} = -\frac{U_{conf}}{\partial \mathbf{s}_i} = -\frac{U_{conf}}{\partial \mathbf{r}_i}\frac{\partial \mathbf{r}_i}{\partial \mathbf{s}_i} = \mathbf{f}_i V^{1/3} \tag{4.71}$$

$$\dot{V} = \frac{\partial V}{\partial t} = \frac{\partial \mathcal{H}}{\partial \mathrm{p}_V} = \frac{\mathrm{p}_V}{Q_p} \tag{4.72}$$

$$\dot{\mathrm{p}}_V = \frac{\partial \mathrm{p}_V}{\partial t} = -\frac{\partial \mathcal{H}}{\partial V} = \frac{1}{3}\frac{1}{V^{5/3}}\sum_{i=1}^{N}\frac{\mathbf{p}_{s,i}^2}{m_i} - \frac{\partial U_{conf}}{\partial V} - p. \tag{4.73}$$

When the momentum $\mathbf{p}_{s_i}$ conjugated with the scaled coordinated $\mathbf{s}_i$ in Eq. 4.73 is replaced by the momentum $\mathbf{p}_i$, it can be seen that the first two terms in Eq. 4.73 represent the ideal gas (kinetic) and residual contribution to the instantaneous pressure $\wp$ of the system (see Eqs. 7.9 and 7.4, and Sect. 7.1 for more details on the derivation of the virial expression of the pressure). Thus,

$$\dot{\mathrm{p}}_V = \frac{\partial \mathrm{p}_V}{\partial t} = \wp - p. \tag{4.74}$$

In most textbooks, the equations of motion of the Andersen barostat are expressed in terms of the second derivatives of the scaled coordinate and the volume, $\ddot{\mathbf{s}}_i$ and $\ddot{V}$, which can be readily obtained from the Eqs. 4.70 and 4.72

$$\ddot{\mathbf{s}}_i = \frac{\dot{\mathbf{p}}_{s,i}}{m_i V^{2/3}} - \frac{2}{3}\frac{\mathbf{p}_{s,i}}{m_i V^{5/3}}\dot{V} = \frac{\mathbf{f}_i}{m_i V^{1/3}} - \frac{2}{3}\dot{\mathbf{s}}_i\frac{\dot{V}}{V} \tag{4.75}$$

$$\ddot{V} = \frac{1}{Q_p}\frac{\partial \mathrm{p}_V}{\partial t} = \frac{\wp - p}{Q_p}. \tag{4.76}$$

It was shown that the Andersen barostat correctly samples the isobaric-isenthalpic *NpH*-ensemble [1]. Nosé [21] combined the Andersen barostat with his approach of thermostatting (see Sect. 4.3.2) to enable simulations in the *NpT*-ensemble. Thus, the real variables of the system are derived from the virtual parameters by both scaling of the coordinates by $V^{1/3}$ and scaling of the timestep δt by the parameter s, which represents the thermal heat bath.

Hoover [18] proposed an alternative approach to decribe the uniform dilation of the system with time by using the strain rate $\dot{\varepsilon}$

$$\dot{\varepsilon} = \frac{\dot{V}}{V} \tag{4.77}$$

as dynamic variable instead of the volume. The equation of motion of the strain rate is given by [18]

$$\ddot{\varepsilon} = \frac{1}{k_b T \tau_p^2} V(t)(\wp - p), \tag{4.78}$$

wherein τ_p is the relaxation time of the pressure fluctuation with $\tau_p^2 = Q_p / N_f k_B T$. In the modification of the Hoover barostat by Melchionna et al. [20], which is for instance implemented in the *DL_POLY_2* MD simulation package [27], a friction or scaling factor χ is used as dynamic variable instead of the strain rate. When coupled with the Nosé-Hoover thermostat discussed in Sect. 4.3.2, the equations of motions are given by [27]

$$\frac{d\mathbf{r}_i(t)}{dt} = \mathbf{v}_i + \chi\left(\mathbf{r}_i - \mathbf{R}_{CoM}\right) \tag{4.79}$$

$$\frac{d\mathbf{v}_i(t)}{dt} = \frac{\mathbf{f}_i}{m_i} - \left[\zeta(t) + \chi(t)\right]\mathbf{v}_i(t) \tag{4.80}$$

$$\frac{d\zeta}{dt} = \frac{N_f k_B}{Q_T}[\mathcal{T}(t) - T] + \frac{1}{Q_T}\left(Q_p\chi(t)^2 - k_B T\right) \tag{4.81}$$

$$\frac{d\chi}{dt} = \frac{3}{Q_p}V(t)[\wp(t) - p] - \chi(t)\zeta(t) \tag{4.82}$$

$$\frac{dV(t)}{dt} = 3\chi(t)V(t). \tag{4.83}$$

Therein $\mathbf{R}_{CoM}$ is the center of mass of the system. The conserved quantity is the Gibbs free energy of the system G.

Parinello and Rahmen [23] extended the approach by Andersen to enable control of both pressure and stress. With this, the simulation box is allowed to also change its shape, which might be helpful in simulation studies of solids [1]. The shape of the simulation cell is described by three cell vectors $\mathbf{h}_1$, $\mathbf{h}_2$, $\mathbf{h}_3$ that define the cell matrix $\boldsymbol{H}$. The potential energy term U_V and the kinetic energy term $E_{kin,V}$ associated with the change of the box size or shape are given by [1]

$$U_V = pV = p\left(\mathbf{h}_1 \cdot \mathbf{h}_2 \times \mathbf{h}_3\right), \quad E_{kin,V} = \frac{Q_p}{2}\sum_\alpha\sum_\beta \dot{H}^2_{\alpha\beta} \tag{4.84}$$

The equations of motion for the scaled variable $\mathbf{s}_i$ and the cell matrix $\boldsymbol{H}$ then become

$$m_i\ddot{\mathbf{s}}_i = \boldsymbol{H}^{-1}\mathbf{f}_i - m_i\boldsymbol{G}^{-1}\dot{\boldsymbol{G}}\dot{\mathbf{s}}_i \tag{4.85}$$

$$Q_p\ddot{\boldsymbol{H}} = (\wp - \mathbf{1}p)\,V(t)\left(\boldsymbol{H}^{-1}\right)^T \tag{4.86}$$

$$\text{with} \quad \boldsymbol{G} = \boldsymbol{H}^T\boldsymbol{H}. \tag{4.87}$$

4.4.2 Gauss Barostat

In a similar way as in the Gauss thermostat, a constraint friction force can be introduced into the equation of motion to generate constant instantaneous pressure $\wp$. The barostatting method proposed by Evans and Morris [8] is based on the Doll's tensor Hamiltonian

$$\mathcal{H} = \sum_{i=1}^{N} \frac{\mathbf{p}_i^2}{2m_i} + U_{conf}\left(\mathbf{r}^N\right) + \chi \sum_{i=1}^{N} \mathbf{r}_i \mathbf{p}_i. \tag{4.88}$$

Therein the last term is the potential energy of the piston that acts on the system to impose the pressure. The constraint represented by the piston is introduced using a Lagrange multiplier χ, so that applying the Gauss principle of least constraint, the deviation between the Newtonian and the constraint trajectory is minimized. The Lagrange multiplier χ can thereby be interpreted as dilation rate. The equations of motion for the coordinates and momenta can then readily be obtained from the Hamiltonian

$$\dot{\mathbf{r}}_i = \frac{\partial \mathbf{r}_i}{\partial t} = \frac{\partial \mathcal{H}}{\partial \mathbf{p}_i} = \frac{\mathbf{p}_i}{m_i} + \chi(t)\mathbf{r}_i \tag{4.89}$$

$$\dot{\mathbf{p}}_i = \frac{\partial \mathbf{p}_i}{\partial t} = -\frac{\partial \mathcal{H}}{\partial \mathbf{r}_i} = \mathbf{f}_i - \chi(t)\mathbf{p}_i. \tag{4.90}$$

These need to be supplemented by an equation of motion for the volume, which is again given by

$$\dot{V} = \frac{dV(t)}{dt} = 3\chi(t)V(t). \tag{4.91}$$

The multiplier χ is determined to constrain the pressure to the imposed value p, i.e. to results in $\dot{\wp} = 0$. To derive the condition equation for the parameter χ, we assume for convenience that all particles have the same mass m. The instantaneous pressure of the system is determined by the virial equation (see Sect. 7.1)

$$3\wp V = \sum_{i=1}^{N} \frac{\mathbf{p}_i^2}{m} + \sum_{i=1}^{N} \sum_{j>i}^{N} \mathbf{r}_{ij}\mathbf{f}_{ij}. \tag{4.92}$$

Thus, the time derivative is given by

$$3\dot{\wp} V + 3p\dot{V} = \sum_{i=1}^{N} \frac{2\mathbf{p}_i\dot{\mathbf{p}}_i}{m} + \sum_{i=1}^{N} \sum_{j>i}^{N} \dot{\mathbf{r}}_{ij}\mathbf{f}_{ij} + \sum_{i=1}^{N} \sum_{j>i}^{N} \mathbf{r}_{ij}\dot{\mathbf{f}}_{ij}. \tag{4.93}$$

Inserting the equations of motion from Eqs. 4.89 to 4.91 yields

$$\begin{aligned} 3\dot{\wp} V + 3p\left(3\chi V\right) = & \sum_{i=1}^{N} \frac{2\mathbf{p}_i\left(\mathbf{f}_i - \chi(t)\mathbf{p}_i\right)}{m} + \sum_{i=1}^{N} \sum_{j>i}^{N} \left(\frac{\mathbf{p}_{ij}}{m} + \chi(t)\mathbf{r}_{ij}\right)\mathbf{f}_{ij} \\ & + \sum_{i=1}^{N} \sum_{j>i}^{N} \mathbf{r}_{ij}\dot{\mathbf{f}}_{ij}. \end{aligned} \tag{4.94}$$

The time derivative of the force $\mathbf{f}_{ij}$ can be obtained from

$$\dot{\mathbf{f}}_{ij} = -\dot{\mathbf{r}}_{ij}\left(\frac{\partial^2 U_{conf}}{\partial \mathbf{r}_{ij}^2}\right) = -\left(\frac{\mathbf{p}_i - \mathbf{p}_j}{m} + \chi(t)\mathbf{r}_{ij}\right)\left(\frac{\partial^2 U_{conf}}{\partial \mathbf{r}_{ij}^2}\right). \tag{4.95}$$

We then introduce the abbreviations

$$\varphi' = \frac{\partial U_{conf}}{\partial \mathbf{r}_{ij}} = -\mathbf{f}_{ij}, \qquad \varphi'' = \frac{\partial^2 U_{conf}}{\partial \mathbf{r}_{ij}^2} \qquad \mathbf{p}_{ij} = \mathbf{p}_i - \mathbf{p}_j. \tag{4.96}$$

With this, the time derivative of the instantaneous pressure can be expressed as

$$\dot{\wp} = \left(-9\chi pV + \sum_{i=1}^{N} \frac{2\mathbf{p}_i\mathbf{f}_i}{m} - \chi \sum_{i=1}^{N} \frac{2\mathbf{p}_i^2}{m} - \sum_{i=1}^{N}\sum_{j>i}^{N} \frac{\mathbf{p}_{ij}}{m}\varphi' - \chi \sum_{i=1}^{N}\sum_{j>i}^{N} \mathbf{r}_{ij}\varphi' \right.$$
$$\left. - \sum_{i=1}^{N}\sum_{j>i}^{N} \frac{\mathbf{r}_{ij}\mathbf{p}_{ij}}{m}\varphi'' - \chi \sum_{i=1}^{N}\sum_{j>i}^{N} \mathbf{r}_{ij}\varphi'' \right) /3V = 0 \tag{4.97}$$

Solving this equation for χ yields the conditional equation for the dilation rate

$$\chi = \frac{\sum_{i=1}^{N} \frac{2\mathbf{p}_i\mathbf{f}_i}{m} - \sum_{i=1}^{N}\sum_{j>i}^{N} \frac{\mathbf{r}_{ij}\mathbf{p}_{ij}}{m}\left(\frac{\varphi'}{r_{ij}} + \varphi''\right)}{\sum_{i=1}^{N} \frac{2\mathbf{p}_i^2}{m} + \sum_{i=1}^{N}\sum_{j>i}^{N} \mathbf{r}_{ij}^2\left(\frac{\varphi'}{r_{ij}} + \varphi''\right) + 9pV}. \tag{4.98}$$

For simulations in the *NpT*-ensemble, both constraint forces—to generate constant kinetic temperature dynamics and a constant instantaneous pressure—are introduced into the Newton's equation of motion simultaneously. This yields

$$\dot{\mathbf{r}}_i = \frac{\mathbf{p}_i}{m_i} + \chi(t)\mathbf{r}_i \tag{4.99}$$

$$\dot{\mathbf{p}}_i = \mathbf{f}_i - (\zeta(t) + \chi(t))\,\mathbf{p}_i \tag{4.100}$$

As before, the friction parameter ζ as Lagrange multiplier of the constraint thermostatting force is determined to ensure $\dot{\mathcal{T}} = 0$. This now results in a slightly modified conditional equation compared to the pure thermostat application in Eq. 4.58

$$\zeta = \frac{\sum_{i=1}^{N} \mathbf{f}_i(t)\cdot \mathbf{v}_i(t)}{\sum_{i=1}^{N} m_i \mathbf{v}_i^2(t)} - \chi. \tag{4.101}$$

The multiplier χ of the barostat is derived to yield $\dot{\wp} = 0$. According to the derivations from Eqs. 4.94–4.96, the time derivative of the instantaneous pressure can be expressed as

$$3\dot{\wp}V + 3p\,(3\chi V) = \sum_{i=1}^{N} \frac{2\mathbf{p}_i\dot{\mathbf{p}}_i}{m} - \sum_{i=1}^{N}\sum_{j>i}^{N} \left(\frac{\mathbf{p}_{ij}}{m} + \chi(t)\mathbf{r}_{ij}\right)\varphi' - \sum_{i=1}^{N}\sum_{j>i}^{N} \mathbf{r}_{ij}\varphi'' \left(\frac{\mathbf{p}_{ij}}{m} + \chi\mathbf{r}_{ij}\right). \tag{4.102}$$

However, in this case, the first term on the right side is canceled to comply with the constraint of constant kinetic temperature. This yields a simplified equation for the parameter χ

$$\chi = \frac{-\sum_{i=1}^{N}\sum_{j>i}^{N} \frac{\mathbf{r}_{ij}\mathbf{p}_{ij}}{m}\left(\frac{\varphi'}{r_{ij}} + \varphi''\right)}{\sum_{i=1}^{N}\sum_{j>i}^{N} \mathbf{r}_{ij}^2\left(\frac{\varphi'}{r_{ij}} + \varphi''\right) + 9pV}. \tag{4.103}$$

4.5 Simulation of Molecules

In Sects. 4.1 and 4.2 we have discussed the fundamental equations of motion and finite-difference methods for their numerical integration for the simplified case of spherical particles that only perform translational motions. When we now consider molecules composed of several atoms, we also have to take into account that molecules rotate, that we will observe torsions within the molecule, or in general, changes in the molecular structure due to inter- and intramolecular interactions. The forces acting on the single atoms of the molecules then not only arise from intermolecular interactions but also from bonding forces within the molecules. Chapter 6 provides an overview on analytical potential energy functions to account for bond stretching, angle bending and internal rotations (torsions) around dihedral angles. Analytical expressions to derive the forces acting on the atoms from different intramolecular interactions can be found, for instance, in the manual of the *DL_POLY_2* MD simulation package [27]. Thus, when we consider molecules as fully flexible, we treat their atoms as free bodies that experience different forces from inter- and intramolecular interactions as shown in Fig. 4.5. The problem is then again reduced to formulate the equations of motion for the translational motion of the atoms. Solving the equations of motion for the different atoms will generate the system's trajectory that allows for following how the position of the molecules (center of mass) and the molecular structure evolve in time. The time step used in the integration of the equation of motion should be shorter by a factor of 10 than the period of the highest frequency

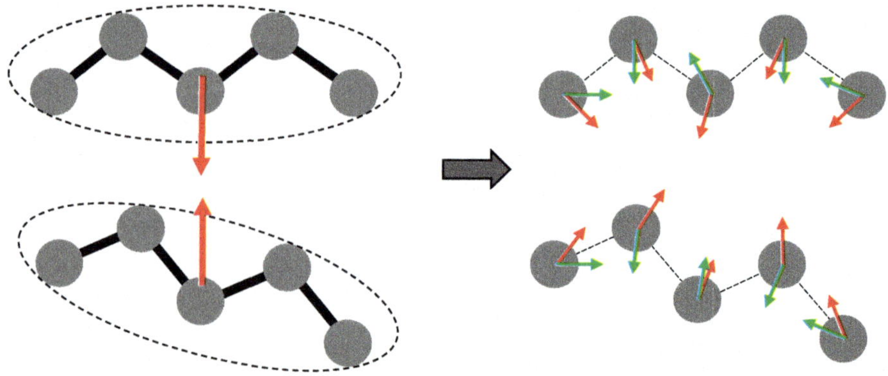

Fig. 4.5 Cutting molecules into atoms moving as free bodies under the influence of intermolecular forces (*red*) and intramolecular force (*green*) due to bonding interactions, i.e. bond stretching, angle bending and torsion

motion [10]. Intramolecular motions, especially bond stretching, are high frequent vibrations, and the highest frequency is observed in vibrations of bonds to hydrogen atoms: the wavenumber of $O - H$ stretching is approximately 3600 cm^{-1}, that of $C - H$ is ca. 3000 cm^{-1} [29], which yields periods of (9.3–11) fs. This then sets an upper limit to the time step to round about 1 fs. Thus, especially bonds to hydrogen atoms are often treated as rigid to allow for longer time steps in the integration. The following section will provide an introduction to constraint dynamics that are used to 'freeze' intramolecular degrees of freedom. However, small molecules such as N_2, CO_2, etc. are often treated as single rigid units, and their motion is divided into a translational motion of the molecular center of mass (CoM) and a rotation about the CoM. Thus, Sect. 4.5.2 deals with the formulation and solution of the rotational equation of motion of rigid bodies.

4.5.1 Constraint Dynamics

In Sect. 4.3.3 we have already introduced the concept of constraint dynamics that force the system to satisfy an imposed constraint by solving the equation of motion. In this case, the equations of motion of the atoms are solved under the constraint of maintaining constant bond lengths, bond angles etc. As an example, we consider the bond between two atoms i and j that should be fixed to the equilibrium bond length r_0. Thus, the system is subject to the constraint

$$\sigma\left(\mathbf{r}_i, \mathbf{r}_j\right) = \left(\mathbf{r}_j(t) - \mathbf{r}_i(t)\right)^2 - r_0^2 = 0. \tag{4.104}$$

Thus, a constraint force along the bond vector $\mathbf{r}_{ij}(t) = \mathbf{r}_j(t) - \mathbf{r}_i(t)$ is introduced into the equations of motion of the two atoms by means of a Lagrange multiplier λ

$$m_i\mathbf{a}_i = \mathbf{f}_i + \lambda\mathbf{r}_{ij} \tag{4.105}$$

$$m_j\mathbf{a}_j = \mathbf{f}_j - \lambda\mathbf{r}_{ij}. \tag{4.106}$$

When the Verlet algorithm (see Sect. 4.2) is employed, the positions of the two atoms are forwarded by

$$\mathbf{r}_i(t+\delta t) = 2\mathbf{r}_i(t) - \mathbf{r}_i(t-\delta t) + \left(\frac{\mathbf{f}_i(t)}{m_i} + \frac{\lambda(t)}{m_i}\mathbf{r}_{ij}(t)\right)\delta t^2 \tag{4.107}$$

$$\mathbf{r}_j(t+\delta t) = 2\mathbf{r}_j(t) - \mathbf{r}_j(t-\delta t) + \left(\frac{\mathbf{f}_j(t)}{m_i} - \frac{\lambda(t)}{m_j}\mathbf{r}_{ij}(t)\right)\delta t^2. \tag{4.108}$$

The first three terms on the right side represent the equation for the positions without accounting for the constraint force, and we denote the predicted position without constraint force $\mathbf{r}^*(t+\delta t)$. Equation 4.107 can then be expressed as

$$\mathbf{r}_i(t+\delta t) = \mathbf{r}_i^*(t+\delta t) + \frac{\lambda(t)}{m_i}\mathbf{r}_{ij}(t)\delta t^2 \tag{4.109}$$

$$\mathbf{r}_j(t+\delta t) = \mathbf{r}_j^*(t+\delta t) - \frac{\lambda(t)}{m_j}\mathbf{r}_{ij}(t)\delta t^2. \tag{4.110}$$

Constraining the distance between the two atoms to the equilibrium bond length r_0 then yields

$$\begin{aligned} \left|\mathbf{r}_j(t+\delta t) - \mathbf{r}_i(t+\delta t)\right|^2 &= r_0^2 \\ &= \left|\mathbf{r}_j^*(t+\delta t) - \mathbf{r}_i^*(t+\delta t) - \lambda(t)\mathbf{r}_{ij}(t)\left(\frac{1}{m_j} + \frac{1}{m_i}\right)\delta t^2\right|^2. \end{aligned} \tag{4.111}$$

This represents a quadratic equation for the Lagrange multiplier that needs to be solved in order to obtain the value for λ. This value is then used to correct the positions $\mathbf{r}_{i/j}^*(t+\delta t)$ to the $\mathbf{r}_{i/j}(t+\delta t)$ that yield the correct bond length. This approach is accordingly applied to every constrained bond length in the system. Bond angles can be fixed by constraining the distances between all pairs of the three atoms involved. When more than one bond to an atom i is frozen, several constraints have to be satisfied simultaneously. When, for example, the *C–H* bonds in a methane molecule are treated as rigid, four constraints have to be implemented in the equation of motion of the carbon atom. The constraint force $\mathbf{g}_i$ acting on the atom i due to l constraints σ is expressed as [26]

$$\mathbf{g}_i = -\sum_{k=1}^{l} \lambda_k(t)\nabla_i\sigma_k. \tag{4.112}$$

The calculation of all λ_l multipliers then requires solving a system of l coupled quadratic constraint equations, which becomes computationally expensive.

Ryckaert et al. [26] have proposed an iterative scheme to satisfy the constraints in succession, wherein one λ is computed while all other multipliers are fixed. As atoms are often involved in more that one constraint, the different λ_k are coupled so that their successive calculation has to be repeated until convergence. Ryckaert et al. used their procedure with the Verlet integration, which is known as *SKAKE* algorithm. The *RATTLE* algorithm by Andersen [3] is the formulation of the iterative solution for the Lagrange multiplier with the Velocity Verlet integrator.

4.5.2 Equation of Motion of Rigid Bodies

As stated before in the introduction of Sect. 4.5, the motion of a rigid body in space can be decomposed into a translational motion of its center of mass (CoM) and a rotation about the CoM. The translational dynamics is governed by the total force acting on the molecule, and can be described by the Newton's equation of motion as discussed before in Sect. 4.1. Thus, this section focuses on the formulation of the rotational equation of motion, which is determined by the total torque applied to the molecule. When a molecule i is divided into its atoms a (or more general, interaction sites), and each experiences a total force $\mathbf{f}_{i,a}$, the resulting torque $\mathbf{M}_i$ of the molecule is given by,

$$\mathbf{M}_i = \sum_a \left(\mathbf{r}_{i,a} - \mathbf{r}_{i,CoM}\right) \times \mathbf{f}_{i,a} = \sum_a \mathbf{d}_{i,a} \times \mathbf{f}_{i,a}, \tag{4.113}$$

with $\mathbf{d}_{i,a}$ being the distance vector between the atom a and the molecule's center of mass. It is more convenient to express the rotation of the molecule not by space-fixed (index s), but by body-fixed coordinates (index b). Thereby, the origin of the body-fixed system is located in the center of mass of the molecule, and the principal axes are orientated in such a way that the inertia tensor $\mathbf{I}$ becomes diagonal

$$\mathbf{I}_i = \begin{pmatrix} I_{xx} & 0 & 0 \\ 0 & I_{yy} & 0 \\ 0 & 0 & I_{zz} \end{pmatrix}, \tag{4.114}$$

with I_{xx}, I_{yy} and I_{zz} being the three principal moments of inertia of the molecule. The orientation of the molecule in space is specified by the relation of the body-fixed to the space-fixed coordinate system, which can be defined by the Euler angles Θ, Φ and Ψ as depicted in Fig. 4.6.

The rotation about the original z-axis is described by the Euler angle Φ, and the corresponding elementary rotation matrix A_Φ is given by

$$A_\Phi = \begin{pmatrix} \cos\Phi & -\sin\Phi & 0 \\ \sin\Phi & \cos\Phi & 0 \\ 0 & 0 & 1 \end{pmatrix}. \tag{4.115}$$

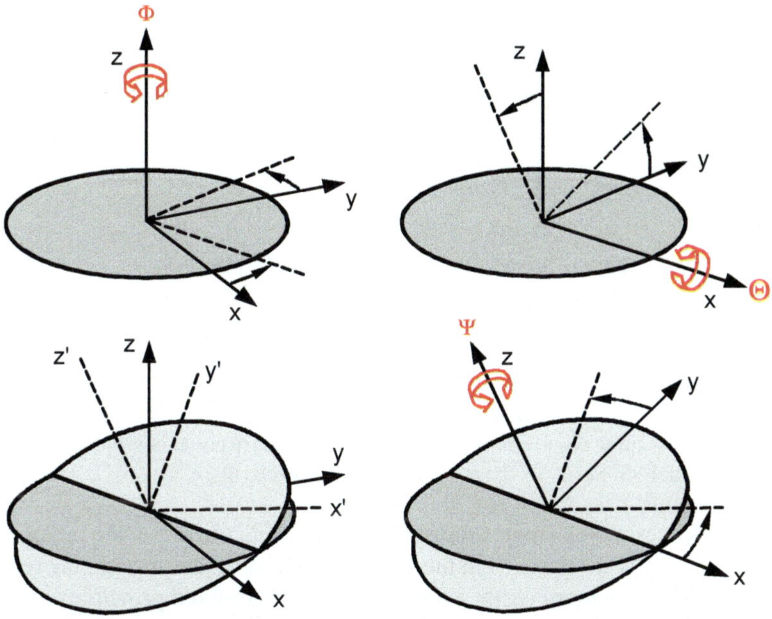

Fig. 4.6 Illustration of Euler angles to present the orientation and rotation on an object in three-dimensional space (after [1, 13])

The Euler angle Θ then describes the rotation about the 'new' x-axis, with

$$A_\Theta = \begin{pmatrix} 1 & 0 & 0 \\ 0 & \cos\Theta & -\sin\Theta \\ 0 & \sin\Theta & \cos\Theta \end{pmatrix}, \tag{4.116}$$

and the Euler angle Ψ finally the rotation about the then newly orientated z-axis

$$A_\Psi = \begin{pmatrix} \cos\Psi & -\sin\Psi & 0 \\ \sin\Psi & \cos\Psi & 0 \\ 0 & 0 & 1 \end{pmatrix}. \tag{4.117}$$

Thus, a rotation in space can be expressed by executing these three elementary rotations in series. The multiplication of the three elementary rotation matrices yields

$$\begin{bmatrix} \cos\Phi\cos\Psi - \sin\Phi\cos\Theta\sin\Psi & -\cos\Phi\sin\Psi - \sin\Phi\cos\Theta\cos\Psi & \sin\Phi\sin\Theta \\ \sin\Phi\cos\Psi + \cos\Phi\cos\Theta\sin\Psi & -\sin\Phi\sin\Psi + \cos\Phi\cos\Theta\cos\Psi & -\cos\Phi\sin\Theta \\ \sin\Theta\sin\Psi & \sin\Theta\cos\Psi & \cos\Theta \end{bmatrix}$$

$= A^{-1} = A^T.$

This is the inverse matrix or transpose of the rotation matrix $\mathbf{A}$ that relates the body-fixed to the space-fixed vectors. Thus, the transition of the coordinates, the angular velocity $\boldsymbol{\omega}$ and the torque between the body-fixed and the space-fixed axis system is given by

$$\mathbf{r}_{ia}^{b} = \mathbf{A}\cdot\mathbf{r}_{ia}^{s} \rightarrow \mathbf{r}_{ia}^{s} = \mathbf{A}^{-1}\cdot\mathbf{r}_{ia}^{b} = \mathbf{A}^{T}\cdot\mathbf{r}_{ia}^{b} \tag{4.118}$$

$$\boldsymbol{\omega}_{i}^{b} = \mathbf{A}\cdot\boldsymbol{\omega}_{i}^{s} \rightarrow \boldsymbol{\omega}_{i}^{s} = \mathbf{A}^{-1}\cdot\boldsymbol{\omega}_{i}^{b} = \mathbf{A}^{T}\cdot\boldsymbol{\omega}_{i}^{b} \tag{4.119}$$

$$\mathbf{M}_{i}^{b} = \mathbf{A}\cdot\mathbf{M}_{i}^{s} \rightarrow \mathbf{M}_{i}^{s} = \mathbf{A}^{-1}\cdot\mathbf{M}_{i}^{b} = \mathbf{A}^{T}\cdot\mathbf{M}_{i}^{b} \tag{4.120}$$

with the rotation matrix
$A =$

$$\begin{bmatrix} \cos\Phi\cos\Psi - \sin\Phi\cos\Theta\sin\Psi & \sin\Phi\cos\Psi + \cos\Phi\cos\Theta\sin\Psi & \sin\Psi\sin\Theta \\ -\cos\Phi\sin\Psi - \sin\Phi\cos\Theta\cos\Psi & -\sin\Phi\sin\Psi + \cos\Phi\cos\Theta\cos\Psi & \cos\Psi\sin\Theta \\ \sin\Phi\sin\Theta & -\sin\Theta\cos\Phi & \cos\Theta \end{bmatrix}.$$

whereas the forces on the single atoms and the resulting torque on the molecule (see Eq. 4.113) are easily determined in the space-fixed coordinate system, the rotational equation of motion is more easily expressed in the body-fixed principle axis system due to its diagonal inertia tensor $\mathbf{I}$ [1]. The rotational equation of motion in the body-fixed frame is given by

$$\dot{\mathbf{j}}^{b} + \boldsymbol{\omega}^{b}\times\mathbf{j} = \mathbf{M}^{b}, \tag{4.121}$$

with the angular momentum $\mathbf{j}$

$$\mathbf{j}^{b} = \begin{pmatrix} j_x \\ j_y \\ j_z \end{pmatrix}^{b} = \begin{pmatrix} I_{xx}\omega_x \\ I_{yy}\omega_y \\ I_{zz}\omega_z \end{pmatrix}^{b} \rightarrow \dot{\mathbf{j}}^{b} = \begin{pmatrix} \dot{j}_x \\ \dot{j}_y \\ \dot{j}_z \end{pmatrix}^{b} = \begin{pmatrix} I_{xx}\dot{\omega}_x \\ I_{yy}\dot{\omega}_y \\ I_{zz}\dot{\omega}_z \end{pmatrix}^{b}. \tag{4.122}$$

This yields the first order differential equations for the components of the angular velocity $\boldsymbol{\omega}$ in the body-fixed axis system [1]

$$\dot{\omega}_{x}^{b} = \frac{M_{x}^{b}}{I_{xx}} + \left(\frac{I_{yy}-I_{zz}}{I_{xx}}\right)\omega_{y}^{b}\omega_{z}^{b} \tag{4.123}$$

$$\dot{\omega}_{y}^{b} = \frac{M_{y}^{b}}{I_{yy}} + \left(\frac{I_{zz}-I_{xx}}{I_{yy}}\right)\omega_{x}^{b}\omega_{z}^{b} \tag{4.124}$$

$$\dot{\omega}_{z}^{b} = \frac{M_{z}^{b}}{I_{zz}} + \left(\frac{I_{xx}-I_{yy}}{I_{zz}}\right)\omega_{x}^{b}\omega_{y}^{b}. \tag{4.125}$$

Additionally, equations of motion of the molecular orientation itself, i.e. the Euler angles are required. In the body-fixed frame, the angular velocity is related to the Euler rotational velocity by

$$\begin{pmatrix} \omega_x \\ \omega_y \\ \omega_z \end{pmatrix}^b = \begin{pmatrix} \sin\Theta\sin\Psi & cos\Psi & 0 \\ \sin\Theta\cos\Psi & -\sin\Psi & 0 \\ cos\Theta & 0 & 1 \end{pmatrix} \cdot \begin{pmatrix} \dot{\Phi} \\ \dot{\Theta} \\ \dot{\Psi} \end{pmatrix}. \tag{4.126}$$

Inversion of the matrix then yields the first order differential equations for the Euler angles

$$\dot{\Phi} = \frac{\sin\Psi}{\sin\Theta}\omega_x^b + \frac{\cos\Psi}{\sin\Theta}\omega_y^b \tag{4.127}$$

$$\dot{\Theta} = \cos\Psi\,\omega_x^b - \sin\Psi\,\omega_y^b \tag{4.128}$$

$$\dot{\Psi} = -\frac{\sin\Psi\cos\Theta}{\sin\Theta}\omega_x^b - \frac{\cos\Psi\cos\Theta}{\sin\Theta}\omega_y^b + \omega_z^b. \tag{4.129}$$

This means that the rotational dynamics have to be evaluated step-by-step:

1. At timestep t, the Euler angles are known, and the rotation matrix $\mathbf{A}$ and its inverse matrix $\mathbf{A}^{-1}$ are given;
2. From the known positions of the atoms in the space-fixed frame, the force $\mathbf{f}_{i,a}$ acting on all atoms can be determined;
3. Equation 4.113 then yields the torque acting on the molecule in the space-fixed axis system;
4. The torque is transformed into the body-fixed frame by Eq. 4.120 and enters the equations of motion of the angular velocities in Eq. 4.123;
5. The equations of motion of the angular velocities are solved using suitable Finite-Difference methods (Gear Predictor-Corrector, see Sect. 4.2.4) to yield the angular velocities at the next time step $t + \delta t$;
6. With the then known angular velocities in the body-fixed frame, the differential equations for the Euler angles (Eq. 4.127) are solved to give the Euler angles at $t + \delta t$, from which the new rotation matrices $\mathbf{A}$ and $\mathbf{A}^{-1}$ are determined.

4.5.2.1 Quaternions

The formulation of the rotational equations of motion in terms of Euler angles becomes problematic when the Euler angle Θ approaches 0 or π. Then the $\sin\Theta$ term approaches 0, and divergence occurs in the differential equations for the Euler angles (Eq. 4.127), and the matrix in Eq. 4.126 becomes singular. A more convenient and stable notation for presenting the orientation and rotation of the rigid molecules is provided by the mathematical concept of quaternions that extend the complex numbers. For describing and calculating three-dimensional rotations, a set of four quaternions is used

$$Q = (q_1, q_2, q_3, q_4) \tag{4.130}$$

that satisfies the constraint [1]

$$q_1^2 + q_2^2 + q_3^2 + q_4^2 = 1. \tag{4.131}$$

The quaternions can thereby be expressed in different ways. Following the definition of the quaternions by Rapaport [25]

$$\begin{aligned}
q_1 &= \sin\left(\frac{\Theta}{2}\right)\cos\left(\frac{\Phi-\Psi}{2}\right)\\
q_2 &= \sin\left(\frac{\Theta}{2}\right)\sin\left(\frac{\Phi-\Psi}{2}\right)\\
q_3 &= \cos\left(\frac{\Theta}{2}\right)\cos\left(\frac{\Phi+\Psi}{2}\right)\\
q_4 &= \cos\left(\frac{\Theta}{2}\right)\sin\left(\frac{\Phi+\Psi}{2}\right),
\end{aligned} \tag{4.132}$$

the sin and cos terms of the Euler angles in the rotational matrix can be expressed in terms of quaternions by [25]

$$\begin{aligned}
\sin\Theta &= \sqrt{2(q_1^2+q_2^2)(1-q_1^2-q_2^2)}\\
\cos\Theta &= 1-2(q_1^2+q_2^2)\\
\sin\Phi &= 2(q_1q_3+q_2q_4)/\sin\Theta\\
\cos\Phi &= 2(q_1q_4-q_2q_3)/\sin\Theta\\
\sin\Psi &= 2(q_1q_3-q_2q_4)/\sin\Theta\\
\cos\Psi &= 2(q_1q_4+q_2q_3)/\sin\Theta.
\end{aligned} \tag{4.133}$$

This then yields the rotational matrix $\mathbf{A}$ in quaternion notation

$$\mathbf{A} = 2\begin{pmatrix} q_1^2+q_4^2-\frac{1}{2} & q_1q_2+q_3q_4 & q_1q_3-q_2q_4\\ q_1q_2-q_3q_4 & q_2^2+q_4^2-\frac{1}{2} & q_2q_3+q_1q_4\\ q_1q_3+q_2q_4 & q_2q_3-q_1q_4 & q_3^2+q_4^2-\frac{1}{2} \end{pmatrix}. \tag{4.134}$$

The angular velocity in the body-fixed frame is expressed in terms of quaternion velocities by [25]

$$\begin{pmatrix}\omega_x\\ \omega_y\\ \omega_z\\ 0\end{pmatrix}^b = 2\mathbf{W}\begin{pmatrix}\dot{q}_1\\ \dot{q}_2\\ \dot{q}_3\\ \dot{q}_4\end{pmatrix} \quad \mathbf{W} = \begin{pmatrix} q_4 & q_3 & -q_2 & -q_1\\ -q_3 & q_4 & q_1 & -q_2\\ q_2 & -q_1 & q_4 & -q_3\\ q_1 & q_2 & q_3 & q_4 \end{pmatrix}. \tag{4.135}$$

Rapaport [25] also derived equations of motion for the quaternions

$$\begin{pmatrix} \ddot{q}_1 \\ \ddot{q}_2 \\ \ddot{q}_3 \\ \ddot{q}_4 \end{pmatrix} = \frac{1}{2}\mathbf{W}^T \begin{pmatrix} \dot{w}_x^b \\ \dot{w}_y^b \\ \dot{w}_z^b \\ -2(\dot{q}_1^2 + \dot{q}_2^2 + \dot{q}_3^2 + \dot{q}_4^2) \end{pmatrix}. \tag{4.136}$$

Therein, the derivatives of the components of angular velocity can be replaced by Eq. 4.123. The components of $\boldsymbol{\omega}$, which then appear in Eq. 4.136, are in turn replaced by Eq. 4.135. By doing so, a set of second-order differential equations of motion for the quaternions is derived that do not involve the angular velocity $\boldsymbol{\omega}$. This has the advantage that both the rotational and the translation equations of motion can be solved using the same Finite-Difference method (see Sect. 4.2) [24].

4.5.2.2 Linear Molecules

Linear molecules allow for a simplified representation of the rotational equations of motion as the moment of inertia for the axis of symmetry vanishes. Furthermore, the orientation of the molecule in space can be solely specified by the vector along the molecular axis. When $\mathbf{e}_i^s$ is the unit vector along the molecular axis in the space-fixed coordinate system, Eq. 4.113 for calculating the torque can be rewritten as [9]

$$\mathbf{M}_i = \mathbf{e}_i^s \times \sum_a \mathbf{d}_{i,a} \cdot \mathbf{f}_{i,a} = \mathbf{e}_i^s \times \mathbf{f}_i^{rot}, \tag{4.137}$$

wherein $\mathbf{f}_i^{rot}$ is the resulting 'turning' force from intermolecular interactions. Only the component of $\mathbf{f}_i^{rot,\perp}$ perpendicular to the molecular axis has to be taken into account as the parallel components have no effect on the rotation. The parallel components of $\mathbf{f}_i^{rot}$ are subtracted out by [9]

$$\mathbf{f}_i^{rot,\perp} = \mathbf{f}_i^{rot} - \mathbf{e}_i^s \left(\mathbf{f}_i^{rot} \cdot \mathbf{e}_i^s\right). \tag{4.138}$$

When the linear molecule can be regarded as symmetric rotor, the remaining two principle moments of inertia become equal, denoted as I. The rotational equations of motion are then given by [9]

$$\mathbf{j}_i = I\boldsymbol{\omega}_i^s \tag{4.139}$$

$$\frac{d\boldsymbol{\omega}_i^s}{dt} = \frac{\mathbf{M}_i}{I} \tag{4.140}$$

$$\frac{\mathbf{e}_i^s}{dt} = \boldsymbol{\omega}_i^s \times \mathbf{e}_i^s = \mathbf{u}_i^s. \tag{4.141}$$

It is often more convenient to express the rotation dynamics in terms of the velocity $\mathbf{u}_i^s$ of the unit axis vector instead of the angular velocity $\boldsymbol{\omega}_i^s$. The equation of motion of $\mathbf{u}_i^s$ can be derived from [9]

$$\frac{\mathbf{u}_i^s}{dt} = \frac{d\boldsymbol{\omega}_i^s}{dt} \times \mathbf{e}_i^s + \boldsymbol{\omega}_i^s \times \frac{d\mathbf{e}_i^s}{dt} \tag{4.142}$$

$$= \frac{\mathbf{f}_i^{rot,\perp}}{I} - u^2\mathbf{e}_i^s. \tag{4.143}$$

The second term in Eq. 4.143 can be regarded as a centripetal acceleration to constrain the length of $\mathbf{e}_i^s$ to unity [9]. Thus, u^2 is often replaced by a Lagrange multiplier λ that is determined to normalize $\mathbf{e}_i^s$.

References

1. Allen M, Tildesley DJ (1987) Computer simulation of liquids. Oxford Science Publication
2. Andersen HC (1980) Molecular dynamics simulations at constant pressure and/or temperature. J Chem Phys 72:2384–2393
3. Andersen HC (1983) Rattle: a 'velocity' version of the shake algorithm for molecular dynamics calculation. J Comput Phys 52:24–34
4. Basconi JE, Shirts MR (2013) Effects of temperature control algorithms on transport properties and kinetics in molecular dynamics simulations. J Chem Theory Comput 9:2887–2899
5. Berendsen HJC, Postma JPM, Van Gunsteren WF, Di Nola A, Haak JR (1984) Molecular dynamics with coupling to an external bath. J Chem Phys 81:3684–3690
6. Brown D, Clarke JHR (1984) A comparison of constant energy, constant temperature and constant pressure ensembles in molecular dynamics simulations of atomic liquids. Mol Phys 51:1243–1252
7. Evans DJ (1983) Computer experiment for nonlinear thermodynamics of Couette flow. J Chem Phys 77:3297–3302
8. Evans DJ, Morris G (1983) Isothermal-Isobaric molecular dynamics. Chem Phys 77:63–66
9. Fincham D (1993) Leap frog rotational algorithm for linear molecules. Mol Sim 11:79–89
10. Frenkel D, Smit B (1996) Understanding molecular simulation. Academic Press
11. Gear CW (1966) The numerical integration of ordinary differential equations of various orders. Report ANL 7126, Argonne National Laboratory
12. Gear CW (1971) Numerical initial value problems in ordinary differential equations. Prentice-Hall, Englewood Cliff, New York
13. Goldstein H, Poole C, Safko J (2001) Classical mechanics, 3rd edn. Addison Wesley
14. Haile JM (1997) Molecular dynamics simulation. Wiley Professional Paperback Series
15. Hünenberger PH (2005) Thermostat algorithms for molecular dynamics simulations. Adv Polym Sci 173:105–149
16. Hockney RW (1970) The potential calculation and some applications. Methods Comput Phys 9:136–211
17. Hoover WG (1985) Canonical dynamics: equilibrium phase-space distributions. Phys Rev A 31(3):1695
18. Hoover WG (1986) Constant-pressure equation of motion. Phys Rev A 34:2499–2500
19. Hoover WG, Ladd AJC, Moran B (1982) High strain rate plastic flow studied via nonequilibrium molecular dynamics. Phys Rev Lett 48:1818–1820
20. Melchionna S, Ciccotti G, Holian BL (1993) Hoover NPT dynamics for systems varying in shape and size. Mol Phys 78:533–544
21. Nosé, (1984) A unified formulation of the constant temperature molecular dynamics method. J Chem Phys 81:511–519
22. Nosé, (1984) A molecular dynamics method for simulations in the canonical ensemble. Mol Phys 52(2):255–268

23. Parrinello M, Rahman A (1980) Crystal structure and pair potentials: a molecular dynamics study. Phys Rev Lett 45:1196–1199
24. Rapaport DC (1985) Molecular dynamics simulation using quaternions. J Comput Phys 60:306–314
25. Rapaport DC (2004) The art of molecular dynamics simulation. Cambridge University Press
26. Ryckaert JP, Ciccotti G, Berendsen HJC (1977) Numerical integration of the cartesian equation of motion of a system with constraints: molecular dynamics of n-alkane. J Comput Phys 23:327–341
27. Smith TR, Forester TR, Todorov IT DL_POLY classic user manual, http://www.ccp5.ac.uk/DL_POLY_CLASSIC. Daresbury Laboratory, UK
28. Swope WC, Andersen HC, Berens PH, Wilson KR (1982) A computer simulation method for the calculation of equilibrium constants for the formation of physical cluster of molecules: application to small water clusters. J Chem Phys 76:637–649
29. Tasumi M (2015) Introduction to experimental infrared spectroscopy fundamentals and practical methods. Wiley

Chapter 5
Running Molecular Simulations

The preceding chapters have provided an introduction into the theoretical background of molecular simulations given by the statistical thermodynamics, and into the general principles of the Monte Carlo and molecular dynamics simulation techniques. This chapter now will discuss strategies for performing molecular simulations, i.e. setting up the simulation, treatment of boundaries, techniques to speed up the simulation, handling of long-range electrostatic interactions etc. A more comprehensive discussion of these practical aspects of molecular simulation studies is, for instance, given by the standard textbooks by Allen and Tildesley [1], or Frenkel and Smit [10].

5.1 Setting-Up of the Simulation

Setting up a Monte Carlo or molecular dynamics simulation first of all requires selecting reasonable initial positions of the molecules within the simulation box. When performing MD simulations, we additionally have to assign initial velocities to the molecules. Thus, the following sections provide an introduction to initialization strategies for both the configuration and the velocity distribution.

5.1.1 *Initial Configuration*

For an ergodic system, simulation results for the thermophysical properties at equilibrium should be insensitive to the initial conditions [10]. Thus, an initial configuration only has to be reasonable—meaning that it should not contain non-physical overlaps between molecules that result in high-energy interactions, and that it should allow the system to relax to the equilibrium structure of the state point aimed to simulate. Thus, when simulating solids, the initial configuration is prepared according

G. Raabe, *Molecular Simulation Studies on Thermophysical Properties*,
Molecular Modeling and Simulation, DOI 10.1007/978-981-10-3545-6_5

to the experimentally determined crystal structure of the component, for instance using information on the structure from X-ray experiments. For liquids or gases, the molecules could be placed at random positions within the simulation box, though in doing so energy 'hot spots' [16] have to be avoided: if molecules are positioned too close to each other, large short range repulsive interactions arise that may cause a disruption of the simulation. Thus, an algorithm to randomly position molecules in the box needs to be combined with an optimization algorithm that ensures that the minimum distance between particles is greater or equal to a predefined tolerance. This approach of an automated generation of initial configurations with a packing optimization is for instance realized in the package 'Packmol' [18, 19].

A common approach to set up an initial configuration for a liquid while avoiding overlaps is to place the particles on sites of a regular lattice, which then usually melts rapidly into a fluid structure. Convenient lattice structures are thereby 'simple cubic' or 'face-centered cubic' (fcc) as shown in Fig. 5.1. The lattice size is chosen to give an appropriate density. For a molecular system though, its atoms have to be positioned in such a way that all bond length, bond angles and dihedrals agree with their equilibrium or reference values. Thus, a coordinate file or connectivity map for each type of molecule has to be provided that defines the relative positions of the atoms within the molecule. Additionally, an orientation has to be assigned to each molecule on the lattice. In general, all molecules may be positioned with the same orientation as shown in Fig. 5.2.

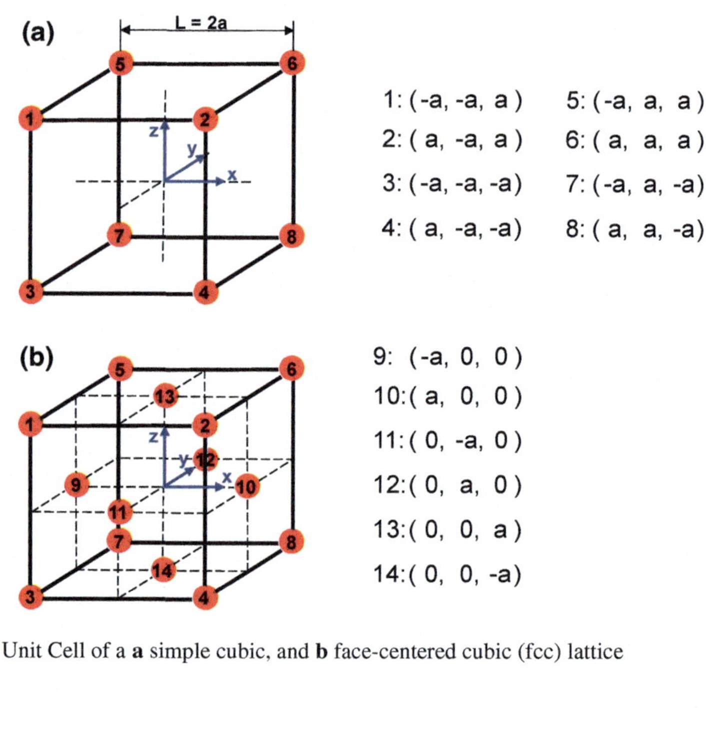

Fig. 5.1 Unit Cell of a **a** simple cubic, and **b** face-centered cubic (fcc) lattice

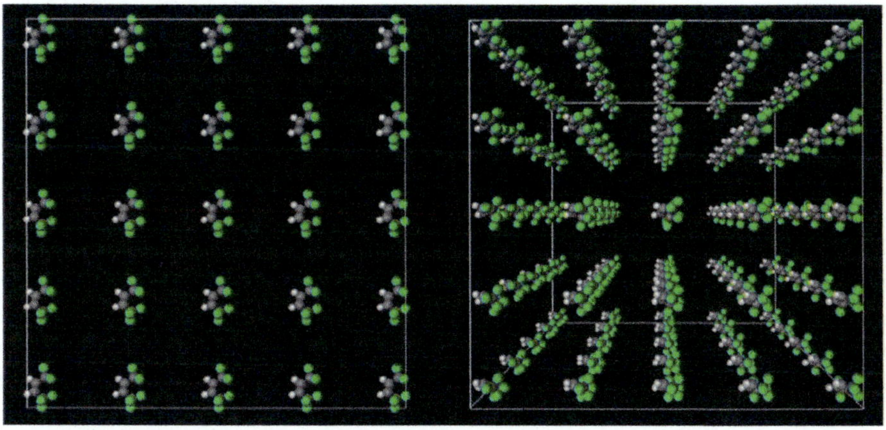

Fig. 5.2 Positioning of cis133mzz molecules on a simple cubic lattice with same orientation for all molecules (initial configuration generated by TOWHEE [17], visualized with Ovito [25])

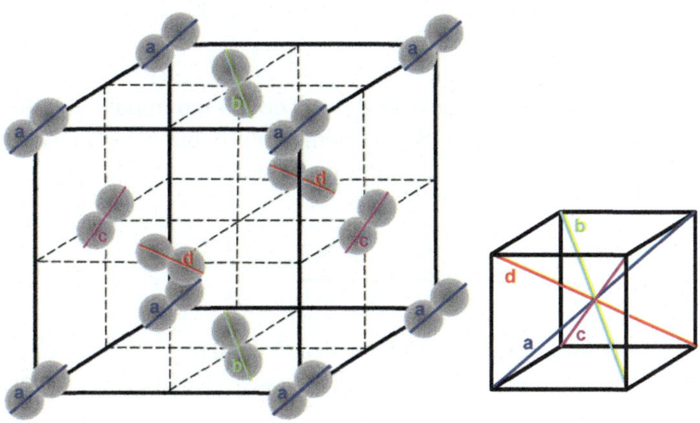

Fig. 5.3 Orientation of molecules along the diagonals of the unit cells

Though, for long chain molecules there is the risk of overlaps between tails of neighboring molecules. Thus, a random orientation may be assigned to each molecule, or the molecules may be orientated along the diagonals of the unit cells [16] as illustrated by Fig. 5.3. However, for large molecules it becomes necessary to check for overlap between molecules on neighboring lattice sites, or the lattice size has to be increased accordingly.

When not starting simulations for a component from the scratch, it is common practice to use an equilibrated configuration of an earlier simulation as starting configuration for the subsequent simulation.

5.1.2 Initial Velocities

In molecular dynamics studies, the set up of the simulation also comprises the assignment of initial velocities to all particles. Thereby the resulting kinetic energy of the system should yield an instantaneous temperature that agrees with the desired value according to Eq. 4.33. Thus, the velocities may be chosen from the Maxwell-Boltzmann distribution for the desired temperature (see Sect. 2.4), with the direction of the velocity being selected randomly. Alternatively, each velocity component of every particle may be assigned a randomly chosen value from a uniform distribution in the interval $[-0.5; 0.5]$ [10]. Then, however the velocities need to be scaled to adjust the temperature. The required scaling factor λ is determined by the square root of the relation of the desired temperature to the instantaneous temperature resulting from the unscaled velocities (see Eq. 4.35). A simple approach to assign initial velocities is to attribute the same value to each velocity component of every particle, the required value then results from Eq. 2.123, i.e.

$$|v_{i,x}| = |v_{i,y}| = |v_{i,z}| = \sqrt{\frac{k_B T}{m}} \tag{5.1}$$

with randomly assigned algebraic sign $+/-$. Though the initial velocities not only need to be selected subject to the desired temperature, but also to ensure that there is no resulting total (linear) momentum of the system

$$\mathbf{p}_{total} = \sum_{i=1}^{N} \mathbf{p}_i = \sum_{i=1}^{N} m_i \mathbf{v}_i = 0. \tag{5.2}$$

Therefore,the velocities again have to be scaled to yield $\mathbf{p}_{total} = 0$, i.e. to eliminate an overall flow [23]. The scaling is performed for each velocity component according to

$$v_{i,x}^{new} = v_{i,x}^{old} - \frac{1}{N} \sum_{i=1}^{N} v_{i,x}^{old}, \; v_{i,y}^{new} = \ldots \tag{5.3}$$

5.2 Periodic Boundary Conditions

In the previous section we have discussed strategies to position the molecules into the simulation box. If our simulation cell was bounded by real physical walls, we would have to consider interactions of the particles with the walls, which may extend up 10 molecular diameters into the fluid [16]. In typical simulation systems that contain some 1000 particles, this means that a considerable fraction of particles would experience different interactions from particles in the bulk. Consequently,

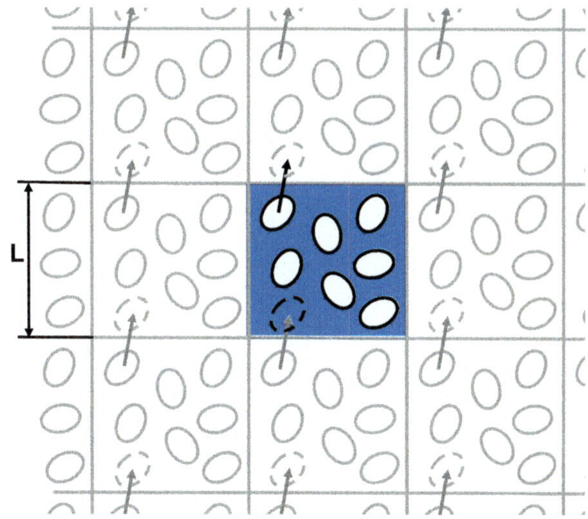

Fig. 5.4 2-D visualization of the periodic boundary condition (PBC) (after Allen and Tildesley [1])

the properties of the molecules subject of these boundary effects differ from the thermophysical properties in the bulk that we intend to determine. In order to enable simulations in small systems of some 1000 particles without distorting the result by boundary or surface effects, the system has to be bounded without physical walls. This is realized by employing periodic boundary conditions (PBC) that mimic the presence of a bulk by surrounding the simulation box imaginarily by an infinite number of identical copies as illustrated in Fig. 5.4.

In the course of the simulation, the image particles in the periodic copies move exactly the same way as the particles in the central box. When a molecule then leaves the central simulation box, one of its images enters the box from the opposite side, so that the number of particles in the central box is conserved. This 'wraparound effect' [23] for the molecular movement has to be included when determining new positions either by displacements in MC or by integrating the equation of motion in MD studies. For a cubic box with a boxlength L and the origin of coordinates located in the center of the box, the PBC is applied to the components of the position vector by

$$\begin{aligned} &\text{if } r_{i,x} > L/2 \ \ r_{i,x}^{new} = r_{i,x} - L \\ &\text{if } r_{i,x} < -L/2 \ r_{i,x}^{new} = r_{i,x} + L, \end{aligned}$$

accordingly for the $r_{i,y}$ and $r_{i,z}$ coordinates. It should be noted that molecular simulation packages handle the application of PBC to molecules differently. Whereas some simulation codes apply the PBC to each atom that leaves the central box, other simulation packages only apply it to the molecule as a whole when its center of mass crosses the cell boundary.

The cubic box is the simplest periodic system and therefore the most widely used cell shape for a simulation box. Though, also other cell shapes may be

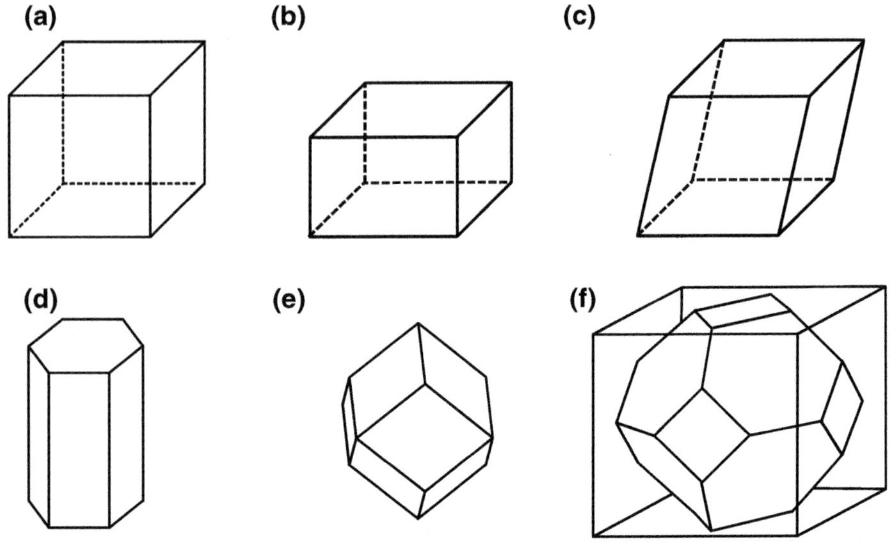

Fig. 5.5 Possible cell shapes for periodic systems: **a** cubic, **b** orthorhombic, **c** parallelepiped, triclinic, **d** hexagonal prism, **e** rhombic dodecahedron, and **f** truncated octahedron

employed—provided that its replication results in a space filling array. Possible periodic cell shapes are shown in Fig. 5.5.

The orthorhombic, parallelpiped or triclinic cell are used for simulations of crystalline materials with corresponding shapes and dimensions of their unit cells. The hexagonal prism represents an appropriate shape for simulation studies on elongated structures such as strand, fibres, tubes or polymer chains. The rhombic dodecahedron and truncated octahedron are the shapes that best approximate a spherical cell. They might therefore be suitable for simulations on spherical structures or large molecules in solution. Compared with a cubic cell of the same number density, distances between image particles are larger for the rhombic dodecahedron and truncated octahedron. This is advantageous for calculating distribution functions [1] (see Sect. 7.8.1), or allows for simulations with fewer particles (solvent molecules) [16, 24]. Finally it should be mentioned that the use of PBC may lead to falsified effects for long range interactions, correlations and fluctuations. For long-range interactions that outreach the dimensionality of the simulation box, accounting for interactions between particles and its own images imposes a symmetry on the system and leads to spurious correlations that do not exist in reality [1]. On the other hand, despite using PBC, finite size effect on the simulation results can still be observed when correlation lengths exceed the system size. Furthermore, the application of PBC suppresses density fluctuations with wavelengths λ outranging the box length. Density fluctuations close to the gas-liquid critical point have long wave length of some 100 nm, clearly illustrated by the phenomenon of critical opalescence, which results from the fact that the size of density fluctuations become comparable to the wavelength of visible light, so that the light is scattered. The correct representation of these long wavelength fluctuations would require box length $L \geq \lambda$, which is general beyond the accessible system size in molecular simulations. It is therefore not possible to perform simulations in the vicinity of the gas-liquid critical point [1].

5.2.1 Minimum Image Convention and Cut-Off Radius

The heart of molecular simulations is the calculation of intermolecular interactions between the particles in the system to derive the configurational energy, and additionally for MD studies, the resulting forces on the particles. Applying PBC, i.e. surrounding the simulation box by an infinite number of identical periodic copies theoretically entail calculations of interactions between all periodic images. In many classical potential energy functions (force fields), the Pauli repulsion at short ranges and the attractive dispersive van-der Waals interactions at larger distances between two particles i and j are modeled by the (12-6)-Lennard-Jones (LJ) potential (see Sect. 6.1)

$$u_{ij}^{LJ} = 4\varepsilon_{ij}\left[\left(\frac{\sigma_{,ij}}{r_{ij}}\right)^{12} - \left(\frac{\sigma_{ij}}{r_{ij}}\right)^{6}\right]. \tag{5.4}$$

Therein, ε_{ij} is the depth of the potential well, and σ_{ij} is the distance at which the intermolecular potential is zero, as shown in Fig. 5.6.

The depiction of the interaction energy between two argon atoms in Fig. 5.6 illustrates that dispersive interactions drop quite rapidly with the intermolecular distance. Thus, for the LJ potential—or for short-range pair potential energy functions in

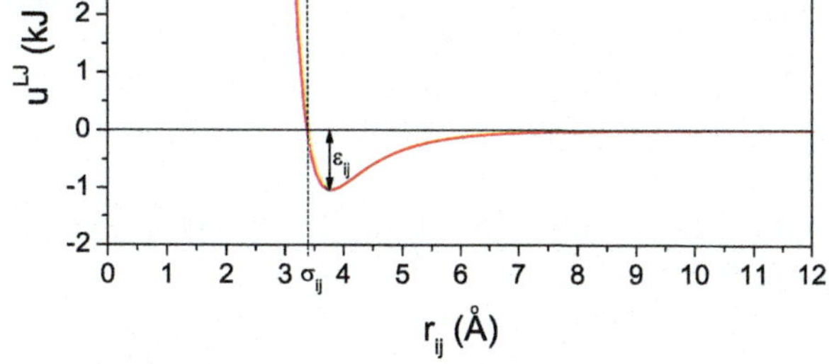

Fig. 5.6 Interaction between two argon atoms, described by the 12-6-Lennard-Jones potential with parameters from White [31]

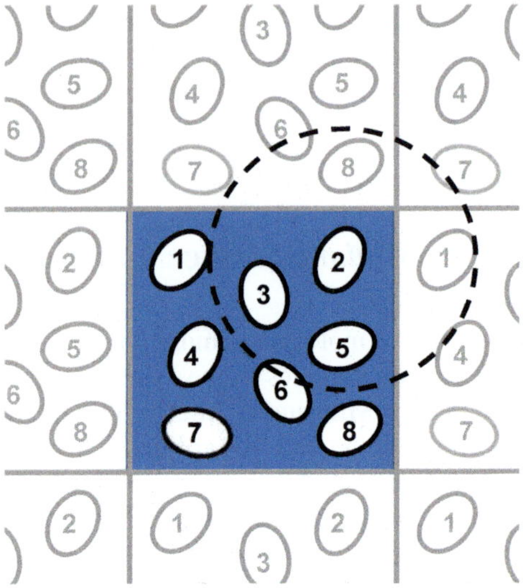

Fig. 5.7 2-D visualization of the minimum image convention (MIC)

general—the largest contribution to the configurational energy and to the forces on a tagged molecule i arise from its nearest neighbors, so that interactions with distant periodic image particles can be neglected. Consequently, the Minimum Image Convention (MIC) can be applied, which states that only interactions with the nearest periodic image of a particle needs to be considered. From this follows that we only consider interactions with (image) particles that are located within a spherical radius of $r_{cut,MIC} = L/2$ around our particle i of interest when we intend to calculate its configurational energy or the force acting on it. Setting the so called cut-off radius to $r_{cut,MIC} = L/2$ secures not only that a particle interacts at most with one image of the other particles, but also that it never interacts with one of its own periodic images. The application of the MIC is illustrated by Fig. 5.7 for the calculation of the interactions of particle #2. The particles #1 and #8 in the central box are located outside $r_{cut,MIC}$, whereas their periodic images in the right and upper periodic cell are within. Thus, only the interactions with the images of particles #1 and #8 are considered when determining the energy of particle #2.

The MIC is realized by employing a wraparound effect to the calculation of the pair separations r_{ij} similar to that discussed for the positions in Eq. 5.2. The implementation of the MIC for a cubic simulation cell with boxlength L is given by

$$\begin{aligned} &\text{if } \; r_{ij,x} > L/2 \quad r_{ij,x}^{new} = r_{ij,x} - L \\ &\text{if } \; r_{ij,x} < -L/2 \; r_{ij,x}^{new} = r_{ij,x} + L, \end{aligned}$$

and is accordingly applied to the $r_{ij,y}$ and $r_{ij,z}$ components of the distance vector $\mathbf{r}_{ij} = \mathbf{r}_j - \mathbf{r}_i$.

For large simulation systems and short-range energy functions, particles located within $r_{cut,MIC} = L/2$ may still be so distant from the particle i that their contribution to the configurational energy or force on i is negligible. Then, the calculation of interaction energies is restricted to the nearest neighbors by applying a spherical cut-off of $r_{cut} < r_{cut,MIC} = L/2$. In the following, we assume that the contribution of intermolecular interactions to the configurational energy of the system U_{conf} is determined by the sum of all interaction energies between molecular pairs, which is described by a pair potential $u_{ij}(r_{ij})$. The implementation of a spherical cut-off in the simulation is then realized by ignoring pair interactions u_{ij} between two particles i and j for which the length of their distance vector $\mathbf{r}_{ij}$ exceeds r_{cut}

$$u_{ij}(r_{ij}) = \{ \begin{matrix} u_{ij}(r_{ij}) & r_{ij} \leq r_{cut} \\ 0 & r_{ij} > r_{cut} \end{matrix} . \tag{5.5}$$

For each single pair interaction energy, the contribution for $r_{ij} > r_{cut}$ might be negligible small, though the number of neighboring particles increases rapidly with the pair separation r_{ij} [10]. Thus, the sum of all neglected interaction energies for $r_{ij} > r_{cut}$ might result in a significant error in the calculation of the configurational energy of the system or of its particles. The error increases with decreasing value of r_{cut}, and can be minimized by choosing an adequate cut-off radius. Additionally, the error arising from ignored interaction energies at larger distances may be corrected by applying so called long-range corrections (LRC). This is based on the relation that for a given pair potential $u_{ij}(r_{ij})$, the total interaction energy of a system with N particles can be determined from

$$U^{total} = \frac{N\varrho}{2} \int_{r_{ij}=0}^{\infty} 4\pi r_{ij}^2 g_{ij}(r_{ij}) u_{ij}(r_{ij}) dr_{ij}. \tag{5.6}$$

Therein, ϱ is the system density, and $g_{ij}(r_{ij})$ is the pair distribution function, which gives the probability of finding the molecule j at distance r_{ij} apart from molecule i, relative to the expected probability for a uniform distribution (see Sect. 7.8.1). For large distances, i.e. $r_{ij} > r_{cut}$, it can be assumed that the distribution of particles is only weakly correlated so that $g_{ij}(r_{ij}) = 1$. With this, the long range (tail) correction to the pair interaction energy can be estimated from

$$U^{LRC} = \frac{N\varrho}{2} \int_{r_{cut}}^{\infty} 4\pi r_{ij}^2 u_{ij}(r_{ij}) dr_{ij}. \tag{5.7}$$

For simulations in the NVT-ensemble with a fixed value for r_{cut}, U^{LRC} is a constant value, which only has to be determined once in the course of the simulation. For simulations in the NpT-ensemble, though, the long range correction needs to be recalculated every time the density changes. Tail corrections are not only added to the interaction energy but also to the resulting thermophysical properties of the system such as its pressure, or the chemical potential of the fluid. The expressions for the LRC of these properties are provided in Chap. 7.

The Minimum Image Convention or even shorter cut-off radii can only be applied when the resulting systematic error in the calculation of the interaction energy stays small. This requires that the tail correction to the interaction energy U^{LRC} takes a finite value, which demands in turn that the pair potential $u_{ij}(r_{ij})$ in Eq. 5.7 decays more rapidly than r_{ij}^{-3} [10]. This condition is satisfied for dispersive, van-der Waals interactions (see Eq. 5.4 or Sect. 6.1), but not for electrostatic interactions (see Eq. 5.11 or Sect. 6.1). Thus, in the calculation of long-range electrostatic interactions, all periodic images have to be considered explicitly, and the proper handling of these interactions requires special methods such as the Ewald sum that is discussed in Sect. 5.4.

5.3 Neighbor Lists

The heart and most time consuming part of each molecular simulation is the calculation of the configurational energy and of the resulting forces on the particles of the system. Assuming pairwise additivity of interaction energies, the evaluation of the contribution of intermolecular interactions to the configurational energy of a system of N interaction sites (e.g. atoms), requires the consideration of $\frac{1}{2}N(N-1)$ pairs. For each pair of particles i and j, the evaluation of their interaction consists of two steps

1. Compute their distance $r_{ij}^2 = \left(r_{i,x} - r_{j,x}\right)^2 + \left(r_{i,y} - r_{j,y}\right)^2 + \left(r_{i,z} - r_{j,z}\right)^2$, subject to Periodic Boundary Conditions (see Sect. 5.2).
2. For $r_{ij} \leq r_{cut}$, evaluate the pair potential to derive the contribution to the configurational energies of i and j, and to the resulting forces acting on them.

The introduction of a spherical cut-off reduces the number of interaction pairs for which the pair potential has to be evaluated in step 2. Though it still requires the computation of the distances between every interaction pair to decide whether their interaction may be ignored or not. The computation time that is wasted for the calculation of all distances $r_{ij} > r_{cut}$ could be saved by employing book-keeping schemes of 'relevant' neighbors. Verlet [29] first proposed a book-keeping technique referred to as Verlet list or Verlet neighbor list. This scheme is discussed in the following Sect. 5.3.1. Section 5.3.2 than provides an introduction to the cell list technique. Though it should be noted that any neighbor list technique involves additional computational overhead for generating and updating the lists. For small systems, the additional computational effort of maintaining the lists might outweigh any efficiency gains in the calculation of pair interactions. Thus, there is always a minimum system size below which the implementation of the list technique doesn't pay off.

5.3.1 Verlet Neighbor List

The Verlet list method [29] generates and updates a list of all particles j that lie within a distance

$$r_{List} = r_{cut} + \Delta r$$

of an atom i, as illustrated by Fig. 5.8.

The list is then used to directly identify the nearest neighbors within r_{cut} of each tagged atom i. The "skin" Δr as reservoir ensures that no atom j initially outside r_{List}, i.e. not listed as neighbor of i, approaches the atoms i closer than r_{cut} before the Verlet list is updated. The use of the Verlet list can noticeable reduce the computational effort for the evaluation of the intermolecular interactions when the number of particles stored in the list, N_{List}, is significantly less than the total number of particles in the system N, i.e.

$$N_{List} = \frac{4}{3}\pi \varrho r_{List}^3 \ll N, \tag{5.8}$$

with ϱ being the systems density. This would suggest to choose the value for Δr as small as possible. Though, the Verlet list has to be updated each time the sum of distances covered by any two atoms exceeds Δr [16]. Therefore, a small Δr results in a high update frequency, and the method might become inefficient due to the computing time taken to update the Verlet list. Thus, the efficiency of the Verlet list scheme clearly depends on an appropriate value for Δr, and it is commonly chosen in such a way that the list only has to be updated every 10 to 20 steps [10]. Sutmann and Stegailov [26] proposed a model to determine the optimal skin radius for the most efficient performance on the Verlet list.

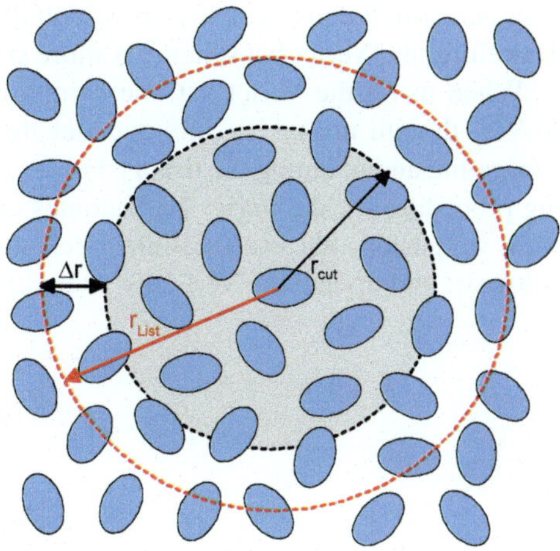

Fig. 5.8 Illustration of the Verlet list: it stores all particles within r_{List} of atoms i although only interaction with neighbors within r_{cut} are calculated

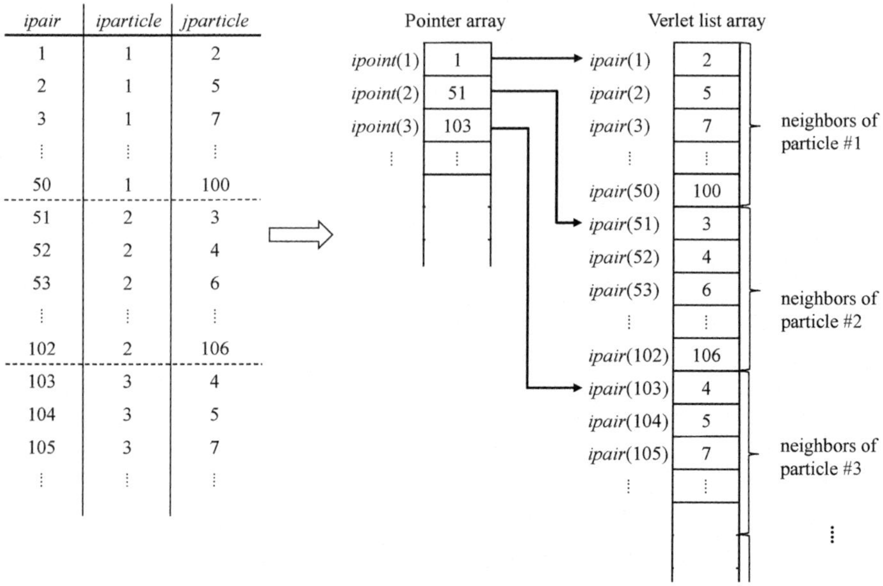

Fig. 5.9 Verlet list book-keeping scheme using list and pointer array

As mechanism to identify neighbors of a specific particle i, two arrays are used: one array is the Verlet list that stores all neighbors, starting with listing all neighbors of particle #1, then those of particle #2 etc. The second array then is a pointer array *ipoint*(i) that indicates the first particle in the list of neighbors of particle i as demonstrated in Fig. 5.9.

In general, the Verlet list method can be applied to both Monte Carlo and molecular dynamics simulations. Though, in MC simulations, it limits the maximal possible displacement δr_{max} (see Sect. 3.1), so that it might only be efficient in simulations at large enough densities [10], for which the δr_{max} value is generally small anyway. Therefore, the Verlet list scheme is more commonly used in molecular dynamics.

When using the Verlet list, the computing time scales roughly $\propto N$, instead of $\propto N^2$—though not taking into account the additional computational overhead for generating and updating the list [1]. For large systems with many particles, the updating procedure for the Verlet list becomes computationally costly. Thus, approaches were suggested that use the linked-cell methods described in the next section to construct and update the Verlet list more efficiently (see for instance [2, 10]).

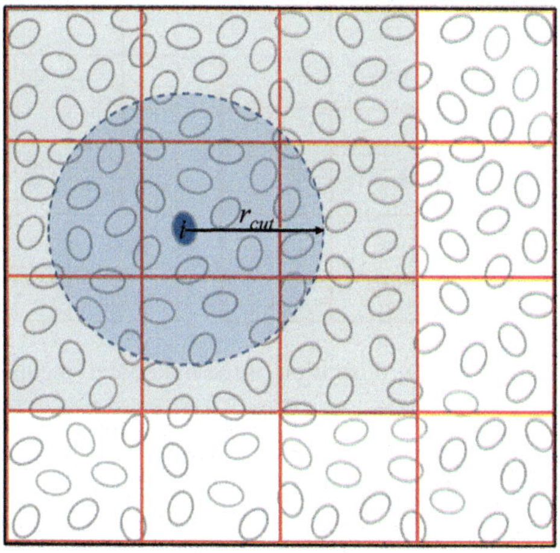

Fig. 5.10 Illustration of the cell list

5.3.2 *Cell List and Linked-Cell Methods*

In the Cell list scheme [15, 22], the simulation box is subdivided into a regular lattice of cells with side length $L_{Cell} \geq r_{cut}$ as illustrated in Fig. 5.10.

The particles are then sorted in these cells according to their positions. Thereby, a linked-list array is created that links together particles belonging to the same cell. For computing the intermolecular interactions of particle i, only molecules in the same and the nearest neighbor cells are considered. In a 3-dimensional cubic box, this means, that 27 cell lists are evaluated, and the number of particles for which the distance to particle i has to be determined is

$$N_{Cell} = 27\varrho L_{Cell}^3. \tag{5.9}$$

Setting $L_{Cell} = r_{cut}$ and $r_{List} = 1.1 \cdot r_{cut}$ gives a ratio of particles considered in the Verlet List (see Eq. 5.8) and the Cell list of

$$\frac{N_{List}}{N_{Cell}} = \frac{\frac{4}{3}\pi\varrho 1.1^3 r_{cut}^3}{27\varrho r_{cut}^3} \approx \frac{1}{5}. \tag{5.10}$$

This suggests that the Verlet list scheme is much more efficient than the Cell list, as considerably less distances r_{ij} need to be evaluated. However, for systems with a large number of particles, the Verlet list is also large, and the updating procedure becomes computationally costly as it requires the recalculation of all distances between pairs of particles [16]. The number of particles listed for each single cell is then significantly lower than the number of particles stored in a Verlet list. Therefore, the updating procedure, but also the searching for relevant neighbors can become more efficient for the Cell list scheme than the Verlet scheme. Additionally, the Cell list method can also be used in MC simulations with random displacements of the particles [10].

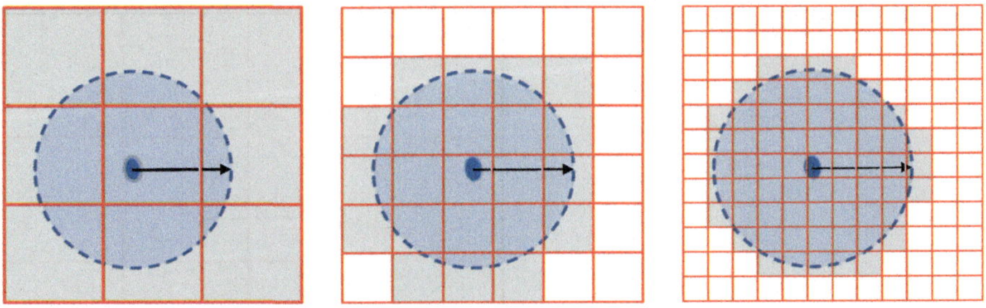

Fig. 5.11 2D-illustration of the linked-cell method with cell sizes $L < r_{cut}$

However, when relevant interaction partners of a tagged particle i are found within an interaction sphere of radius r_{cut}, checking all particles lying in the 27 cells with $L \geq r_{cut}$ for their relative distances to i represents an immense overhead of unnecessary distance calculations. Sutmann and Stegailov [26] determined that even in the best-case of $L = r_{cut}$, only 15% of the calculated particle distances are relevant. Thus, different approaches were suggested to improve the efficiency of the linked-cell method. One idea that goes back to Allen and Tildesley [1] is to reduce the cell size L, which then means, that not only next neighboring cells are considered in the check for interaction partners, but all cells within r_{cut} of particle i, as illustrated by Fig. 5.11.

In the extreme example, the cell size is so far reduced that no more than one particle occupies a cell [20]. This results in a better approximation of the volume of the interaction sphere and reduces significantly the number of unnecessary interparticle distance calculations. However, this gain in efficiency might be offset by the increase in computational effort and memory to maintain the list of neighboring cells [20] and for the inspection of the large number of cells [12].

An alternative approach to increase the efficiency of cell-linked lists was proposed by Gonnet [12]. In his algorithm, the particles are first sorted along their projection onto the axis that connects the centers of neighboring cells. Then only interactions are considered when the interparticle distance along the axis is $\leq r_{cut}$. In [32], Willing and Germano have investigated the efficiency of different linked-cell algorithms for a wide range of simulation set-ups.

5.4 Treatment of Electrostatic Interaction

In classical analytical potential energy functions (force fields), electrostatic interactions between molecules are in general modeled by locating static fractional charges, point charges q_i, at their atomic nuclei (see Sect. 6.1). The interaction between two point charges q_i and q_j is then described by the Coulomb term

$$u^{qq} = \sum \frac{q_i q_j}{4\pi\varepsilon_0 r_{ij}}, \tag{5.11}$$

wherein $\varepsilon_0 = 8.854 \cdot 10^{-12}\,\mathrm{C^2N^{-1}m^{-2}}$ is the permittivity or dielectric constant in vacuo. This means that interactions between two charges decay with r^{-1} and with this, their range exceeds $L/2$. Thus, it is not valid to employ spherical cut-off radii of $r_{cut} \leq L/2$ for the long-range electrostatic interaction. In fact the proper handling of the long-range electrostatic interaction requires not only to consider interactions with all other particles in the (central) simulation box, but also with the infinite number of their periodic images. With this, the total electrostatic interaction due to all pairs of charges is given by

$$U^{el} = \frac{1}{2}\sum_{n}^{\prime}\sum_{i=1}^{N}\sum_{j=1}^{N} \frac{q_i q_j}{4\pi\varepsilon_0 \left(\mathbf{r}_{ij} + \mathbf{n}L\right)}, \quad \mathbf{n} = \begin{pmatrix} n_x \\ n_y \\ n_z \end{pmatrix}, n_{x,y,z} = 0, 1, \ldots \tag{5.12}$$

when assuming a simple cubic cell shape with boxlength L. For the central simulation box ($\mathbf{n} = 0$) the interaction between q_i and q_j for $i = j$ has to be omitted, which is indicated by the prime symbol $'$ in Eq. 5.12 [1].

The sum in Eq. 5.12 is 'conditionally convergent', meaning that it does not converge absolutely but that the result depends on the order of the summation. The most accurate but also computationally most expensive approach to handle the summation in Eq. 5.12 is the Ewald summation [7] that is discussed in the following Sect. 5.4.1. The Smooth Particle Mesh method introduced in Sect. 5.4.2 then represents a modification to the original Ewald sum to reduce the computational costs for the evaluation of electrostatic interactions for large system sizes. A simple though quite accurate truncation method to evaluate Eq. 5.12 was proposed by Wolf et al. [33]. The Wolf damped Coulomb potential and its modifications are discussed in Sect. 5.4.3. A review on non-Ewald methods to calculate electrostatic interactions is for instance provided by Fukuda and Nakamura [11].

5.4.1 Ewald Summation

The trick of the Ewald method to transform the conditionally convergent sum over all pairs of charges in Eq. 5.12 into a rapidly and absolutely converging summation is to convert it into two series. In the first step, each charge q_i is surrounded by a neutralizing diffuse charge cloud—i.e. the total charge of the cloud has the same magnitude as q_i but is of opposite sign, and it dispreads symmetrically from the charge as illustrated by Fig. 5.12.

Usually, a Gaussian charge distribution is used to represent the diffuse cloud surrounding the charge q_i. The charge density $\varrho_{i,Gauss}$ of the Gaussian charge distribution is given by [10]

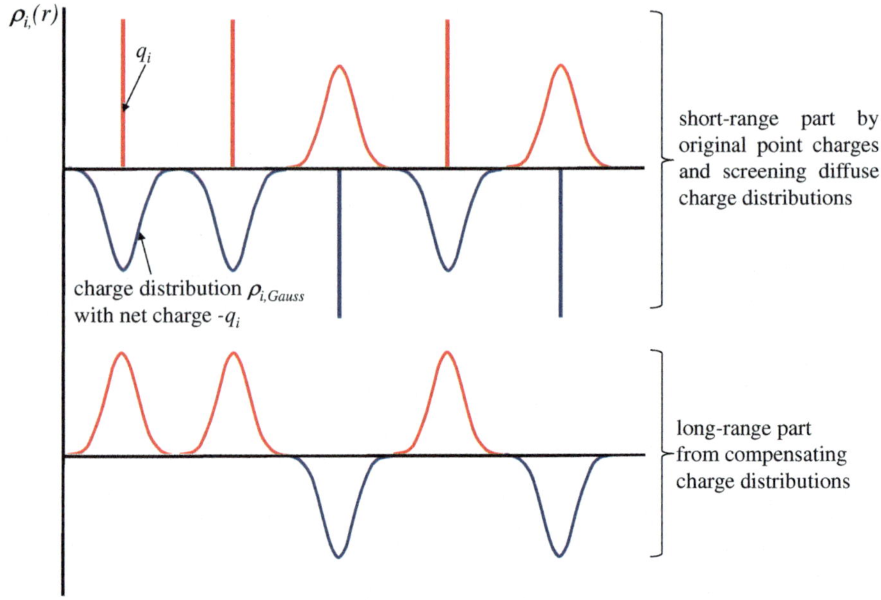

Fig. 5.12 Charge distribution in the Ewald summation [1]

$$\varrho_{i,Gauss}(r) = q_i \left(\frac{\alpha}{\pi}\right)^{\frac{3}{2}} \exp\left(-\alpha r^2\right), \tag{5.13}$$

with the parameter α (or rather $\sqrt{2/\alpha}$) defining the width of the distribution. The electrostatic potential field $\phi_{i,Gauss}$ generated by the charge distribution can be derived from the Poisson's equation. In SI units the relation is given by

$$-\nabla^2 \phi_{i,Gauss}(r) = \frac{\varrho_{i,Gauss}(r)}{\varepsilon_0}. \tag{5.14}$$

Though, in electrostatics, Gaussian (CGS) units are more commonly used, so that the Ewald summation is also usually derived in Gaussian notation. With this, the Poisson's equation of electrostatics becomes

$$-\nabla^2 \phi_{i,Gauss}(r) = 4\pi \varrho_{i,Gauss}(r), \qquad \nabla^2 = \frac{1}{r^2}\frac{\partial}{\partial r}\left(r^2 \frac{\partial}{\partial r}\right). \tag{5.15}$$

Taking into account the spherical symmetry of the Gaussian charge distribution, the Poisson's equation can be expressed by

$$-\frac{1}{r}\frac{\partial^2 r \phi_{i,Gauss}(r)}{\partial r^2} = 4\pi \varrho_{i,Gauss}(r). \tag{5.16}$$

Two successive partial integrations then yield

$$\phi_{i,Gauss}(r) = \frac{q_i}{r} erf\left(\sqrt{\alpha} r\right) \tag{5.17}$$

with

$$erf\left(\sqrt{\alpha} r\right) = 2\sqrt{\frac{\alpha}{\pi}} \int_0^r \exp\left(-\alpha r^2\right) dr \tag{5.18}$$

(see for instance the Appendix B of [10] for more details).

The total electrostatic potential of the point charge q_i and its screening Gaussian charge cloud with opposite net charge $-q_i$ is given by

$$\begin{aligned}\phi_{i,short-range}(r) &= \frac{q_i}{r} + \phi_{i,Gauss}(r) = \frac{q_i}{r} - \frac{q_i}{r} erf\left(\sqrt{\alpha} r\right) \\ &= \frac{q_i}{r} erfc\left(\sqrt{\alpha} r\right).\end{aligned} \tag{5.19}$$

The electrostatic potential of the screened charge now rapidly decays with the inter-charge distance r. When the value for α is chosen large enough, the interactions are limited to a short range so that it does not require summation beyond the central simulation box. The contribution to the configurational energy from the so called short-range interactions between all screened charges in the central box ($\mathbf{n} = 0$) is given by

$$\begin{aligned}U_{short-range} &= \frac{1}{2}\sum_{i=1}^{N}\sum_{j=1}^{N} q_i \phi_{j,short-range} \\ &= \frac{1}{2}\sum_{i=1}^{N}\sum_{j=1}^{N} \frac{q_i q_j}{r_{ij}} erfc\left(\sqrt{\alpha} r_{ij}\right).\end{aligned} \tag{5.20}$$

Due to the error function in Eq. 5.20, the summation now rapidly converges.

To correct for the induced screening charge distribution, a second compensating charge distribution with the same shape as $\varrho_{i,Gauss}$ but opposite sign has to be added that counteract the first distribution (see Fig. 5.12). Then however, also interactions of the second compensating charge distribution have to be taken into account. The electrostatic potential of this second compensating charge distribution is long-ranged so that also interactions with its periodic images have to be considered. The summation of the compensating charge distribution is therefore performed in the reciprocal space where it rapidly converges. The charge distribution at a point r_i as sum of all periodic Gaussian (see Eq. 5.13) is given by [10]

$$\varrho_{long-range}(r_i) = \sum_n \sum_{j=1}^{N} q_j \left(\frac{\alpha}{\pi}\right)^{\frac{3}{2}} \exp\left[-\alpha \left|\mathbf{r}_i - \left(\mathbf{r}_j + \mathbf{n}L\right)\right|^2\right]. \tag{5.21}$$

The Fourier transformation of the charge distribution $\varrho_{long-range}(r_i)$ is given by (see for instance [10] for more details)

$$\begin{aligned}\varrho_{long-range}(\mathbf{k}) &= \int_V \exp\left(-i\mathbf{k}\cdot\mathbf{r}\right) \varrho_{long-range}(r) dr \\ &= \sum_{j=1}^{N} q_j \exp\left(-i\mathbf{k}\cdot\mathbf{r}_j\right) \exp\left(-k^2/4\alpha\right),\end{aligned} \tag{5.22}$$

with the reciprocal vector $\mathbf{k} = 2\pi\mathbf{n}/L^2$ [1]. In the Fourier form, the Poisson's equation of electrostatics in Gaussian notation is given by

$$k^2 \phi_{i,Gauss}(k) = 4\pi \varrho_{i,Gauss}(k). \tag{5.23}$$

Thus, by multiplying Eq. 5.22 by $4\pi/k^2$, we obtain the electrostatic potential due to the sum of periodic Gaussians in the Fourier form

$$\phi_{long-range}(\mathbf{k}) = \frac{4\pi}{k^2} \sum_{j=1}^{N} q_j \exp\left(-i\mathbf{k}\cdot\mathbf{r}_j\right) \exp\left(-k^2/4\alpha\right). \tag{5.24}$$

Applying Inverse Fourier transformation gives

$$\begin{aligned}\phi_{long-range}(\mathbf{r}_i) &= \frac{1}{V} \sum_{k\neq 0} \phi_{long-range}(\mathbf{k}) \exp\left(-i\mathbf{k}\cdot\mathbf{r}\right) \\ &= \frac{1}{V} \sum_{k\neq 0} \sum_{j=1}^{N} \frac{4\pi}{k^2} q_j \exp\left[-i\mathbf{k}\cdot\left(\mathbf{r}_i - \mathbf{r}_j\right)\right] \exp\left(-k^2/4\alpha\right).\end{aligned} \tag{5.25}$$

With this, the so called long-range (or reciprocal sum) contribution to the configurational energy due to interactions of the compensating charge distribution can be derived as

$$\begin{aligned}U_{long-range} &= \frac{1}{2} \sum_{i=1}^{N} \sum_{j=1}^{N} q_i \phi_{long-range}(r_i) \\ &= \frac{1}{2} \sum_{k\neq 0} \sum_{i=1}^{N} \sum_{j=1}^{N} \frac{4\pi}{Vk^2} q_i q_j \exp\left[-i\mathbf{k}\cdot\left(\mathbf{r}_i - \mathbf{r}_j\right)\right] \exp\left(-k^2/4\alpha\right).\end{aligned} \tag{5.26}$$

Though, the long-range contribution to the configurational energy in Eq. 5.26 includes a spurious interaction of each Gaussian with itself. To correct for it, we have to subtract the configurational energy at the origin of each Gaussian charge distribution. The electrostatic potential $\phi_{i,Gauss}(r = 0)$ can be derived from Eq. 5.17

$$\phi_{self}(r_i) = \phi_{i,Gauss}(r = 0) = 2q_i\sqrt{\frac{\alpha}{\pi}}. \tag{5.27}$$

With this the contribution to the configurational energy from the self interaction can be expressed by

$$U_{self} = \frac{1}{2}\sum_{i=1}^{N} q_i\phi_{self}(r_i) = \sqrt{\frac{\alpha}{\pi}}\sum_{i=1}^{N} q_i^2. \tag{5.28}$$

It should be noted that the contribution U_{self} does not depend on the particle positions and with this is a constant term in the course of the simulation.

For molecular systems with partial charges located on atomics sites, an additional correction term is required to subtract contributions from intramolecular interactions of a charge on atom a, $q_{i,a}$, and the distributed charges centered at the other atoms within the same molecule i [1]. The correction term is given by

$$\begin{aligned} U_{corr} &= \frac{1}{2}\sum_{i=1}^{N}\sum_{a=1}^{N_{i,atoms}}\sum_{b=1}^{N_{i,atoms}} q_{i,a}\phi_{ib,Gauss} \\ &= \frac{1}{2}\sum_{i=1}^{N}\sum_{a=1}^{N_{i,atoms}}\sum_{b=1}^{N_{i,atoms}} \frac{q_{i,a}q_{i,b}\, erf\,(\alpha\, r_{iab})}{r_{i,ab}}, \end{aligned} \tag{5.29}$$

therein $N_{i,atoms}$ is the number of atoms within the molecule i, and $r_{i,ab}$ is the intramolecular distance between the two atoms a and b in i.

The total contribution of electrostatic interactions to the configurational energy of the system is then described by

$$U^{el} = U_{short-range} + U_{long-range} - U_{self} - U_{corr}. \tag{5.30}$$

In order to determine the electrostatic interactions in SI units, the terms in the Eqs. 5.20, 5.26, 5.28 and 5.29 have to be multiplied by $(4\pi\varepsilon_0)^{-1}$. With this the Ewald sum becomes

$$
\begin{aligned}
U^{el} &= \frac{1}{2}\sum_{i=1}^{N}\sum_{j=1}^{N}\frac{q_i q_j}{4\pi\varepsilon_0 r_{ij}} erfc\left(\sqrt{\alpha} r_{ij}\right) \\
&+ \frac{1}{2}\sum_{k\neq 0}\sum_{i=1}^{N}\sum_{j=1}^{N}\frac{1}{\varepsilon_0 V k^2} q_i q_j \exp\left[-i\mathbf{k}\cdot\left(\mathbf{r}_i - \mathbf{r}_j\right)\right]\exp\left(-k^2/4\alpha^2\right) \\
&+ \sqrt{\frac{\alpha}{\pi}}\sum_{i=1}^{N}\frac{q_i^2}{4\pi\varepsilon_0} \\
&+ \frac{1}{2}\sum_{i=1}^{N}\sum_{a=1}^{N_{i,atoms}}\sum_{b=1}^{N_{i,atoms}}\frac{q_{i,a} q_{i,b}\, erf\left(\alpha r_{i,ab}\right)}{4\pi\varepsilon_0 r_{i,ab}}.
\end{aligned}
\tag{5.31}
$$

The accuracy of the Ewald sum to reproduce the exact coulombic energy and its convergence behavior depends on the parameter α, the real space cut-off r_{cut} and the largest reciprocal space vector $\mathbf{k}_{max}$ that determines the number of cells included in the reciprocal sum. Several studies in literature have addressed the proper choice of the Ewald parameters with regard to both accuracy and efficiency (see for instance [28, 30]). A large value for α yields a narrow Gaussian distribution and results in a fast converging real-space sum in $U_{short-range}$ (Eq. 5.20). Though the number of terms that have to be included in the Fourier sum in $U_{long-range}$ (Eq. 5.26) increases with α so that the summation in reciprocal sum faster converges for small α values. Thus, real space and reciprocal summation need to be balanced. Recommended value for α and k_{max} are for instance [24, 28]

$$
\alpha = \frac{3.2}{r_{cut}} \ldots \frac{3.5}{r_{cut}}, \qquad k_{max} > \frac{3.2L}{r_{cut}}. \tag{5.32}
$$

Though, some simulation programs such as DL_POLY classic [24] provide estimates for α and k_{max} for a user defined required accuracy of the Ewald sum.

Although the Ewald summation represent the most accurate method to handle long-range electrostatic interactions, it induces an artificial periodicity in the system which might have an impact on simulation results for inherent non-periodic systems. Hünenberg and McCamman [14] for instance have studied the influence of this effect in simulations of biomolecules with explicit solvent. The large system size of biomolecular simulations have also motivated modifications to the classical Ewald summation such as the Smooth Particle Mesh Ewald (SPME) [6] method that is discussed in the next section.

5.4.2 *Smooth Particle Mesh Ewald (SPME)*

The Particle Mesh method represents an approach to improve the efficiency for the computation of the reciprocal part of the Ewald sum. It is based on the fact that the

Poisson equation (Eq. 5.23) can be solved more efficiently when the charge distribution is approximated by a gridded distribution with charges attributed to mesh points of a three dimensional rectangular grid. This yields a discretized Poisson equation that can then be solved by Fast Fourier Transform (3DFFT). This approach was first introduced by Hockney and Eastwood [15] in their particle-particle particle-mesh (PPPM) algorithm. Several modifications of this method have been published until then, and the interested reader is referred to reviews on the topic, for instance by Deserno and Holm [5]. This section only covers the Smooth Particle Mesh Ewald method (SPME) by Essmann et al. [6], which is nowadays widely used, and implemented in popular MD simulation packages such as NAMD [21], DL_POLY_4 [27], or AMBER [3].

In the SPME method, complex B-spline functions are used as basis functions to interpolate the structure factor $S(\mathbf{k})$ of the reciprocal space onto a regular grid. The use of continuous B-spline functions allows to obtain forces by analytical differentiation of the electrostatic energies. The forces themselves are then also smooth continuous functions of the particle positions.

The SPME method is based on an approximation to the reciprocal space sum by approximating its structure factor $S(\mathbf{k})$, which is defined by [6]

$$S(\mathbf{k}) = \sum_{j=1}^{N} q_i \exp\left(2\pi i \mathbf{k}\mathbf{r}_j\right). \tag{5.33}$$

With this, the long-range or reciprocal term of the Ewald sum in Eq. 5.26 can be rewritten

$$U_{long-range} = \frac{1}{2\pi V} \sum_{k \neq 0} \frac{\exp\left(-\pi^2 \mathbf{k}^2/\alpha^2\right)}{\mathbf{k}^2} S(\mathbf{k}) S(-\mathbf{k}). \tag{5.34}$$

When K_1, K_2, K_3 are the total number of grid points in each direction, the coordinates $\mathbf{r}$ can be replaced by scaled coordinates $\mathbf{u}$ so that the exponential term in Eq. 5.33 can be expressed by

$$\exp\left(2\pi i \mathbf{k}\mathbf{r}_j\right) = \exp\left(2\pi i \frac{k_1 u_{j,1}}{K_1}\right) \cdot \exp\left(2\pi i \frac{k_2 u_{j,2}}{K_2}\right) \cdot \exp\left(2\pi i \frac{k_3 u_{j,3}}{K_3}\right). \tag{5.35}$$

Therein, the k_1, k_2, k_3 are the integer indices of the reciprocal vector $\mathbf{k}$ for the three principal directions [24]. Cardinal B-splines $M_n(u)$ are then applied for the interpolation of the complex exponential terms in Eq. 5.35. The interpolation term for one dimension is given by

$$\exp\left(2\pi i\,\frac{k_1 u_{j,1}}{K_1}\right) \approx b(k_1)\sum_{l=-\infty}^{\infty} M_n(u_{j,1}-l)\exp\left(2\pi i\frac{k_1 l}{K_1}\right), \tag{5.36}$$

and for the other two directions accordingly. The coefficients $M_n(u)$ are B-splines of order n, and the constant $b(k)$ is defined by [24]

$$b(k) = \frac{\exp\left(2\pi i\frac{(n-1)k}{K}\right)}{\sum_{l=0}^{n-2} M_n(l+1)\exp\left(2\pi i\,\frac{lk}{K}\right)}. \tag{5.37}$$

With this, the structure factor in Eq. 5.33 is approximated by

$$S(\mathbf{k}) \approx b_1(k_1)b_2(k_2)b_3(k_3)F(Q)(k_1,k_2,k_3), \tag{5.38}$$

wherein $F(Q)$ is the discrete Fourier transform of the charge array Q, defined as

$$\begin{aligned} Q(l_1,l_2,l_3) = \sum_{j=1}^{N} q_j \sum_{n_1,n_2,n_3} & M_n(u_{j,1}-l_1-n_1K_1)\cdot M_n(u_{j,2}-l_2-n_2K_2) \\ & \cdot M_n(u_{j,3}-l_3-n_3K_3). \end{aligned} \tag{5.39}$$

The sum over n_1, n_2, n_3 accounts for contributions from relevant periodic cells. With the approximated structure factor defined in Eq. 5.38, the approximated long-range (reciprocal space) contribution to the electrostatic energy can be evaluated

$$\begin{aligned} U_{long-range} = \frac{1}{2\pi V}\sum_{k\neq 0} & \frac{\exp\left(-\pi^2\mathbf{k}^2/\alpha^2\right)}{\mathbf{k}^2}B(k_1,k_2,k_3) \\ & \cdot F(Q)(k_1,k_2,k_3)F(Q)(-k_1,-k_2,-k_3), \end{aligned} \tag{5.40}$$

with

$$B(k_1,k_2,k_3) = |b_1(k_1)|^2\cdot|b_2(k_2)|^2\cdot|b_3(k_3)|^2. \tag{5.41}$$

For MD simulations, the force is determined as exact derivative of the approximated energy. More detailed derivations of the equations, background information on the discrete Fourier transform and B-spline interpolation, as well as a discussion on the accuracy and efficiency of the method is provided in the original paper by Essmann et al. [6]. They also state that the SPME method is more efficient than the conventional Ewald summation for systems containing more than 10.000 atoms.

5.4.3 The Wolf Damped Coulomb Potential

Wolf et al. [33] have observed that the effective Coulomb interactions in condensed ionic systems are actually short ranged. From this, they have concluded that the error in calculating the electrostatic interaction of an ion i from the direct pairwise evaluation

$$u_i^{el} = \sum_{\substack{j \neq i \\ r_{ij} \leq r_{cut}}}^{N} \frac{q_i q_j}{4\pi\varepsilon_0 r_{ij}} \tag{5.42}$$

is due to the fact that the spherical truncation of the Coulomb pair potential at r_{cut} always results in a net charge $\Delta q_i(r_{cut}) = q(N^+ - N^-)$ in the spherical volume around the ion i. They also stated that the error in calculating the correct coulombic interaction energy of ion i when truncating the interaction at r_{cut} is proportional to the resulting net charge $\Delta q_i(r_{cut})$. With this, the correct coulombic energy of ion i can be estimated by charge neutralization, i.e. by simply subtracting the interactions due to the net charge from the energy calculated by Eq. 5.42

$$u_i^{el} = \sum_{\substack{j \neq i \\ r_{ij} \leq r_{cut}}}^{N} \frac{q_i q_j}{4\pi\varepsilon_0 r_{ij}} - u_i^{neutral}(r_{cut}). \tag{5.43}$$

For determining the charge-neutralization potential $u_i^{neutral}(r_{cut})$, Wolf et al. grouped ions A^+ and counterions B^- to dipolar molecules, with their maximum ion distance being b. Then all dipoles separated by the cutoff sphere can be found in a distance $r_{cut} - b < r_{ij} < r_{cut} + b$ from ion i. Assuming that b is small compared to r_{cut}, the net charge can be located at the surface of the spherical volume at r_{cut}, instead of considering the charge distribution in the spherical shell. $u_i^{neutral}(r_{cut})$ is then given by the Coulombic interaction of ion i with the net charge on the surface at r_{cut}

$$u_i^{neutral}(r_{cut}) = \frac{q_i \Delta q_i(r_{cut})}{4\pi\varepsilon_0 r_{cut}}. \tag{5.44}$$

As the net charge Δq_i in the spherical truncated volume is simply the sum of all included charges

$$\Delta q_i(r_{cut}) = \sum_{\substack{j=1 \\ r_{ij} \leq r_{cut}}}^{N} q_j, \tag{5.45}$$

the charge-neutralization potential can also be expressed by

$$u_i^{neutral}(r_{cut}) = \sum_{\substack{j=1 \\ r_{ij} \leq r_{cut}}}^{N} \frac{q_i q_j}{4\pi\varepsilon_0 r_{cut}}. \tag{5.46}$$

When separating the self term ($j = i$) in Eq. 5.46, the total electrostatic energy of the system can finally be determined by

$$U^{el} = \frac{1}{2}\sum_{i=1}^{N}\sum_{\substack{j \neq i \\ r_{ij} \leq r_{cut}}}^{N} \frac{q_i q_j}{4\pi\varepsilon_0}\left(\frac{1}{r_{ij}} - \frac{1}{r_{cut}}\right) - \frac{1}{2r_{cut}}\sum_{i=1}^{N}\frac{q_i^2}{4\pi\varepsilon_0}. \tag{5.47}$$

The first term is interpreted as shifted pair potential, in which charge neutralization is achieved by placing for every charge q_j at distance r_{ij} an image charge of opposite sign on the surface of the truncated spherical volume at r_{cut} [33]. Though the calculated energy of the charge neutralized system as function of r_{cut} approaches the correct electrostatic energy of the system in an oscillatory manner and exhibits slow convergence [11]. In order to faster flatten out the oscillations, Wolf et al. introduced a damping function to the Coulombic potential, and opted for damping via the complementary error function with the damping parameter α. The total electrostatic energy of the system is then given by [11, 34]

$$U^{el} = \frac{1}{2}\sum_{i=1}^{N}\sum_{\substack{j \neq i \\ r_{ij} \leq r_{cut}}}^{N}\left(\frac{q_i q_j}{4\pi\varepsilon_0 r_{ij}} erfc\left(\alpha r_{ij}\right) - \frac{q_i q_j}{4\pi\varepsilon_0 r_{cut}} erfc\left(\alpha r_{cut}\right)\right)$$
$$- \left(\frac{1}{2r_{cut}} erfc\left(\alpha r_{cut}\right) + \frac{\alpha}{\sqrt{\pi}}\right)\sum_{i=1}^{N}\frac{q_i^2}{4\pi\varepsilon_0}. \tag{5.48}$$

The accuracy of the method depends on a suitable choice of the parameters α and r_{cut}. Zahn et al. [34] proposed a two-step strategy to determine the parameters. Demontis et al. [4] proposed

$$\alpha = \frac{2}{r_{cut}}$$

as reasonable choice.

In [33] Wolf et al. derived the forces directly from the damped Coulomb potential. Though Zahn et al. [34] argue that when deriving forces from a shifted potential that converges to zero at r_{cut}, it does not cause the force to be zero at r_{cut} as well. This results in an inconsistency in the calculation of energies and forces, and in a poor energy conservation. Thus, they proposed an alternative approach by first deriving an expression for the shifted force and then deducing a shifted-force potential by integration. Their expression for the electrostatic pair potential is given by [34]

$$u^{el}(r_{ij}) = \frac{q_i q_j}{4\pi\varepsilon_0}\left[\frac{erfc\left(\alpha r_{ij}\right)}{r_{ij}} - \left(\frac{erfc\left(\alpha r_{cut}\right)}{r_{cut}^2} + \frac{2\alpha}{\sqrt{\pi}}\frac{exp\left(-\alpha^2 r_{cut}^2\right)}{r_{cut}}\right)\left(r_{ij} - r_{cut}\right)\right]. \quad (5.49)$$

Though Zahn et al. admit that their approach to derive a force-shifted potential leads to the loss of the intuitive physical meaning of charge neutralization.

Alternative expressions for pairwise electrostatic interaction potentials extended from the Wolf method are for instance discussed by Fennell and Gezelter [8]. They also compare the results of the Wolf variants regarding the reproduction of energies and dynamics to that of the Smooth Particle Mesh method. A brief review on variants of the Wolf method is also provided by Fukuda and Nakamura [11].

5.5 Stages of a Simulation

A molecular simulation consists of four stages as illustrated by Fig. 5.13

1. initialization
2. equilibration
3. production
4. analysis

The first stage, the initialization of molecular simulations is discussed in Sect. 5.1. The setting up of the simulation thereby comprises the selection of initial positions of the molecules within the simulation box, and additionally the assignment of initial velocities to the molecules in case of MD studies. However, these initial positions and velocities are not representative for the conditions of the state we intend to study in the simulation, and the system has first of all to be equilibrated. Therefore, the simulation runs for a period in which the system evolves to configurations that are representative for the state condition of interest. The equilibration can be monitored by recording instantaneous values of various properties such as configurational energy, density (in *NpT*), pressure (*NVT* in gas phase) etc. Figure 5.14 shows exemplarily the equilibration period of a GEMC simulation in the mixture *R*-32 + CO_2. Initially,

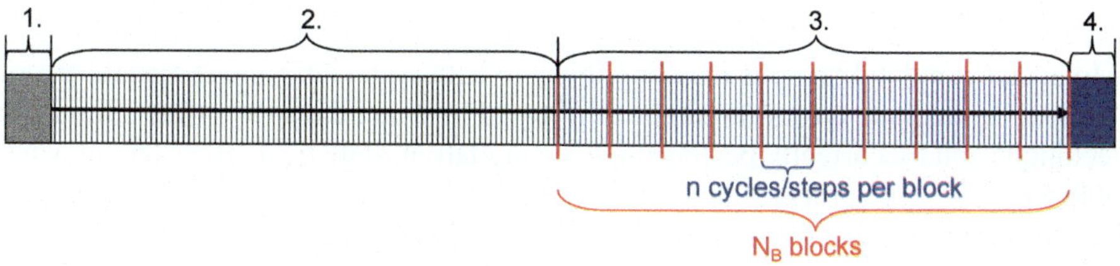

Fig. 5.13 Stages of a molecular simulation, and illustration of block averaging technique

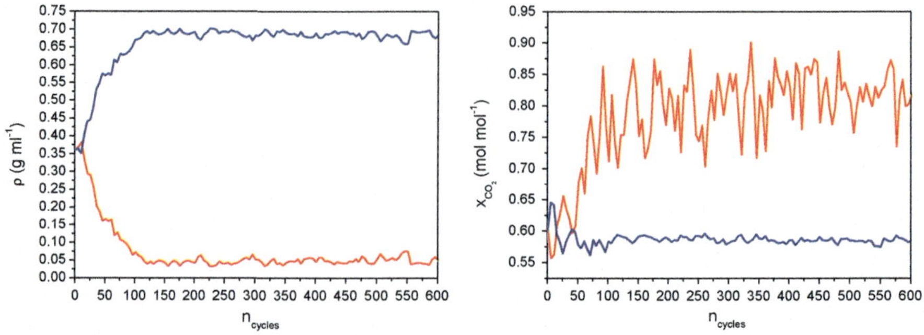

Fig. 5.14 Equilibration period of a GEMC simulation in the mixture R-32 and CO_2 at $T = 273\,K$ and $p = 2$ MPa with $N = 400$ molecules. Each cycle consists of $N = 400$ attempted moves. Due to the smaller number of molecules in the gas phase, its composition is subject of larger fluctuations

the molecules were placed on sites of a cubic lattice, with equal densities and compositions for both boxes (phases). During the equilibration period, one box evolves to the liquid phase (blue) and the other to the vapor phase (red) with the corresponding densities and compositions. The equilibration period has to be extended until the monitored quantities no longer exhibit a systematic drift but start to oscillate about a steady mean value [1].

When equilibration is guaranteed, the production period of the simulation starts, in which the output data are produced to derive ensemble or time averages for the properties of interest. Various properties can be evaluated during the production run as discussed in Chap. 7. Averages are either accumulated during the simulation, or configurations and instantaneous values of the properties of interest are frequently stored during the production run to allow for an analysis of the data after the simulation is completed. In both cases, it is recommended to store configurations and data during the production period and at the end of the simulation to provide sufficient information to reanalyse the simulation results and to restart the simulation. An important aspect of the accumulation of averages is the evaluation of their statistical uncertainties. The next section therefore deals with the widely used block averaging technique to estimated standard errors of ensemble or time averages.

5.5.1 Quantifying Uncertainties by Block Averaging

Molecular simulation results may be subject to both systematic and statistical errors. Whereas systematic errors should be eliminated by using the appropriate simulation technique and by correctly performing the simulation, statistical errors are unavoidable and have to be properly quantified. The widely used block averaging technique was first reported by Flyvbjerg and Petersen [9]. Thereby, the production run consisting of N_{total} cycles (MC) or timesteps (MD) is divided into N_B blocks, each covering

n cycles or steps. Then the average of any property of interest A is calculated for each block

$$\bar{A}_b = \frac{1}{n}\sum_{i=1}^{n} A_i \tag{5.50}$$

giving N_B block averages $\bar{A}_b$. These block averages are then used to determine the ensemble average $\langle A \rangle$

$$\langle A \rangle = \frac{1}{N_B}\sum_{b=1}^{N_B} \bar{A}_b. \tag{5.51}$$

The standard deviation among the block averages, $\sigma(\bar{A})$, can be estimated from [1]

$$\sigma^2(\bar{A}) = \langle A^2 \rangle - \langle A \rangle^2 \approx \frac{1}{N_B}\sum_{b=1}^{N_B} \left[\bar{A}_b - \langle A \rangle\right]^2. \tag{5.52}$$

This is then used to calculate a running estimate of the overall standard errors *BSE* [13]

$$BSE = \frac{\sigma(\bar{A})}{\sqrt{N_B - 1}}. \tag{5.53}$$

In the original approach by Flyvbjerg and Petersen, the size n of the blocks is systematically increased until consecutive block averages are uncorrelated, and the *BSE* cease to vary with n. Then the running estimate of the overall standard error *BSE* asymptotes the true variance of the ensemble average $\sigma\left(\langle A \rangle\right)$.

In practice, the block averaging technique is generally applied in a much simpler way, i.e. by dividing the production run into N_B blocks of n cycles or time steps without variation of the block size (see. Fig. 5.13). The ensemble average of any property $\langle A \rangle$ is derived from the block averages $\bar{A}_b$ as described by the Eqs. 5.51 and 5.50. The standard deviation of the ensemble average is determined by Eq. 5.53 by setting $\sigma\left(\langle A \rangle\right) = BSE$, i.e.

$$\sigma\left(\langle A \rangle\right) = \frac{\sigma(\bar{A})}{\sqrt{N_B - 1}}. \tag{5.54}$$

This simplified procedure though requires that the block size n is chosen large enough to yield independent consecutive block averages. For studying dynamic properties in MD studies, this means that the block length has to be significantly larger than the correlation time (see Sect. 7.2.1).

References

1. Allen M, Tildesley DJ (1987) Computer simulation of liquids. Oxford Science Publication
2. Auerbach DJ, Paul W, Lutz C, Bakker AF, Rudge WE, Abraham FF (1987) A special purpose parallel computer for molecular dynamics: motivation, design, implementation, and application. J Phys Chem 91:4881–4890
3. Case DA, Pearlman DA, Caldwell JW, Cheatham TE III, Ross WS, Simmerling CL, Darden TA, Merz KM, Stanton RV, Cheng AL, Vincent JJ, Crowley M, Tsui V, Radmer RJ, Duan Y, Pitera J, Massova I, Seibel GL, Singh UC, Weiner PK, A KP, (1999) AMBER 6. University of California, San Francisco
4. Demontis P, Spanu S, Suffriti GB (2001) Application of the wolf method for the evaluation of coulombic interactions to complex condensed matter systems aluminosilicates and water. J Chem Phys 114:7980–7988
5. Deserno M, Holm C (1998) How to mesh up Ewald sums. I. A theoretical and numerical comparison of various particle mesh routines. J Chem Phys 109:7678
6. Essmann U, Perera L, Berkowitz ML, Darden T, Lee H, Pedersen LG (1995) A smooth particle mesh Ewald method. J Chem Phys 103:8577
7. Ewald P (1921) Die Berechnung optischer und elektrostatischer Gitterpotentiale. Ann Phys 64:253–287
8. Fennell CJ, Gezelter J (2006) Is the Ewald summation still necessary? Pairwise alternatives to the accepted standard for long-range electrostatics. J Chem Phys 124:234104
9. Flyvbjerg H, Petersen AG (1989) Errors estimates on averages of correlated data. J Chem Phys 91:461–466
10. Frenkel D, Smit B (1996) Understanding molecular simulation. Academic Press
11. Fukuda I, Nakamura H (2012) Non-Ewald methods: theory and applications to molecular systems. Biophys Rev 4:161–170
12. Gonnet P (2007) A simple algorithm to accelerate the computaion of non-bonded interactions in cell-besed molecular dynamics simulations. J Comput Chem 28(2):580–593
13. Grossfield A, Zuckerman DM (2009) Quantifying uncertainties and sampling quality in biomolecular simulations. Ann Rep Comput Chem 5:23–48
14. Hünenberg PH, McCammon JC (1994) Affect of artificial periodicity in simulations of biomolecules under Ewand boundary conditions: a continuum electrostatic study. Biophys Chem 78:69–86
15. Hockney RW, Eastwood JW (1981) Computer simulations using particles. McGraw-Hill, New York
16. Leach A (1996) Molecular modelling: principles and applications, 1st edn. Prentice Hall
17. Martin MG (2013) MCCCS Towhee: a tool for Monte Carlo molecular simulation. Mol Simulat 39:1212–1222
18. Martínez JM, Martínez L, (2003) Packing optimization for automated generation of complex system's initial configurations for molecular dynamics and docking. J Comput Chem 24:819–825
19. Martínez L, Andrade R, Birgin EG, Martínez JM, (2009) Packmol: A package for building initial configurations for molecular dynamics simulations. J Comput Chem 30:2157–2164
20. Mattson W, Rice BM (1999) Near neighbor calculations using a modified cell-linked list method. Comput Phys Commun 119:135–148
21. Phillips JC, Braun R, Wang W, Gumbart J, Tajkhorshid E, Villa E, C C, Skeel RD, Kalé, Schulten K, (2005) Scalable molecular dynamics with NAMD. J Comp Chem 26
22. Quentrec B, Brot C (1975) New method for searching for neighbors in molecular dynamics computations. J Comput Phys 13:430–432
23. Rapaport DC (2004) The art of molecular dynamics simulation. Cambridge University Press
24. Smith TR, Forester TR, Todorov IT DL_POLY Classic User Manual, http://www.ccp5.ac.uk/DL_POLY_CLASSIC. Daresbury Laboratory, UK
25. Stukowski A (2009) Visualization and analysis of atomistic simulation data with OVITO—the open visualization tool. Model Simul Mater Sci Eng 18:015012

26. Sutmann G, Stegailov V (2006) Optimization of neighbor list techniques in liquid matter simulations. J Mol Liquids 125:197–203
27. Todorov IT, Smith W, Trachenko K, Dove MT (2006) DL_POLY3: new dimensions in molecular dynamics simultions via massive parallelism. J Mater Chem 1911–1918
28. Toukmaji AY, Board JJA (1996) Ewald summation techniques in perspective: a survey. Comput Phys Commun 95:73–92
29. Verlet L (1967) Computer "experiments" on classical fluid, I. Thermodynamical properties of Lennard-Jones molecules. Phys Rev 159:98–103
30. Wang Z, Holm C (2001) Estimate of the cutoff erros in the Ewald summation for dipolar systems. J Chem Phys 115:6351
31. White JA (1999) Lennard-Jones as a model for argon and test of extended renormalization group calculations. J Chem Phys 111:9352–9356
32. Willig U, Germano G (2011) Efficiency of linked cell algorithm. Comput Phys Commun 182(3):611–615
33. Wolf D, Keblinski P, Phillpot SR, Eggebrecht J (1999) Exact method for the simulation of coulombic systems by spherically truncated pairwise r^{-1} summation. J Chem Phys 110:8254–8282
34. Zahn D, Schilling B, Kast SM (2002) Enhancement of the Wolf damped coulomb potential: static, dynamic, and dielectric properties of liquid water from molecular simulation. J Phys Chem B 106:10725–10732

Chapter 6
Molecular Models (Force Fields)

Molecular simulation studies require the accurate calculation of the potential energy of the system as function of its configuration and the structures of the molecules. Theoretically, ab initio quantum mechanical calculations can be employed to determine these properties. However, ab initio simulations are still restricted to systems of some 100 atoms and cannot be used to calculate properties in the condensed phase. Thus, molecular simulations are primarily performed using simplified analytical potential energy functions, so called force fields, that relate the chemical structure of the system to its conformational energy. In this chapter, we will mainly focus on classical force fields for fluid phases and their parametrization. The classical molecular models (first and second generation) use fixed partial charges to describe the charge distribution within the molecules. However, in Sect. 6.2 we will also discuss approaches to introduce polarization into the force field (third generation). Section 6.3 provides a short overview on other force field types such as potentials for metals or reactive force fields. In Sect. 6.4, we will finally discuss specific aspects on molecular modeling, i.e. approaches to account for many-body effects and the influence of intermolecular flexibility on the description of thermophysical properties of water in Sect. 6.4.
The discussion on force fields in this chapter concentrates on so called all-atoms models that consider all atoms including hydrogen. Though most conclusion also apply for united-atom models in which hydrogens bonded to carbons are combined with them to single interaction sides. However, coarse graining models in which larger groups of atoms are presented by pseudo atoms are not within the scope of this discussion.

6.1 Classical Force Fields for Condensed Phases

For most force fields used in condensed phase simulations, the configurational energy U_{conf} is described by the following standard functional form [77]

G. Raabe, *Molecular Simulation Studies on Thermophysical Properties*,
Molecular Modeling and Simulation, DOI 10.1007/978-981-10-3545-6_6

$$
\begin{aligned}
U_{conf} = & \sum_{bonds} k_r(r - r_0)^2 + \sum_{angles} k_\theta(\theta - \theta_0)^2 \\
& + \sum_{dihedrals} k_\phi \left[1 + cos(n\phi - \delta)\right] \\
& + \sum_{improper} k_\omega(\omega - \omega_0)^2 \\
& + \sum_{electrostatics} \left(\frac{q_i q_j}{4\pi\varepsilon_0 r_{ij}}\right) + \sum_{vdW} \varepsilon_{ij} \left[\left(\frac{R_{min,ij}}{r_{ij}}\right)^{12} - 2\left(\frac{R_{min,ij}}{r_{ij}}\right)^{6}\right].
\end{aligned}
\tag{6.1}
$$

It accounts for additive interactions between the atoms (or sites) of the system's molecules, with the atoms treated as point masses centered on their nuclei. Molecular models following this simple functional form are referred to as class I force fields. This class includes the AMBER [22], OPLS [60–62], CHARMM [35, 78] or the GROMOS [113] force fields that are widely used in studies on biological and organic molecules such as proteins or nucleic acids. Beyond that, many force fields to compute thermophysical properties in chemical engineering, amongst others for components such as hydrocarbons [18, 83, 86], refrigerants [93, 95, 98] or ionic liquids [10, 14, 80] were developed in the framework of the functional form given by Eq. 6.1. A common feature shared by class I (first generation) force fields is that they only employ a simple harmonic representation by the Hooke's law for the bond stretching and angle bending (given by the first line in Eq. 6.1). There, r_0 and θ_0 are the equilibrium bond length and bond angle, k_r and k_θ the corresponding force constants. The energies arising from internal rotations (torsions) around dihedral angles are generally expressed as Fourier cosine series expression, with n representing the multiplicity, δ the phase of the dihedral term and k_ϕ the force constant. To enhance the flexibility of the dihedral term and with this the accuracy in reproducing the torsion profile, each dihedral angle in a molecule can be treated by a sum of dihedral terms. The improper term describes the so called out-of-plane bending of trigonal groups, and is used e.g. for aromatic rings to help maintain their planarity. The terms for the bond stretching, angle bending, torsion and out-of-plane bending are referred to as internal or intramolecular terms. The non bonded or intermolecular terms generally comprise van der Waals, dispersive interactions as well as electrostatic interactions, given by the last line of Eq. 6.1. In most current force fields, the repulsive interactions at short distances and attractive dispersive interactions are treated with Lennard-Jones (LJ) 12-6 site-site terms and interaction centers on the atomic sites. The potential parameters are the depth of the potential well, ε, and the distance at the minimum of the potential R_{min}. In alternative formulations of the 12-6 Lennard-Jones term, the distance σ at which the intermolecular potential is zero is used instead of R_{min}, with $R_{min} = \sigma \cdot 2^{1/6}$. The Lennard-Jones parameters are typically obtained not for single atoms but rather for atom types that are defined by their element type and their chemical environment, i.e. hybridization, aromaticity etc. [53, 135]. The Lennard-Jones parameters for the heteroatomic interaction between two unlike atom types i and j are often determined by the Lorentz-Berthelot combining rule

$$\varepsilon_{ij} = \sqrt{\varepsilon_{ii} \cdot \varepsilon_{jj}}, \quad \text{geometric mean} \tag{6.2}$$

$$\sigma_{ij} = \frac{\sigma_{ii} + \sigma_{jj}}{2}, \quad \text{arithmetic mean} \tag{6.3}$$

In some force fields, such as OPLS, the σ_{ij} are also obtained via the geometric mean. The 12-6 Lennard Jones potential combines the attractive dispersive London force at long ranges, described by the r^{-6} to account for interactions between induced dipoles, with a repulsive r^{-12} part at short ranges.

A more general description of the van der Waals interactions $u_{ij}(r_{ij})$ is given by the Mie-potential [84]

$$u_{ij}(r_{ij}) = \left(\frac{n}{n-m}\right)\left(\frac{n}{m}\right)^{m/(n-m)} \varepsilon_{ij} \left[\left(\frac{\sigma_{ij}}{r_{ij}}\right)^{n} - \left(\frac{\sigma_{ij}}{r_{ij}}\right)^{m}\right]. \tag{6.4}$$

Therein, the exponents n and m can be varied for a more flexible modeling of the steepness of the repulsive part of the potential and of the attractive parameter, respectively. Setting $n = 12$ and $m = 6$ then yields the 12-6-Lennard-Jones potential.

The use of the r^{-12} in the LJ potential arise from its computational efficiency although it lacks theoretical justification and yields a too steep rise of repulsion. Thus, it is sometimes replaced with an r^{-9} or r^{-10} term. In some force fields [31], the vdW interactions are described by the Buckingham exponential-6 instead of the Lennard-Jones potential, as it employs a softer repulsive exponential term that is in better accordance with theory, but is computational more expensive

$$U_{vdW}(r_{ij}) = \frac{\varepsilon_{ij}}{1 - 6/\alpha}\left[\frac{6}{\alpha}\exp\left(\alpha\left[1 - \frac{r_{ij}}{R_{min,ij}}\right]\right) - \left(\frac{R_{min,ij}}{r_{ij}}\right)^{6}\right] \tag{6.5}$$
$$\text{for} \quad r_{ij} > r_{max}.$$

The electrostatic interactions are often included by using the Coulombic term, as shown in Eq. 6.1, with static partial (point) charges q_i located at the atomic nuclei. The majority of force fields excludes dispersive vdW and electrostatic interactions between atoms separated by only one or two covalent bonds, whereas external interactions between atoms separated by three covalent bonds, the so called 1–4 interactions, are often considered. However, they are generally scaled to decrease the magnitude of the interactions. Typical scaling factors are 0.5 for the dispersive and 0.5 (OPLS) or 0.8333 (AMBER) for the electrostatic interactions. For smaller molecules, molecular models have been proposed (i.e. [119]) that account for electrostatic interactions by point dipole or point quadrupolar pair potentials. Though these approaches will not be further discussed here.

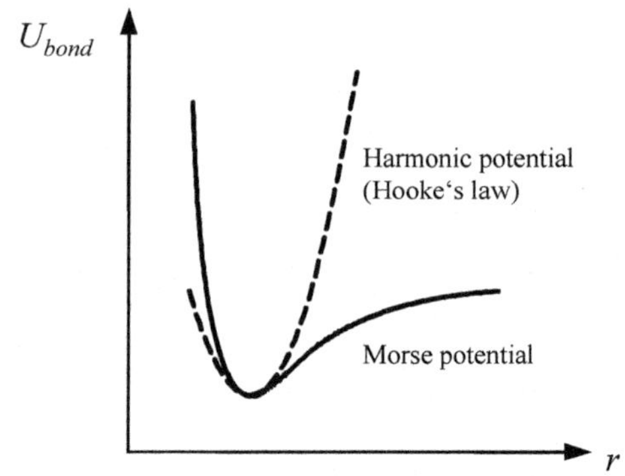

Fig. 6.1 Comparison of the bond stretching energy U_{bond} described by the harmonic model (Hooke's law) and the Morse potential

Despite their simplified functional form, class I force fields have shown to yield accurate predictions for thermodynamic and transport properties in bulk phases. However, extended potential energy functions, termed class II or second generation force fields, are required to accurately determine subtleties of molecular geometries or vibrational frequencies. To provide a better reproduction of vibrational data and a more realistic description of the bond stretching energy at higher stretching (see Fig. 6.1), some class II force fields replace the Hooke's law representation by the anharmonic Morse potential [53]

$$U_{bond} = D_e \{1 - \exp[-\beta(r - r_0)]\}^2 , \tag{6.6}$$

wherein D_e is the well-depth, and β an adjustable parameter.

Another approach to improve the representation of the bond stretching energy is to include higher order polynomial terms. This is then also applied to the description of the angle bending energy. Additionally, class II force fields include cross terms to account for the coupling between internal coordinates, as exemplarily shown by Eq. 6.7. The class II force fields include for instance the Merck Molecular Force Field (MMFF94, [44]) that was developed for simulations on conformational energies, structures and vibrational frequencies of small organic molecules to enable screening of functional groups and compounds in pharmaceutical drug design. Other class II force fields are the Mechanical Mechanics force fields (MM_n) by Allinger [3], the Universal Force Field, UFF [106] or the COMPASS force field [138]. As class II force fields are rarely used for molecular simulations studies on thermophysical properties of bulk phases, which are the main focus of this work, we will in the following limit our discussion to class I force fields.

$$
\begin{aligned}
U^{intra} &= \sum_{bonds} \left[k_{r_2}(r-r_0)^2 + k_{r_3}(r-r_0)^3 + k_{r_4}(r-r_0)^4\right] \\
&+ \sum_{angles} \left[k_{\theta_2}(\theta-\theta_0)^2 + k_{\theta_3}(\theta-\theta_0)^3 + k_{\theta_4}(\theta-\theta_0)^4\right] \\
&+ \sum_{dihedrals} \left[k_{\phi_1}(1-cos\phi) + k_{\phi_2}(1-cos2\phi) + k_{\phi_3}(1-cos3\phi)\right] \\
&+ \sum_{improper} k_{\omega}\omega^2 \\
&+ \sum_{bonds}\sum_{bonds'} k_{rr'}(r-r_0)(r'-r'_0) \\
&+ \sum_{angles}\sum_{angles'} k_{\theta\theta'}(\theta-\theta_0)(\theta'-\theta'_0) \\
&+ \sum_{bonds}\sum_{angles} k_{r\theta}(r-r_0)(\theta-\theta_0) \\
&+ \sum_{bonds}\sum_{dihedrals} (r-r_0)\left[k_{\phi,r_1}\cos\phi + k_{\phi,r_2}\cos 2\phi + k_{\phi,r_3}\cos 3\phi\right] \\
&+ \sum_{bonds'}\sum_{dihedrals} (r'-r'_0)\left[k_{\phi,r'_1}\cos\phi + k_{\phi,r'_2}\cos 2\phi + k_{\phi,r'_3}\cos 3\phi\right] \\
&+ \sum_{angles}\sum_{dihedrals} (\theta-\theta_0)\left[k_{\phi,\theta_1}\cos\phi + k_{\phi,\theta_2}\cos 2\phi + k_{\phi,\theta_3}\cos 3\phi\right] \\
&+ \sum_{angles}\sum_{angles'}\sum_{dihedrals} (\theta-\theta_0)(\theta'-\theta'_0)\,cos\phi.
\end{aligned}
\tag{6.7}
$$

6.1.1 *Parametrization of Force Fields*

A force field is not only defined by its functional form, but also by its parameters. In fact, the success and usefulness of a force field model is considerably determined by its parametrization. Details on the parametrization of force fields, i.e. the optimization of their parameters, have already been presented by a number of authors, for instance, in the reviews by A. D. MacKerrell Jr. [77, 79], or in the review on the parametrization strategies for the AMBER force field [36, 134, 135]. In this section, an overview is provided over the strategies used to determine the parameters of the intramolecular terms for bond stretching, angle bending and the dihedral potential, and for the external non-bonded interactions described by the Lennard-Jones parameters and the partial charges. However, it should be noted that the Lennard-Jones parameters and partial atomic charges are highly correlated, and the same goes for the different intramolecular degrees of freedom. Furthermore, if 1–4 interactions are considered, the internal parameters of the torsion potential will also depend on the non bonded parameters. Thus, the parametrization of a force field is generally an iterative procedure and a laborious task that requires a gradually refinement of the parameters.

6.1.1.1 Parameters for Bond Stretching and Angle Bending

The nominal bond lengths r_0 and bond angles θ_0 are usually obtained from quantum mechanical calculations in which the energy minimized structure of an isolated molecule is determined. However, in a bulk phase, the internal geometry of a molecule might differ from its structure as isolated molecule that is determined in the ab initio calculations. Thus, experimental information on equilibrium bond lengths and angles e.g. from X-ray and neutron diffraction studies on crystal structures are also used as target data to obtain geometry parameters that include condensed phase effects [135]. There exist several well-known databases for molecular structures such as the Cambridge Structural Database (CSD) [40] for crystal structures data of organic and metal-organic compounds, or the Brookhaven Protein Data Bank (PDB) for biological macromolecular structural data [53]. The corresponding harmonic force constants k_r and k_θ are also often derived from quantum mechanics QM simulations via a normal-mode analysis. Thus, the Hessian matrix holding the second partial derivatives of the potential energy with respect to the displacement of the atoms is calculated from both ab initio and molecular mechanics (MM) simulations using the force field. The force constants of the force field are then usually optimized such that the MM vibrational spectra, represented by the eigenvalues of the Hessian, reproduce best the ab initio data. However, it is often tedious to identify all normal modes and to relate them to the vibration of a specific degree of freedom. Alternatively, the force constants can be derived by comparing all matrix elements of the MM Hessian to the corresponding QM Hessian [63]. It is important to note that a normal-mode analysis is only valid for an optimized structure. That means that in the QM simulations, the structure of the molecule of interest not only has to be optimized at first, but also that the frequency calculation has to be performed for the optimized structure on the same level of theory and with the same basis sets. This is required as the Hessian matrix only yields meaningful results when the first derivatives of the potential energy with respect to the displacement of the atoms are zero.

Moreover, the target data from ab initio simulations are also often obtained on the HF/6-31G* level of theory that tends to overestimate the frequencies. Thus, the HF/6-31G* frequencies need to be scaled by the factor 0.89 [36]. Scaling factors for ab initio frequencies calculated on other levels of theory and other basis sets are available from the NIST Computational Chemistry Comparison and Benchmark Database [87]. Alternatively, the force constants, k_r and k_θ can be determined to reproduce experimental vibrational spectra. By this, it is also possible to account for bulk phase effects by comparing the MM vibrational spectra to those from experimental condensed phase vibrational studies.

Instead of using a normal mode analysis, the force constants for the bond stretching and angle bending can also be derived from ab initio simulations by perturbating the bond length or angle around their equilibrium value [121]. Thus, QM geometry optimizations are performed for the perturbed geometry in which the degree of freedom of interest is scanned in small intervals while all other internal coordinates are allowed to relax. The k_r and k_θ are then derived from the fitting of the harmonic

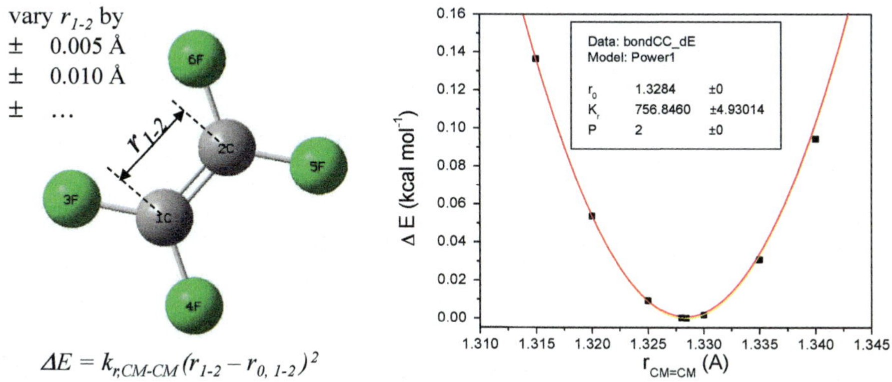

Fig. 6.2 This figure illustrates the procedure of deriving force constants from measuring the related change in the ab initio energy when perturbating the degree of freedom. Here, the bond r_{CM-CM} between the atoms 1 and 2 is changed by 0.005 Å. Now the structure is optimized to allow all other degrees of freedom to relax. When using the ab initio simulation package GAUSSIAN [37], this is realized by using the 'opt=modredundant' keyword. The difference between the energy of this configuration E_{Config} and the energy E_0 of the unperturbed, optimized structure can then be assigned to the change in the bond r_{CM-CM}. Repeating this procedure for different perturbations yields the potential energy curve ΔE versus the change of the bond r_{CM-CM} from which the force constant $k_{r,CM-CM}$ can be derived

potential (or another appropriate expression) to the resulting energy change versus the degree of perturbation, as shown in Fig. 6.2.

For large and complex compounds such as biological or organic molecules, parameters for the bond stretching and angle bending are often established by studying simple, small-sized model molecules containing the bond or angle of interest [36, 77]. In the AMBER force field, parameters for functional groups are determined by selecting a set of model compounds that represent this functional group. The parameters are then optimized to reproduce the target data of the model compounds. This approach shall also ensure and enable the transferability of the force field parameters to other molecules [36]. Thus, in the force field development for new compounds such as ionic liquids, the authors often transfer AMBER parameters for chemical similar bond stretchings or angle bendings to the new compounds. However, the transferability of parameters from literature is only valid as long as the equilibrium value of the bond length or angle of interest not differs significantly from the force field parameter [14], and should in any case be checked.

Force constants for the bond stretching k_r and the angle bending k_θ may also be estimated by empirical rules from tabulated reference data. In the MMFF94, Halgren [43] applied an inverse sixth-power dependence to derive the desired force constants for the stretching of a bond $r_{0,ij}$ between two atoms types i and j from tabulated reference values for this bond type, i.e. $k_{r,ij}^{ref}$ and $r_{0,ij}^{ref}$

$$k_{r,ij} = k_{r,ij}^{ref} \left(\frac{r_{0,ij}^{ref}}{r_{0,ij}} \right)^6 . \tag{6.8}$$

The MMFF94 also uses an empirical rule to estimate the force constant $k_{\theta,ijk}$ for the bending of an angle between atoms types i, j and k

$$k_{\theta,ijk} = \beta Z_i \, C_j \, Z_k \, (r_{0,ij} + r_{0,jk})^{-1} \, \theta_{0,ijk}^{-2} \, exp(-2D) \tag{6.9}$$
$$D = \frac{(r_{0,ij} - r_{0,jk})^2}{(r_{0,ij} + r_{0,jk})^2},$$

where Z and C are tabulated parameters depending on the atom types. A similar approach for estimating force constants is used in the the General AMBER Force Field (GAFF, [135]). There, the force constant for bond stretching is determined from

$$k_{r,ij} = k_{r,ij}^{ref} \left(\frac{1}{r_{0,ij}} \right)^m \tag{6.10}$$
$$k_{r,ij}^{ref} = \frac{k_{r,ii}^{ref} |r_{0,ij}^{ref} - r_{0,jj}^{ref}| + k_{r,jj}^{ref} |r_{0,ij}^{ref} - r_{0,ii}^{ref}|}{|r_{0,ij}^{ref} - r_{0,jj}^{ref}| + |r_{0,ij}^{ref} - r_{0,ii}^{ref}|} .$$

The empirical rule to estimate the force constant for an angle bending was adopted from Halgren with newly fitted reference data for the Z, C and the $r_{0,ij}^{ref}$ and the parameter β set to 143.9. The reference data for different atom types are given in the literature [135].

6.1.1.2 Parameters for Electrostatic Interactions

A large number of studies have addressed the development of reliable methods to determine the partial atomic charges. Today, the most widely used method is the electrostatic potential (ESP) approach. In the ESP approach, QM simulations are used to calculate the molecular electrostatic potential (MEP) on a large number of points around the geometry-optimized molecule of interest. There exist different ESP-charge fitting schemes, that mainly differ on the choice of gridpoints for the MEP calculation. In the CHELP [21] scheme, these points are chosen to be on concentric spherical shells around each atom of the molecule. In the mostly used CHELPG scheme [11], the points are placed on a regularly spaced cubic grid surrounding the molecule, whereas in the MK scheme [117], the gridpoints are located on multiple layers around the molecule. All schemes then employ a least-square fit to derive partial charges that reproduce the molecular electrostatic potential (MEP) as closely as possible. ESP charges can easily be determined using an ab initio simulation package such as GAUSSIAN [37], as shown in Fig. 6.3. However, the fitting procedure generally yields an ambiguous solution. i.e. there might exist significantly different sets

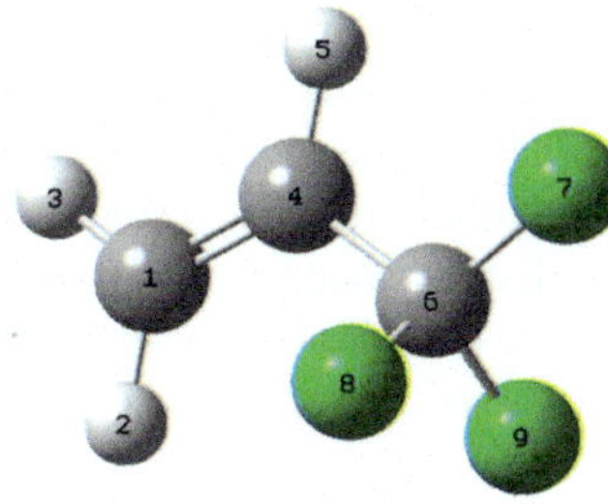

(a)
```
-----------------------
# rhf/6-31g(d) pop=CHELP
-----------------------
596 points will be used for fitting atomic charges
Fitting point charges to eletrostatic potential
Charges from ESP fit, RMS=   0.00308 RRMS=   0.28378:
Charge=   0.00000 Dipole=    -2.7098     0.6758     0.0002
              1
    1  C   -0.162948
    2  H    0.129525
    3  H    0.140516
    4  C   -0.313627
    5  H    0.153296
    6  C    0.866674
    7  F   -0.278714
    8  F   -0.267449
    9  F   -0.267472
----------------------------------------------------------
```

(b)
```
-----------------------
# rhf/6-31g(d) pop=CHELPG
-----------------------
7909 points will be used for fitting atomic charges
Fitting point charges to eletrostatic potential
Charges from ESP fit, RMS=   0.00205 RRMS=   0.14653:
Charge=   0.00000 Dipole=    -2.7731     0.7833    -0.0002
              1
    1  C   -0.237287
    2  H    0.164943
    3  H    0.163339
    4  C   -0.303465
    5  H    0.171624
    6  C    0.849460
    7  F   -0.270408
    8  F   -0.269124
    9  F   -0.269082
----------------------------------------------------------
```

(c)
```
-----------------------
# rhf/6-31g(d) pop=MK
-----------------------
587 points will be used for fitting atomic charges
Fitting point charges to eletrostatic potential
Charges from ESP fit, RMS=   0.00091 RRMS=   0.06711:
Charge=   0.00000 Dipole=    -2.7206     0.7548     0.0000
              1
    1  C   -0.314567
    2  H    0.197366
    3  H    0.192621
    4  C   -0.286990
    5  H    0.192403
    6  C    0.709723
    7  F   -0.235487
    8  F   -0.227541
    9  F   -0.227530
----------------------------------------------------------
```

Fig. 6.3 ESP charges for 3,3,3-trifluoro-1-propene (HFO-1243*zf*) calculated with the **a** CHELP, **b** CHELPG and **c** MK scheme by the ab initio simulation package GAUSSIAN [37]. Shown are only the relevant details of the Gaussian output (*.log) file. The geometry of the molecules was afore optimized on the B3LYP/DGDZVP level of theory. All ESP charges were then determined on the HF/6-31G* level of theory

of charges that all equally well reproduce the electrostatic potential of the molecule. Thus, in order to obtain a well-defined set of charges, additional conditions i.e. constraints have to be applied in the fitting procedure. This approach is called a restrained electrostatic potential fit (RESP) and is used for instance in the parametrization of the traditional AMBER force field [23] and in its more recent version GAFF (General Amber Force Field, [135]). In the two-stage RESP fit, the magnitudes of the partial charges are restrained, and the point charges of equivalent atoms in the molecule are constrained to have the same value to impose necessary molecular symmetries. With this, RESP charges shall enable an improved representation of conformational properties. However, one problem linked with any electrostatic potential fit is, that it is not well-suited for large complex molecules, where innermost atoms are located too far from the gridpoints of the MEP calculation.

Another problem that arises from this approach is that partial charges determined from ab initio simulations on isolated molecules do not reproduce the averaged polarization of the molecules in bulk phases. Thus, the partial charges need to be increased to include an averaged induced dipole moment in order to account for polarizability effects in the condensed phase. Therefore, in many force fields the partial charges are calculated on the HF/6-31G* level of theory (HF/6-31+G* for anions) that tends to overestimate the dipole moment to yield a dipole moment that is typically by 15–20% above the gas phase value.

Early force fields proposed for ionic liquids have also employed this standard approach to derive the partial charges. The static partial charges were determined for the single ions, setting the total ion charges to $+/-1\,e$, respectively. However, these molecular models fail to describe the dynamic properties of the ionic liquids correctly, as they result in a reduced mobility of the ions (see for instance [27, 64]). It was then realized that the correct description of the ion dynamics requires to account for charge transfer and polarization effect, which result in reduced ion net charges [27, 139]. Some force fields for ionic liquids then simply applied a uniform scaling factor for the atomic charges to approximate the effect of mean polarization—without changing the charge distribution within the ions. In general, these scaling factors range from 0.7 (e.g. [139]) to 0.9 (e.g. [85, 114]). Also advanced approaches were proposed that derive the partial charges of ionic liquids for instance by the CHELPG method for different conformers of ion pairs [74], from periodic ab initio MD (AIMD) simulations on the crystal phase of two or four ion pairs [142], or from quantum chemical calculations on snapshots from MD simulations comprising clusters of a number of ion pairs (for instance in [27, 64]). In the CHARMM force field, partial charges are established by a so-called supramolecular approach [79]. Thus, ab initio simulations are performed for water molecules interacting with different groups of the molecule of interest or with model compounds. Corresponding MM simulations are made for the molecule or the model compound interacting with TIP3P [59] water molecules. The partial charges are then optimized to reproduce the minimum interaction energies and distances from the QM simulations. With this, polarization effects are included in the optimization, as the interaction with the water molecules induces charges on the molecules studied. The resulting charges shall therefore be well suited for the study of condensed phase properties. Instead of studying interactions with water molecules, model compound dimers may also be considered to determine the partial charges. Again, the ab initio target data are typically generated on the HF/6-31G* level of theory due to its tendency to overestimate dipole moments. Another approach to determine partial charges is by bond-increment. Thus, the polarity of a bond between two atoms i and j is represented by a dipole or bond charge increment ω_{ij}. The net partial charge q_i of the atom i is then calculated from the summation of the charge increments of all covalent bonds to this atom [43]

$$q_i = \sum_j \omega_{ij} + q_0 \qquad \text{with} \quad \omega_{ij} = -\omega_{ji}. \tag{6.11}$$

The formal charge q_0 of the molecule is usually zero. Thus, in the bond-increment method the charge of an atom is allowed to change in response to the electronegativity of the atoms it is bonded to. This approach is used the Merck Molecular Force Field MMFF94 [43] where the bond charge increment ω_{ij} is determined by partial bond charge increments

$$\omega_{ij} = \delta_i - \delta_j \tag{6.12}$$

that represent the relative electronegativities of the atom types i and j. These partial bond charge increments were optimized by performing ab initio simulations for a set of molecules to derive their CHELPG dipole moments on the HF/6-31G* level of theory. The δ_i were then optimized to reproduce the ab initio dipole moments, enhanced by 10%. Thus, predetermined partial bond charge increments for different atom types are available from the cited reference. However, the applicability of the bond increment method is usually limited to small molecules.
The partial charges may also be considered as empirical parameters that are optimized to reproduce experimental target data. This approach is for instance used in the OPLS force field, where the partial charges are optimized through Monte Carlo simulations computing densities, heats of vaporization and other thermodynamic as well as structural properties [60].

6.1.1.3 Determination of van der Waals Interaction Parameters

The assignment of the non bonded parameters is one of the most difficult but also most important tasks in the force field development, as the proper treatment of the dispersive interactions is essential to accurately reproduce bulk phase properties. Thus, Lennard-Jones parameters for condensed phase simulations cannot be derived easily from ab initio calculations, as they need to incorporate many-body effects in an averaged sense. Thus, once the atomic charges have been assigned and the intramolecular parameters of the force field are determined, molecular dynamics (MD) or Monte Carlo (MC) simulations on pure components are performed to adjust the Lennard-Jones parameters. Usually, first estimates for the Lennard-Jones parameters are adopted from parameters for similar compounds from literature. Then, the parameters are adjusted to fine-tune agreement of the simulated properties with experimental data. Experimental target data usually comprise liquid densities and heats of vaporization, alternatively also heat capacities, free energy of hydration etc.

This procedure to determine the non-bonded interaction parameters in conjunction with Monte Carlo simulations in the liquid phase was first employed in the parametrization of the OPLS force field [60] and then adopted by other authors.

In most cases, Lennard-Jones parameters for several atom types in a molecule need to be established, whereas only a few experimental data are available for the optimization. Thus, Lennard-Jones parameters are often underdetermined, and additional difficulties in the optimization arise from the fact that the Lennard-Jones parameters of different atom types are correlated and often compensate for each other. A general methodology for most force field is, to derive Lennard-Jones parameters for complex molecules from simple model compounds containing the same atom types. Parameters for additional functional groups are then gradually added. This helps to overcome the parameter correlation problem and also provides a greater transferability of the interaction parameters. In addition, some attempts were made to combine the empirical optimization with ab initio simulations. Chen et al. [19] used an approach to derive parameters for the CHARMM force field by determining the relative magnitude of the LJ parameters from minimum ab initio interaction energies and geometries of model compounds interacting with neon or helium. The absolute values of the interaction parameters were then optimized to reproduce experimental heats of vaporization and molar volumes. A similar approach was used by Sum et al. [121] in their force field for triglyceride. Initial values for the Lennard-Jones parameters for their model compound methylacetate were derived from a non linear least square fit to ab initio interaction energies calculated for several hundred configuration of a molecular pair. The interaction parameters were then again fine tuned by condensed phase simulations to improve agreement with experimental data.

6.1.1.4 Fitting of Torsion Parameters

Besides the optimization of the non-bonded interaction parameters, the proper treatment of the rotation around dihedrals is regarded to affect the performance of a force field the most, as it represent the largest structural change in a molecule. The dihedral terms are usually parameterized last, and therefore often cover approximations and simplifications that where made in the optimization of the other force field parameters.

The parameters for the torsion potentials are derived from torsion scanning calculations. Thus, the potential energy surface for the rotation around a given dihedral is determined from ab initio calculations by perturbating the dihedral in intervals of 10–20°. There, the scanned dihedral is constrained at the desired angle, while all other internal coordinates are allowed to relax. This avoids straining of the other intramolecular terms, as otherwise particularly the bond stretching would contribute with a large part to the potential energy. Corresponding molecular mechanics simulations are then performed in which the parameters of the dihedral term which shall be adjusted are set to zero. The dihedral term is then fitted to reproduce the rotational profile, i.e. the energy difference between the QM and MM simulations for different conformations as shown in Fig. 6.4 for the rotation around the dihedral $F - CM - CT - FCT$ in 2,3,3,3-tetra-fluoro-1-propene (HFO-1234*yf*). The optimization requires the initial assignment of appropriate values for the multiplicity n and the phase δ of the dihedral term (see Eq. 6.1). For a rotation around a $Csp3 - Csp3$ single bond, $n = 3$ and $\delta = 0°$ are usually assigned. For a dihedral associated with a $C = C$ double bond ($sp2 - sp2$), $n = 2$ and $\delta = 180°$ is used, or $n = 1$ and $\delta = 0°/180°$ to distinguish between cis and trans configurational isomers. Again, the different dihedral terms in a molecule are determined step by step in such a way that in each torsion scanning, only one torsion term with unknown parameters is involved. Thus, complex molecules are conceptually broken up into a series of smaller compounds, and a 'building-up' procedure is used to gradually optimize all dihedral terms within a molecule.

It is important to mention that the rotational profile for a given dihedral is affected by all force field terms. Especially the 1–4 interactions, i.e. the non bonded interactions between atoms or sites separated by three bonds, will contribute largely to the torsion energy. Thus, the parametrization of the dihedral parameters is iterative and needs to be repeated every time the Lennard-Jones parameters are modified during the fine-tuning of the force field.

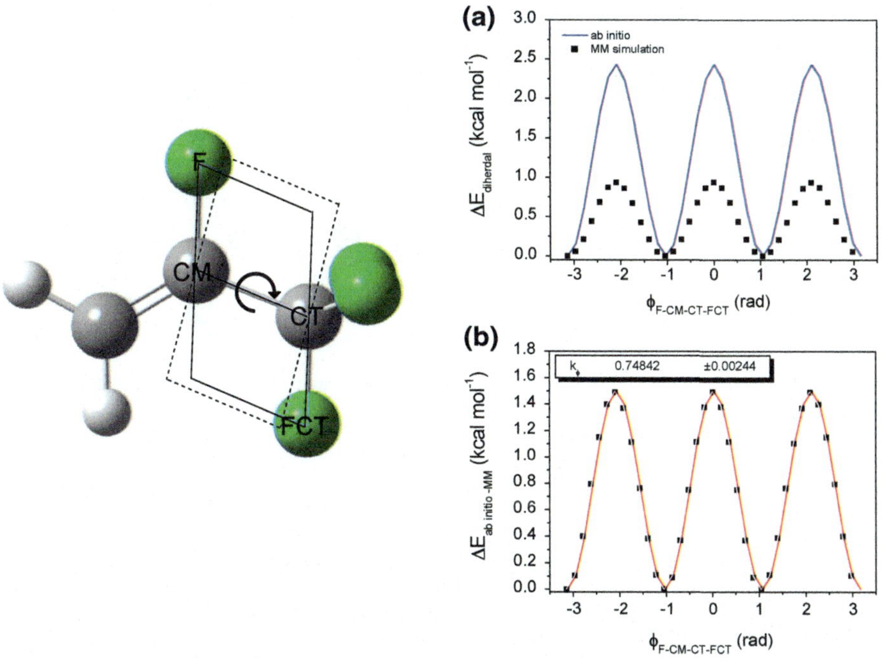

Fig. 6.4 Fit to reproduce the torsion profile for the rotation around the $F - CM - CT - FCT$ dihedral in HFO-1234*yf*: (**a**) shows the resulting change in the ab initio energy that occurs during the rotation in comparison to the MM results, when the dihedral term is set to zero. The energy difference, shown in (**b**), yields the rotational profile to which the torsion term is fitted

6.1.1.5 Example: Transferable Force Field for Hydrofluoroolefines

Different hydrofluoroolefine (HFO) compounds such as 2,3,3,3-tetrafluoro-1-propene (HFO-1234*yf*) or trans-1,3,3,3- tetrafluoro-1-propene (HFO-1234*ze*(*E*)) have been proposed as working fluids for different technical applications (see Chap. 8) as they exhibit a much lower Global warming potentials (GWP) than the hydrofluorocarbons (HFC) refrigerants that are now conventionally employed. Information on the thermophysical properties of HFOs and their mixtures from experiment are still generally rare. Thus, we have introduced a transferable force field for these compounds [95, 98, 99] that enables reliable predictions for their thermophysical and transport properties by molecular simulation studies to complement experimental data. The functional form of the all-atoms force field is based on Eq. 6.1. Figure 6.5 provides an overview over the fluoropropene compounds that were initially covered by the force field, i.e. 3,3,3-trifluoro-1-propene (HFO-1243*zf*), 2,3,3,3-tetrafluoro-1-propene (HFO-1234*yf*), hexafluoro-1-propene (HFO-1216) trans- and cis-1,3,3,3- tetrafluoro-1-propene (HFO-1234*ze*(*E*), HFO-1234*ze*(Z)) and cis-1,2,3,3,3-pentafluoro-1-propene (HFO-1225*ye*(Z)). Also shown is the nomenclature for the different LJ atom types that were used in the molecular

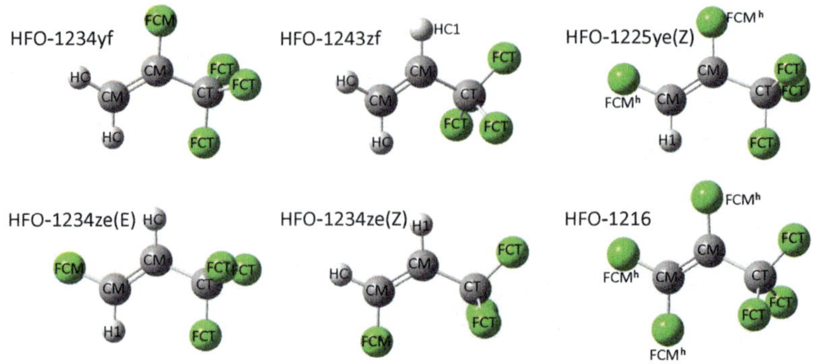

Fig. 6.5 Structure of the fluoropropenes covered by the force field, and nomenclature for the different LJ atoms types

modeling. Following the AMBER methodology, the atom types *CM* and *CT* denote double and single bonded carbons. The hydrogens are of the HC or H1 type. We additionally introduced the atom types FCM and FCT for the fluorines bonded to the CM or CT carbon, respectively

The molecular model is transferable with regard to both the intramolecular terms and the Lennard-Jones parameters, i.e. it employs, for example, the same LJ parameters for the double-bonded carbon *CM* in all compounds, or the same force constant for the $CM - FCM$ bond. The nominal bond lengths r_0 and bond angles θ_0 were obtained from ab initio simulations (B3LYP/DGDZVP, using Gaussian [37]) to determine the energy minimized structures of the isolated molecules. These quantum chemical simulations were performed for all compounds independently, but then the different r_0 and θ_0 values were averaged in order to get a transferable force field with a reduced number of different parameters. The force constants k_r and k_θ were derived from B3LYP/DGDZVP ab initio geometry optimizations for the perturbed geometries in which the degree of freedom of interest was scanned in small intervals, while all other internal coordinates were allowed to relax, as described in Sect. 6.1.1.1. The different force constants have been determined by stages, i.e. that in every perturbation only one degree of freedom with unknown force constant was involved. Thus, most force constants have been derived by studying smaller model compounds, i.e. tetrafluoromethane (CF_4), ethene (C_2H_4) and tetrafluoroethene (C_2F_4), and only the remaining force constants not present in these compounds were then derived from the fluoropropenes. In a similar approach, i.e. by torsion scanning calculations (see Sect. 6.1.1.4), we have determined the parameters for the torsion potentials associated with the $CM - CT$ bond, while we used the AMBER parameters [22] for the torsion associated with the $CM = CM$ double bond. It should be noted that the phase angles δ and multiplies n in the torsion potential (see Eq. 6.1) need to be assigned attentively to distinguish between cis- and trans-isomers, and to account for different equilibrium geometries regarding the rotation of the -CF_3 end group (see Fig. 6.5).

The partial charges were derived for all compounds individually from ab initio simulations by the ESP approach with the CHELPG fitting scheme [11], as described in Sect. 6.1.1.2. Thus, the geometries of the isolated molecules were initially optimized on the B3LYP/DGDZVP level of theory, before their electrostatic potentials have been determined on the HF/6-31G* level.

Once the intramolecular parameters and the atomic charges were assigned, molecular simulations on the pure components were performed to establish the Lennard-Jones parameters for the different atom types CM, CT, FCT, FCM, HC, $HC1$ and FCM^h. The parameters for the CM and FCM atom types were optimized by fine-tuning agreement with experimental data [71] for the model compound C_2F_4, whereas the hydrogen parameters HC and $H1$ were taken from the AMBER force field. Following the AMBER methodology, we use the HC parameters for hydrogens bonded to CM atoms with no electron-withdrawing substituents, and the $H1$ atom type when the hydrogen atom is bonded to the same CM atom as the FCM fluorine (or the chlorine atom). Only the remaining LJ parameters were then optimized to reproduce experimental data for the saturated liquid densities and vapor pressures of HFO-1234*yf* [55, 124], HFO-1243*zf* [24, 116, 141] and HFO-1225*ye*(*Z*) [116]. As no LJ parameters have been adjusted for trans- and cis-1,3,3,3-tetrafluoro-1-propene (HFO-1234*ze*(*E*), HFO-1234*ze*(*Z*)), the simulations for these compounds are purely predictive.

However, we found that the Lennard-Jones parameters we used to model the fluoropropenes with up to four fluorine atoms describe HFO-1225*ye*(*Z*) and HFO-1216 with five and six fluorines as too high boiling compounds [95]. Thus, we derived the slightly modified FCM^h parameters for these 'heavier' fluoropropenes that were then established to reproduce experimental data [116] for the vapor pressures and saturated densities of HFO-1225*ye*(*Z*). These LJ parameters were then also applied for predictive simulation studies on hexafluoro-1-propene (HFO-1216).

In the next stage of development [97], the molecular model was extended to hydrochlorofluoroolefines (HCFO), i.e. cis- and trans-1-chloro-3,3,3-trifluoropropene HCFO-1233*zd* (see Fig. 6.6). The extension of the molecular model requires the adjustment of LJ-parameters for the chlorine atom and the determination of parameters for the $Cl - CM$ bond stretching and the bending of the $Cl - CM - CM$ and $Cl - CM - H1$ angles. The parameters for the intramolecular terms as well as the partial charges of both compounds were derived from ab initio simulations, following the same strategy of parametrization as described above. The LJ parameters for the Cl atom type were then adjusted to finetune agreement between simulation results for the vapor pressure and saturated densities of HCFO-1233*zd*(*E*) and experimental data [56, 123]. The force field for fluoropropenes was then also employed for simulation studies on the fluorinated butenes cis- and trans-1,1,1,4,4,4-hexa-fluoro-2-butene, HFO-1336*mzz* [97]. Therefore, all force fields for the LJ parameters and intramolecular terms were transferred from the original model for fluoropropenes, and only the partial charges were derived from ab initio simulations. Though due to our findings from our earlier work [95] that the modeling of HFO compounds with more than four fluorine atoms requires slightly modified LJ parameters for the fluorine atoms types, we reduced the ε_{FCT} parameter (now denoted as FCT^h) to 0.21784 kJ/mol, i.e. to

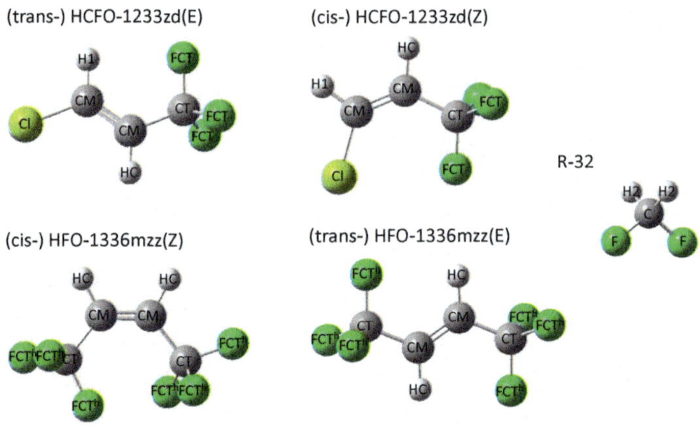

Fig. 6.6 Structure of the HCFO and fluorobutenes compounds covered by the extension of the force field, and nomenclature for the different LJ atoms types. Also shown are the LJ atoms types of the new R-32 model

the same value we used for the ε_{FCM^h} for the heavier fluoropropenes (see. Fig. 6.6). For simulation studies on potential refrigerant blends, we have also derived a new fully-flexible all-atoms force field for difluoromethane R-32, which is compatible with our molecular model for HFO and HCFO compounds [96]. The strategy of parametrization was the same as described before for the HFO/HCFO-model. For the force constants of the $C - F$ bond and the $F - C - F$ angle, the same parameters were used as for the $CT - FCT$ bond and $FCT - CT - FCT$ angle in the HFO/HCFO-model, but the equilibrium bond length and bond angle were derived from ab initio geometry optimizations on R-32. Parameters for the bond stretching and angle bending involving the hydrogen atoms were newly determined from ab initio optimizations on perturbed geometries of R-32. The AMBER $H2$ parameters are used for the hydrogens, whereas the LJ parameters for the carbon and fluorine atoms were established to optimize agreement with REFPROP calculations for the VLE properties of R-32.

A complete list of all force field parameters for the HFO and HCFO compounds, as well as for R-32 is given in the Appendix A. More details on the parametrization are provided in the original papers [94–99].

6.2 Polarizable Force Fields

Regarding current class I force fields, the use of fixed partial charges is considered as a major simplification, as it does not take into account that the electronic distribution of a molecule changes in response to its environment, i.e. the surrounding electrostatic field. Therefore, it is generally believed that the next significant development in force fields (class III) is the explicit inclusion of electronic polarization in molecular modeling in order to obtain molecular models that accurately describe both gas phase and bulk properties. Polarization means the redistribution of charges due to an electrostatic field, and there are different ways to explicitly account for polarization in molecular simulation studies. The most common approach to include electronic polarization into a molecular model is by introducing inducible dipoles, whereas higher order multipole moments are rarely used due to their again higher complexity and computational expense. Widely used methods to generate polarizable force fields are the Drude Oscillators, the fluctuating charge model (charge equilibration) or the induced dipole model. These methods will therefore be discussed briefly in the following subsections, however without considering the presence of external electrostatic potentials. An educational review on the explicit introduction of polarizability into force fields is provided by Yu and van Gunsteren [140]. Applications of polarizable force fields with focus on biological systems are, for instance, reviewed by Halgren and Damm [45] or Ponder and Case [91].

Beyond the high computational expense for the molecular simulations, the explicit introduction of polarization into a force field makes its more complex and therefore the optimization of the parameters more expensive. It has also been shown that the performance of the polarizable force field becomes very sensitive to an inappropriate parametrization [88].

6.2.1 *Drude Oscillator*

The Drude Oscillator, which is also referred to as shell model or Charge-on-Spring method (COS), is probably the simplest method to explicitly account for polarization. It is for instance used in the polarizable CHARMM force field, which is described in a recent review by Vanommeslaeghe and MacKerrel [132]. In the COS approach, the inducible dipole moment of a molecule is described by a Drude particle that is added to the atom's nucleus by a harmonic spring as shown in Fig. 6.7. A Drude particle is in this context a massless, virtual site that carries a partial charge.

The total charge q_i of an atom in a molecule is given by the sum of the partial charge of the nucleus $q_{i,0}$ and the partial charge of the associated Drude particle $q_{i,D}$. The polarizability α_i of the atom i is then described by

$$\alpha_i = \frac{q_{i,D}^2}{k_{i,D}}, \tag{6.13}$$

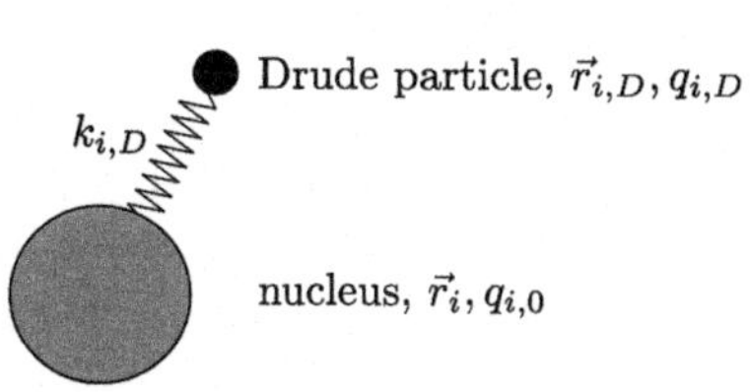

Fig. 6.7 Schematic depiction of a Drude particle added to an atomic nucleus to account for polarizability

where $k_{i,D}$ is the force constant of the harmonic spring between the nucleus and its Drude particle. The force constant of the harmonic spring $k_{i,D}$ and the partial charge of the Drude particle $q_{i,D}$ are treated as adjustable parameters to describe the Drude particle's response to the electrostatic environment. This electrostatic field caused by the partial charges of the atomic nuclei and other Drude particles gives rise to a displacement of the Drude particle, which results in an induced dipole moment given by [38]

$$\boldsymbol{\mu}_i^{ind} = q_{i,D}(\mathbf{r}_{i,D} - \mathbf{r}_i). \tag{6.14}$$

If the displacement is small enough, the induced dipole is nearly ideal. The position of the Drude particle is allowed to relax in the surrounding electrostatic field in such a way that the sum of forces acting on it is zero

$$\mathbf{f}_{i,D} = \mathbf{f}_{i,D}^{restore} + \mathbf{f}_{i,D}^{Coul} = 0. \tag{6.15}$$

The force $\mathbf{f}_{i,D}^{restore}$ is the restoring force of the harmonic spring

$$\mathbf{f}_{i,D}^{restore} = -k_{i,D}(\mathbf{r}_{i,D} - \mathbf{r}_i), \tag{6.16}$$

whereas $\mathbf{f}_{i,D}^{Coul}$ is the force due to the electrostatic field $\mathbf{E}_i$ acting on the Drude particle

$$\mathbf{f}_{i,D}^{Coul} = q_{i,D}\mathbf{E}_i. \tag{6.17}$$

As a result of the mutual dependence of the induced dipoles and the electrostatic field, an iterative self consistent field (SCF) method is necessary to minimize the electric energy of the system with respect to the positions of the Drude particles. Alternatively, a fictitious mass can be assigned to the Drude particle so that the electronic degree of freedom can be treated as dynamic variable in an extended Lagrangian approach. For more details on the Drude Oscillator (Charge-on-the-Spring) method and the resulting expressions for the atomic forces in the simulation, we'd like to refer to the article by Geerke and van Gunsteren [38].

6.2.2 *Fluctuating Charge Model or Charge Equilibration*

In this approach, electronic polarization is modeled by allowing charges to move from an atom or site i to its bonded neighbor j as response to the surrounding electrostatic field. For an isolated atom, the energy required to create an amount of charge q_α is given by [109]

$$U(q_\alpha) = \chi_\alpha q_\alpha + \frac{1}{2} J_\alpha q_\alpha^2 \tag{6.18}$$

where χ_α is the electronegativity of the isolated atom, and J_α(twice) the hardness of the atom, i.e. its resistance to the charge movement. These factors are dependent on the atom type and are generally treated as adjustable parameters. The charges on the sites fluctuate according to the principle of electronegativity, which postulates that in a molecule all atoms should have the same electronegativity. Thus, charges flow between the sites until the instantaneous electronegativities of all sites are equal [45]. However, the charges are not independent due to the charge neutrality constraints, i.e. for a system of neutral molecules, the total sum of charges q_i^{totals} in each molecule has to be zero. Therefore, the Lagrangian approach is usually employed in which the charges on the atoms are treated as dynamic variables, and Lagrange multipliers λ_i are used to include the charge neutrality constraints on each molecule. With this, the contribution of electrostatic interactions U^{el} to the total potential energy for a system of N molecules and $N_{a,i}$ atoms in each molecule i is given by [76]

$$U^{el}(q) = \sum_{i=1}^{N}\sum_{\alpha=1}^{N_{a,i}} \chi_{i\alpha} q_{i\alpha} + \frac{1}{2}\sum_{i=1}^{N}\sum_{j=1}^{N}\sum_{\alpha=1}^{N_{a,i}}\sum_{\beta=1}^{N_{a,j}} J_{i\alpha j\beta} q_{i\alpha} q_{j\beta} \\ + \sum_{i=1}^{N}\lambda_i\left(\sum_{\alpha=1}^{N_{a,i}} q_{i\alpha} - q_i^{totals}\right) \\ + \frac{1}{2}\sum_{i=1}^{N}\sum_{j=1}^{N}\sum_{\alpha=1}^{N_{a,i}}\sum_{\beta=1}^{N_{a,j}} \frac{q_{i\alpha} q_{j\beta}}{4\pi\varepsilon_0 r_{i\alpha j\beta}}. \tag{6.19}$$

The last term of Eq. 6.19 accounts for standard Coulomb interaction between atoms not involved in bonded interactions. The fluctuating charge model has for instance been used in an approach by Patel and Brooks [89] to include polarizability in the CHARMM force field. In [76] Lucas et al. provide a review on charge equilibration force fields.

6.2.3 Induced Point Dipole Model

In the induced point dipole model, each atom i carries not only a partial charge q_i but also an induced point-dipole μ_i^{ind} in response to the surrounding electrostatic field $\mathbf{E}_i$. The induced dipole as vector is given by [53]

$$\boldsymbol{\mu}_i^{ind} = \alpha_i \mathbf{E}_i, \tag{6.20}$$

where $_i$ is the dipole polarizability. For a molecule or a site of the molecule consisting of several bonded atoms, the polarization is in general anisotropic, and the dipole polarizability is a second-rank tensor. The polarization on an atom however is isotropic and α_i therefore becomes a scalar [53, 140]. The induced dipole itself induces other dipoles. Thus, the electric field affecting the atom is represented by the electrostatic field $\mathbf{E}_i^0$ due to the static, permanent point charges q_i, and by the contributions from induced dipoles on the other sites in the system [69]

$$\mathbf{E}_i = \mathbf{E}_i^0 + \sum_{j=1, j\neq i}^{N} \mathbf{T}_{ij}\boldsymbol{\mu}_j^{ind}. \tag{6.21}$$

Thus, $E_i^{ind} = T_{ij}\mu_j^{ind}$ is the dipolar field component and $\mathbf{T}_{ij}$ is the dipole tensor, whose elements are given by [140]

$$T_{ij} = \frac{1}{4\pi\epsilon_0}\frac{1}{r^3}\left[\frac{3r_\alpha r_\beta}{r^2} - 1\right]. \tag{6.22}$$

In the induced dipole model the total electrostatic energy of the system U^{el} is made up of the coulombic interactions between the permanent charges U^{qq} and the electrostatic energy due to the induced dipoles U^{ind}, which in turn represents different types of interactions [140]

$$U^{ind} = U^{stat} + U^{\mu\mu} + U^{self} \tag{6.23}$$

$$= -\sum_{i=1}^{N}\boldsymbol{\mu}_i\mathbf{E}_i^0 - \frac{1}{2}\sum_{i=1}^{N}\sum_{j=1,j\neq i}^{N}\boldsymbol{\mu}_i\mathbf{T}_{ij}\boldsymbol{\mu}_j + \sum_{i=1}^{N}\frac{\mu^2}{2\alpha_i}. \tag{6.24}$$

The terms U^{stat} and $U^{\mu\mu}$ represent the energies of the induced dipoles due to the field of the permanent charges and the field of the other induced dipoles respectively, whereas U^{self} is the self energy that is required to create the induced dipole. It should be noted that at short distances the interactions between the sites may mutually increase their induced dipole moments to infinite size. To avoid this so-called polarization catastrophe, the induced dipoles are smeared according to the Thole's screening method [126]. The induced dipole model is for instance used in the APPLE& P (Atomistic Polarizable Potential for Liquids, Electrolytes & Polymers) force field [7]. Another example of polarizable potentials following this approach is the AMOEBA force field [92]. Though the AMOEBA force field also accounts for polarizable atomic multipoles, i.e. each atomic site carries not only a permanent charge q_i (monopole) but also a permanent dipole and quadrupole moment. In [108], the authors also provide an Ewald summation formulation for the multipole polarization.

6.3 Other Force Field Types

6.3.1 Metal Potentials

Metal force fields are in general based on the concept to describe the bonding of metal atoms in terms of the local electronic density. Examples of density dependent metal potentials are the embedded-atom model (EAM) [34] and its modifications MEAM [5] or 2NN-MEAN [6], but also the Finnes-Sinclair [33] or the Sutton-Chen [122] model. The dependence of the atomic energy on the local environment is included in the potential by an embedding function F that describes the energy associated with placing the atom i in the background electron density $\bar{\varrho}_i$ from contributions of the neighboring atoms. With this, the metal potentials implicitly account for many-body effects. Their general form is given by

$$U = \sum_{i=1}^{N} \left[F\left(\bar{\rho}_i\right) + \frac{1}{2}\phi_{ij}(r_{ij}) \right], \tag{6.25}$$

wherein the potential ϕ_{ij} describes the short range two-body interaction of the atom i with its neighbor j. The aforementioned metal potentials now differ in their formulations of the embedding function $F(\bar{\rho}_i)$, the background electron density $\bar{\rho}_i$ and the pair-potential $\phi_{ij}(r_{ij})$, and with this also in the way they are extended to alloys. We will here only exemplarily describe the widely used EAM model and its modifications MEAM and 2NN-MEAM.
In the EAM model the pair-potential $\phi_{ij}(r_{ij})$ only accounts for the electrostatic interactions between the atom cores with their effective charges Q_i and Q_j

$$\phi_{ij}(r_{ij}) = \frac{Q_i \cdot Q_j}{r_{ij}}. \tag{6.26}$$

The background electron density $\bar{\rho}_i$ at the position of the atom i is approximated by the sum of spherically symmetric (s-orbital) contributions from the neighboring atoms. The embedding function F and the effective charges Q are determined empirically, i.e. by fitting to experimental data for physical constants such as the sublimation energy, the vacancy formation energy, the lattice constant or the elastic constant. The EAM potential though not uses explicit mathematical expressions for these functions but tabulated values.
In the modified embedded-atom model (MEAM, [5]) the embedding function is described by

$$F\left(\bar{\rho}_i\right) = A \cdot E_c \frac{\bar{\rho}_i}{\bar{\rho}_{i,0}} \ln\left(\frac{\bar{\rho}_i}{\bar{\rho}_{i,0}}\right). \tag{6.27}$$

Therein $\bar{\rho}_{i,0}$ is the background electron density in the reference lattice structure at equilibrium, A is an adjustable parameter, and E_c is the sublimation energy. In the MEAM approach, the background electron density $\bar{\rho}_i$ is not only composed of spherically symmetric (s-orbital) contributions from the neighboring atoms, $\rho_i^{(0)}$, but also includes p-, d- and f-orbital symmetries in the terms $\rho_i^{(h)}$, $h = 1, 2, 3$. With this, an angle dependence of the partial electron densities ρ_i is introduced into the model which enables the description of different crystal structures by one single formalism, and also the modeling of covalently bonded systems. The partial electron densities

ρ_i are combined to the total background electron density $\bar{\rho}_i$ by [6]

$$\bar{\rho}_i = \rho_i^{(0)} \cdot G(\Gamma_i), \tag{6.28}$$

with

$$\Gamma_i = \sum_{h=1}^{3} t^{(h)} \left[\frac{\rho_i^{(h)}}{\rho_i^{(0)}} \right]^2 . \tag{6.29}$$

The weighting functions $t^{(h)}$ are adjustable parameters, and different functional forms for $G(\Gamma_i)$ have been proposed. The partial densities $\rho_i^{(h)}$, $h = 0, .., 3$ of the atom i are described as functions of the atomic electronic densities $\rho_j^{a(h)}(r_{ij})$, $h = 0, .., 3$ of the neighboring atoms j at distance r_{ij}. The equations to derive the $\rho_i^{(h)}$, $h = 0, .., 3$ are for instance given in [6]. The atomic electron densities are expressed by

$$\rho^{a(h)} = \exp\left[-\beta^{(h)} \left(\frac{r_{ij}}{r_0} - 1 \right) \right] \tag{6.30}$$

with the adjustable parameters $\beta^{(h)}$ and the nearest-neighbor distance in the reference structure, r_0. In general, there is no functional form given in MEAM for the pair-potential $\phi_{ij}(r_{ij})$, but it is determined by the difference of the energy per atom E^u in the reference structure, given by

$$E^u(r_{ij}) = -E_c \left(1 - a^* + da^{*3}\right) e^{-a^*}, \quad a^* = \alpha \left(\frac{r_{ij}}{r_0} - 1 \right) \tag{6.31}$$

and the embedding energy $F(\bar{\rho}_{i,0})$ at the background electron density in the reference lattice structure. Thus, $\phi_{ij}(r_{ij})$ is derived from

$$\phi_{ij}(r_{ij}) = \frac{2}{Z_1} \left[E^u(r_{ij}) - F(\bar{\rho}_{i,0}) \right]. \tag{6.32}$$

In Eq. 6.31, α and d are adjustable parameters, Z_1 in Eq. 6.32 is the number of nearest-neighbor atoms. This illustrates a major feature of the MEAM model, i.e. that it only accounts for nearest-neighbor interactions. The neglect of interactions with second-nearest neighbors is realized by employing many-body screening functions $S_{ijk}(C, C_{max,ijk}, C_{min,ijk})$. The parameter C defines the length of the minor semi-axis of the ellipse that encloses the position of the atoms i, j and k, and whose length of the major semi-axis is $r_{ij}/2$ (see Fig. 6.8). Thus, if $C > C_{max}$, the atom k has no effect on the interaction between the atoms i and j. The value of the screening functions is then set to $S_{ijk} = 1$. For $C < C_{min}$, the atom k is completely screening the $i - j$ interaction, and $S_{ijk} = 0$. The default values for C_{max} and C_{min} are 2.8 and 2.0 respectively, but in general they can be defined for each triplet $i - j - k$ separately.

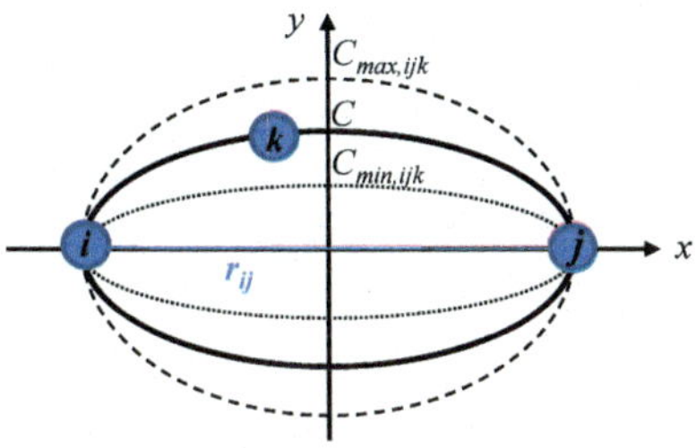

Fig. 6.8 Screening of the $i - j$ interaction by atom k, described by the semi-axis $C_{min,ijk} \leq C \leq C_{max,i,j,k}$ of the ellipse enclosing i, j and k

The resulting screening function S_{ij} for the $i - j$ interaction is then defined as product of the screening factors S_{ijk} from all other neighboring atoms k.

The restriction to only nearest-neighbor interactions in MEAM though results in some stability problems [6]. This motivated the modification of the model to 2NN-MEAM [6], which partially includes second nearest-neighbor interactions by adjusting the screening function, i.e. reducing the $C_{min,ijk}$ value. However, this also requires to extend the computation of pair-interactions to second-nearest neighbors. The pair-potential is now calculated from

$$\phi(r_{ij}) = \phi_{2NN}(r_{ij}) + \sum_{n=1}(-1)^n \left(\frac{S \cdot Z_2}{Z_1}\right)^n \phi_{2NN}(a^n r_{ij}). \tag{6.33}$$

Therein, Z_2 is the number of second nearest-neighbors, and a the ratio between second and first nearest-neighbor distances. The modified pair potential ϕ_{2NN} is again determined from the difference between the energy per atom and the embedding energy in the reference lattice structure, $E^u(r_{ij})$ and $F(\bar{\rho}_{i,0})$, but now by accounting also for second-nearest neighbors.

6.3.1.1 Example: Description of β-Sn

Tin is a widely used material in many commercial applications, most notably its use in soft solders in electronic systems or in corrosion-resistant platings for food-packaging. The molecular modeling of tin though is scientifically interesting as it is borderline between covalent and metallic bonding [107]. Below 286.3 K, α-Sn (grey tin) is the stable phase, in which the atoms are covalently bonded in a diamond cubic structure. Above 286.3 K, tin exist in its metallic form β-Sn (white tin).

In [107], Ravelo and Baskes studied the thermodynamic properties of tin in the α- and β- phase and in the liquid as well as the phase transition temperatures from $\alpha \rightarrow \beta$ and $\beta \rightarrow$ *liquid* by the MEAM potentials. The MEAM parameters they used for the modeling of tin are summarized in Table 6.1.

In [129], Valadez-Huerta optimized parameters for a 2NN-MEAM modeling of β-tin. The optimized values for the parameters A, r_0, $\beta^{(0)}$, C_{min} and C_{max} he derived are also given in Table 6.1. Table 6.2 provides simulation results at 300 K for the

Table 6.1 Parameters for β-Sn for the MEAM and 2NN-MEAM modeling. The parameters $\alpha = 6.2$ and $\beta^{(1)} = \beta^{(2)} = 6.0$ are identical in both models

	E_c (eV)	A	r_0 (Å)	$\beta^{(0)}$	$\beta^{(3)}$	$t^{(1)}$	$t^{(2)}$	$t^{(4)}$	C_{min}	C_{max}
[107]	3.08	1.0	3.44	6.2	6.0	4.5	6.5	−0.183	0.8	2.8
[129]	3.08	0.744	3.2	5.16	5.7	4.5	6.5	−0.183	0.06	0.5

Table 6.2 Comparison of experimental data and MEAM and 2NN-MEAM results for the linear expansion α_l, the lattice constants a and c, and the specific heat capacity c β-Sn

	α_l (10^{-6}/K)	$\alpha_{l,x}$ (10^{-6}/K)	$\alpha_{l,z}$ (10^{-6}/K)	a (Å)	c (Å)	c_p (J/g K)
Experiment	23.5	16.2	32.0	5.83	3.18	0.228
MEAM [107]	18.7	13.7	20.9	5.92	3.23	0.249
2NN-MEAM [129]	20.1	18.7	25.7	5.83	3.12	0.213

coefficient of linear expansion α_l, the lattice constants a and c, and the specific heat capacity c from both MEAM and 2NN-MEAM in comparison with experimental data [20, 25, 70].

For most properties shown in Table 6.2, the 2NN-MEAM extension yields an improved description compared to the MEAM results. However, the linear expansions α_l, $\alpha_{l,x}$ and $\alpha_{l,z}$ still significantly deviate from the experimental values. These results illustrate well that the optimization of the 2NN-MEAM extension is very challenging due to the complex crystal structure (tetragonal) of white tin [129].

6.3.2 *ReaxFF Reactive Force Field*

The classical force fields discussed in Sect. 6.1 are not suited to study chemical reactivity as they are not able to describe bond breaking or formation. Quantum chemical simulation studies on chemical reactions on the other side are limited to system sizes of a few hundred atoms. Thus, so called reactive force fields have been developed to bridge between the system size and time scale of ab initio and classical molecular simulations for studies on chemical reactions. The parameters for the reactive force field are derived from ab initio simulations, for instance for reactions of small molecules, bond dissociation, heats of formation and molecular structures. We will here provide a short introduction to the reactive force field ReaxFF by van Duin et al. [130]. The fundamental assumption of ReaxFF is that there exists a general relationship between the interatomic distance r_{ij} of two atoms i and j and their bond order. The (uncorrected) bond order BO'_{ij} is obtained from the bond lengths by

$$BO'_{ij} = \exp\left[p_{BO,1}\left(\frac{r_{ij}}{r_0^{\sigma}}\right)^{p_{BO,2}}\right] + \exp\left[p_{BO,3}\left(\frac{r_{ij}}{r_0^{\pi}}\right)^{p_{BO,4}}\right] + \exp\left[p_{BO,5}\left(\frac{r_{ij}}{r_0^{\pi\pi}}\right)^{p_{BO,6}}\right]. \tag{6.34}$$

The three exponential terms represent the single (σ), double (π) and triple ($\pi\pi$) bonds with the parameters $p_{BO,i}$, $i = 1, .., 6$. The bond order is then corrected for overcoordination due to residual bonding with second-nearest neighbors in a valence angle ($1 - 3-$interaction).

In the ReaxFF model the total system energy is divided into different contributions

$$E_{system} = E_{bond} + E_{val} + E_{torsion} + E_{under} + E_{over} + E_{conj} + E_{lp} + E_{vdW} + E_{Coul}. \tag{6.35}$$

As in the classical force fields, the terms E_{bond}, E_{val} and $E_{torsion}$ account for energies associated with deviations in the bond length, valence angle and dihedral angle from their reference values. A main feature of ReaxFF is that these connectivity dependent interactions are also described as function of the bond order BO_{ij}, and in such a way that their energy contributions smoothly go to zero upon bond dissociation. Taking the E_{val} valence angle formed by the three atoms i, j, k as an example, this is realized by the following expression [110]

$$E_{val} = \left[1 - \exp(-p_{val,3}BO_{ij}^{p_{val,4}})\right]\left[1 - \exp(-p_{val,3}BO_{jk}^{p_{val,4}})\right] \cdot \left[p_{val,1} - p_{val,1}\exp(-p_{val,2}(\phi_{ijk} - \phi_0)^2)\right], \tag{6.36}$$

with the parameters $p_{val,i}$. The bonding term is described by a distance corrected Morse-potential (see Eq. 6.6), i.e. the well-depth D_e is also multiplied by the bond order BO_{ij} so that the bonding energy also weakens smoothly when the bond between i and j breaks.

The energies E_{over} and E_{under} are multibody terms that present overcoordination penalties to avoid unrealistic high amounts of bond orders on atoms, and under-coordination stabilization, respectively. The term E_{conj} describes contributions of conjugation effects, for instance in aromatics, whereas E_{lp} accounts for energy contributions from lone pairs.

Non-bonded interactions include van der Waals dispersion E_{vdW} and a coulombic term E_{Coul}. In contrast to classical force fields that exclude non-bonded interactions between atoms sharing bonds or valence angles with one another (1–2-, and 1–3-interactions), ReaxFF calculates van der Waals and coulombic interactions between all atoms. However, the interactions are shielded at short distances to avoid excessively high interaction energies. For this purpose, ReaxFF employs a shielding term γ so that the coulombic energies, for instance, are then determined by [110]

$$E_{coul} \propto \frac{q_i q_j}{\left[r_{ij}^3 + (1/\gamma_{ij})^3\right]^{1/3}}. \tag{6.37}$$

The partial charges in ReaxFF are thereby calculated by accounting for polarization within the molecule. The Electron Equilibrium Model (EEM) in ReaxFF is based on the electronegativity of the elements χ and their hardness J. The polarization of an element n is then calculated by [110]

$$\frac{\partial E}{\partial q_n} = \chi_n + 2q_n J_n + C \sum_{j=1}^{n} \frac{q_j}{\left[r_{nj}^3 + (1/\gamma_{nj})^3\right]^{1/3}}, \quad \sum q_i = 0. \tag{6.38}$$

This EEM approach is similar to the CHEQ discussed in Sect. 6.2.2. Further details on the terms to describe the different energy contributions of the ReaxFF model, and on their parametrization are, for instance, provided in [130].
The parametrization of ReaxFF is based on ab initio simulations on the DFT/B3LYP/ G-316** level of theory, which does not account adequately for dispersive van der Waals interactions. To improve the description of long-range dispersive interactions, Liu et al. [73] proposed the modified ReaxFF-lg model, in which an additional interaction term E_{lg} is introduced

$$E_{ReaxFF-lg} = E_{ReaxFF} + E_{lg}. \tag{6.39}$$

The dispersive interaction term E_{lg} is based on a low gradient (lg) model, and is given by [73]

$$E_{lg} = -\sum_{ij,i<j}^{N} \frac{C_{lg,ij}}{r_{ij}^6 + dR_{min,ij}^6}. \tag{6.40}$$

Therein, $R_{min,ij}$ is the equilibrium vdW distance between i and j, $C_{lg,ij}$ is their dispersion energy correction parameter, and d is a scaling factor, which was though set to $d = 1$ by Liu et al.

6.3.2.1 Example: ReaxFF Simulation Studies on Gaseous By-Products in the Silica Reduction

Typically, metallurgical-grade silicon (mg-Si) is industrially produced in an electric-arc furnace above 2.000 °C by the carbothermal reduction of silica according to the simplified (overall) reaction equation

$$SiO_2 + 2C \rightarrow Si + 2CO. \tag{6.41}$$

However, the underlying reaction mechanism is still subject of discussion as it involves a complex system of reactions between solid, liquid and gaseous components with various by-products. Silicon monoxide *SiO* is stated to be one of the most important intermediates in silicon production [111], thus we have tested the ability of ReaxFF to reproduce the formation of gaseous *SiO* [100]. We therefore employed the Silicon parameter for ReaxFF published by van Duin et al. [131] and performed MD simulations with LAMMPS [90].

Figure 6.9 illustrates the structural changes from *SiO* upon heating, starting from a dense, solid-like phase. At round about 1000 K, the dense *SiO* phase decomposes into clusters, whereas free diatomic *SiO*-molecules occur at approx. 1500 K. This is in good agreement with experiment [12]. In [100], we have also studied the reproduction of the densities of the gaseous product *CO* by ReaxFF. Figure 6.10 shows simulation results using the original ReaxFF and the modified ReaxFF-lg models. The ReaxFF model significantly overestimates the *CO* densities over the entire temperature range, whereas the ReaxFF-lg model yields an improved description. This illustrates well the need to adequately account for dispersive van der Waals interactions for this component.

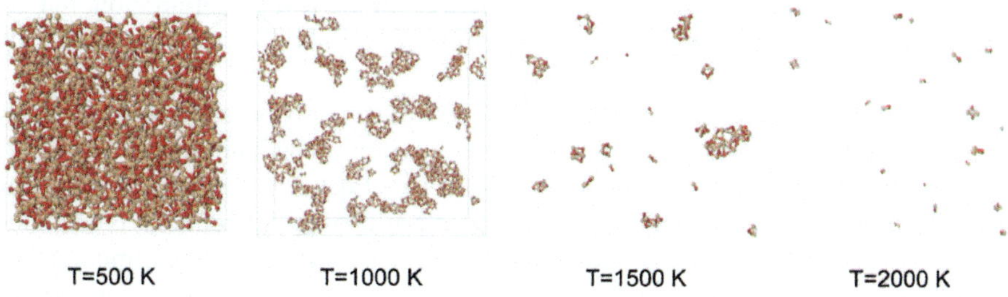

Fig. 6.9 ReaxFF studies on the *SiO* structure upon heating from the dense to the gaseous phase [100]

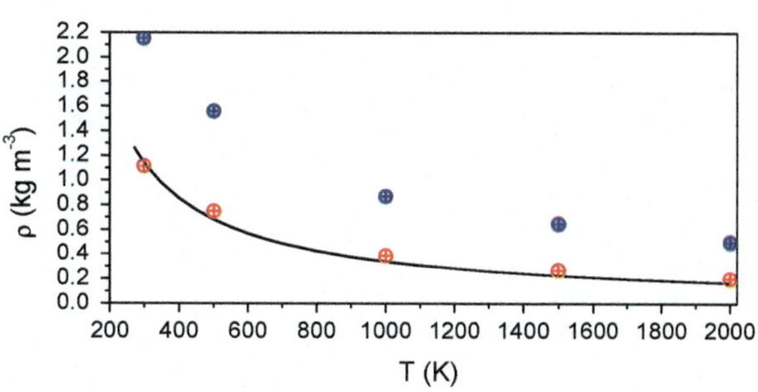

Fig. 6.10 ReaxFF (*blue crossed circles*) and ReaxFF-lg (*red crossed circles*) simulation results for the density of gaseous *CO* at $p = 1$ atm in comparison with calculations based an REFPROP and the ideal gas law at higher temperatures (*black line*)

6.4 Discussion on Specific Aspects of Force Fields

In this section, we discuss specific aspects of the molecular modeling. That includes to the one options to account for multi-body (dispersive) interactions, when the pair interactions are described by purely two-body potentials that were derived from ab initio simulations. To the other, we will discuss for the case study of water the effect of bond flexibility (bond stretching, angle bending) on the reproduction of different thermophysical properties over a wide range of state points.

6.4.1 Accounting for Multi-Body Dispersion

Considerable progress has been made over the past years to develop very accurate ab initio pair potentials for small and medium-sized atoms and molecules such as inert gases [42, 48, 50, 57], nitrogen [46, 68], carbon dioxide [47], methane [49], hydrogen sulfide [51] or mercury [112]. As these pair potentials have been derived from ab initio studies of the potential energy curve of dimers, they represent pure two-body potentials. Thus, they are suitable for simulations in the low density region, but are not able to represent bulk phase properties or phase equilibria, as shown exemplarily in Fig. 6.11 for the VLE of mercury.

This is due to the fact that high density properties can only be described accurately by accounting for the multi-body interactions that occur in bulk phases. Although the Lennard-Jones potential in classical force fields (see Eq. 6.1) also comprehends only pairwise non-bonded interactions, the effect of multi-body interactions is included implicitly by adjusting the Lennard-Jones parameters to represent experimental data of bulk phases (see Sect. 6.1.1.3). Therefore, additional multi-body potentials are required when ab initio two-body potentials shall be applied in simulations on bulk phase properties. For some small compounds such as neon or argon, ab initio

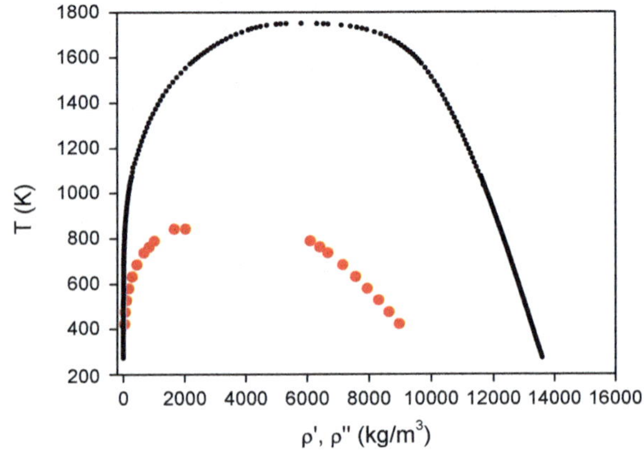

Fig. 6.11 Simulated vapor-liquid coexistence curve of mercury (*red dots*, [101]) by employing an ab initio two-body potential (by Schwerdtfeger et al. [112]) in comparison with experimental data [41, 120, 133] (*black dots*)

three-body potential have also been developed [13, 30]. Recently, Jäger et al. [58] have proposed a new ab initio non-additive three body potential for argon. This was based on an approach by Cencek et al. [15] to derive an analytical expression for the three-body short range and dispersive long-range contribution to the helium interaction potential based on highly accurate ab initio (Couple-Cluster) simulations. However, the development of ab initio three-body potential demands highest levels of theory to accurately calculate the weak dispersive interactions. Furthermore, a large number of single-point calculations are required to establish the three-body potential energy surface. Both aspects make the determination of ab initio three-body potentials computational very expensive [68].

For noble gases [81] and nitrogen [68], vapor-liquid phase equilibria have been successfully predicted in molecular simulation studies by ab initio two-body potentials in connection with the isotropic triple-dipole potential by Axilrod and Teller [4]

$$u_{3,\mu\mu\mu} = \frac{\nu_{\mu\mu\mu}(ijk)(1 + 3\cos\theta_i \cos\theta_j \cos\theta_k)}{(r_{ij} r_{ik} r_{jk})^3} \tag{6.42}$$

to account for three-body non-additive dispersion energies as the leading contribution to multi-body interactions. The $\nu_{\mu\mu\mu}$ in Eq. 6.42 is the non additive coefficient that is related to the polarizability α of the compound and its ionization energy I, $\nu_{\mu\mu\mu} \approx \alpha^3 I$ [68]. The angles and intermolecular separations refer to a triangular configuration of the atoms or sites as shown in Fig. 6.12. The total potential energy is then determined by the contribution of two-body interactions U_2 summed up over for all pairs and of the three-body interactions U_3 from all triples

$$U = U_2 + U_3 = \sum_{i=1}^{N} \sum_{j>i}^{N} u_2(\mathbf{r}_i, \mathbf{r}_j) + \sum_{i=1}^{N} \sum_{j>i}^{N} \sum_{k>j>i}^{N} u_3(\mathbf{r}_i, \mathbf{r}_j, \mathbf{r}_k). \tag{6.43}$$

Due to the additional summation over all triples of atoms or sites, the inclusion of three-body interactions typically requires at least one order of magnitude more computing time than calculations that are only considering pair interactions [82]. Thus, several attempts have been made to incorporate multi-body interactions in a computationally more feasible way. Marcelli and Sadus [82] observed that for noble

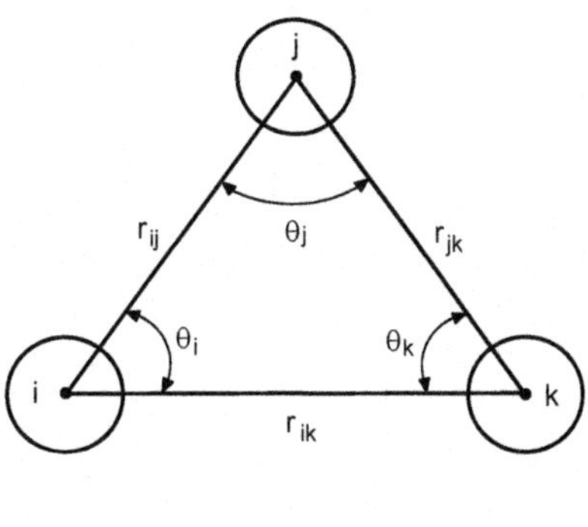

Fig. 6.12 Triangular configurations of three atoms in the Axilrod-Teller triple-dipole potential

gases such as argon, krypton or xenon, the ratio of the two-body and three-body energy is a linear function of the density ϱ. Based on this fact, they developed a method to estimate the energy contribution of three-body interactions U_3 from the two-body interaction energy U_2

$$U_3 = -\frac{2\,\nu_{\mu\mu\mu}\varrho\, U_2}{3\,\varepsilon\sigma^6}, \tag{6.44}$$

where ε is again the well-depth of the two-body potential, and σ the distance at which it is zero. Eskandari Nasrabad [32] has used a semi-empirical approach in which the ab initio two-body interaction potential is scaled by a constant factor to obtain a quantitative agreement for liquid phase properties and the VLE of argon and krypton over a wide range of state points.

For noble gases, the use of a two-body potential alone generates a phase envelope that is always larger than the experimental curve, i.e., the predicted liquid densities are greater than observed experimentally. The inclusion of three-body interactions either by the Axilrod-Teller potential or by the approaches discussed above results in lower predicted liquid densities and therefore in better agreement with experiment. For mercury however, the ab initio pair potential predicts a phase envelope that is inside of the experimental envelope as shown by Fig. 6.11. That indicates that the effect of multi-body interactions in bulk mercury can not be ascribed to non additive dispersion energies as leading contributions. Thus, the Axilrod-Teller potential alone would be insufficient to account for the multi-body interactions, and higher-body potentials for mercury are not available. Following the approach by Silvera and Goldman [118] for a computationally tractable alternative to multi-body calculations, we have proposed a non-additive contributions of multi-body effects by introducing an effective C_9/r^9 term into the ab initio pair potential [101]

$$u_2^{eff} = u_2^{abinitio} + \frac{C_9}{r^9}, \tag{6.45}$$

resulting in a semi-empirical effective potential. The C_9/r^9 term was then adjusted to fine-tune agreement with experimental saturated liquid densities. We were able to reliably predict values of the coexistence properties of mercury (Fig. 6.13) by using for C_9 a simple linear function of temperature

$$C_9 = -161020.46 - 2823.42\left(\frac{k_B T}{\varepsilon}\right). \tag{6.46}$$

Therein, ε is the minimum depth of the Schwerdtfeger ab initio potential, i.e. $\varepsilon/k_B = 525.15\,\mathrm{K}$. Figure 6.13 also shows our molecular simulation results for the vapor pressure p_s of mercury. It should be noted that these simulation results are purely predictive, as no experimental data for p_s were involved in the optimization of the C_9 term. Thus, the good agreement between simulation and experiment [41, 120, 133] is encouraging.

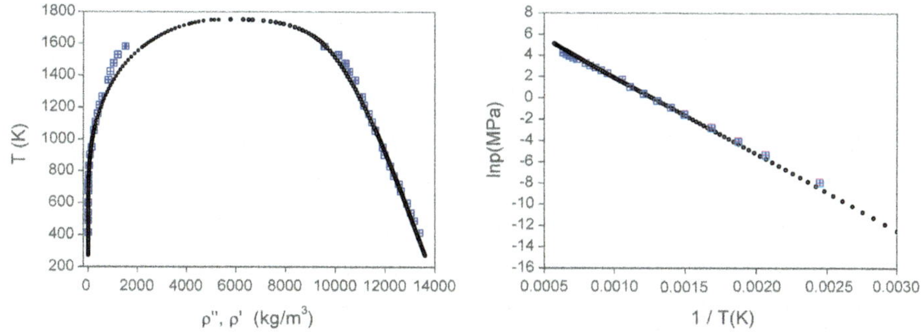

Fig. 6.13 Molecular simulation results (*blue squares*) for the saturation densities (*left*) and the vapor pressure (*right*) of mercury using a semi-empirical effective potential [101] that combines an accurate ab initio two-body potential with an effective C_9/r^9 multi-body term. Also shown as *black dots* are experimental data [41, 120, 133]

In [105], we have investigated the ability of the empirical effective potential to reproduce the transport properties of mercury along its vapor-liquid coexistence curve. We have thereby compared MD simulation results for the viscosity and self-diffusion coefficient with theoretical predictions by an approach by Tippelskirch et al. [127, 128] to describe the transport coefficients of mercury by a modified Enskog theory. Prior to discussing the simulation results, we provide a short introduction to the theoretical background of the Enskog theory. A detailed description is, for instance, given in the textbook by Hirschfelder, Curtiss and Bird [54].

The kinetic theory of gases is expressed by the Boltzmann's equation that represents integro-differential equations for the distribution function $f(r, v, t)$ of the particles in phase space. It describes the change of the distribution with time due to the act of external forces, diffusion of the particles and their collision. It is thereby assumed that only two-particle collisions occur, and that the collisions are not correlated, referred to as principle of molecular chaos. For a system at thermodynamic equilibrium, the distribution is independent of the particles positions r and time t, and therefore reduces to the Maxwell-Boltzmann velocity distribution that we have introduced in Sect. 2.4. For small deviations from equilibrium, i.e. small gradients, Enskog and Chapmann obtained approximate solutions of the Boltzmann equation from perturbation theory. This yields expressions for the transport coefficients in terms of so-called collision integrals, which in turn depend upon the modeling of the intermolecular interactions (see Chapmann and Cawling [16]). For a dilute gas (index 0), in which the effective molecular diameter σ_{eff} is small compared to the average distance between the particles, the Chapmann-Enskog solution of the Boltzmann equation yields the following expression for the shear viscosity of hard-sphere particles [127]

$$\eta_0 = \frac{5}{16}\frac{\sqrt{\pi m k_B T}}{\pi \sigma_{eff}^2}, \tag{6.47}$$

with the mass of a molecule m. The self-diffusion D_0 is given by [28]

$$D_0 = \frac{0.425 V_m}{N_a \sigma_{eff}^2} \sqrt{\frac{N_a k_B T M}{\pi}}. \tag{6.48}$$

Therein, V_m is the molar volume and M the molar mass of the compound.

Enskog advanced the kinetic theory for dense gases of hard-sphere particles by assuming that the deviation from the dilute gas solution can be attributed to a higher collision frequency. He thereby stated that the higher rate of collision in the dense fluids is due to the fact that the average distance that particles travel between collision is decreased by their effective diameter σ_{eff}. With this, the collision frequency is by a factor χ higher than in gases consisting of point particles [28, 54]. The correction factor χ is proportional to the hard-sphere radial distribution function at σ_{eff}, and can be related to the equation of state. The resulting expressions for the viscosity η_E and self-diffusion coefficient D_E of the Enskog theory are given by [28]

$$\eta_E = \eta_0 \left(\frac{B}{V_m}\right) \left[\left(\frac{\chi B}{V_m}\right)^{-1} + 0.800 + 0.761 \left(\frac{\chi B}{V_m}\right) \right], \tag{6.49}$$

$$D^V = D_0 / \chi. \tag{6.50}$$

The quantity B in Eq. 6.49 is the second virial coefficient of a hardsphere fluid

$$B = \frac{2\pi}{3} N_a \sigma_{eff}^3. \tag{6.51}$$

The Enskog theory though still assumes that the particle collisions are not correlated. For the modeling of transport coefficients in the liquid phase, correction factors F are employed to account for the correlated molecular motion

$$\eta^L = \eta_E \cdot F_\eta, \quad D^L = D_E \cdot F_D. \tag{6.52}$$

These correction factors were derived by Alder et al. [2] from MD studies by comparing their simulation results for the autocorrelation functions of different transport coefficients with the predictions by the Enskog theory.

For the theoretical description of the transport coefficients of mercury along the vapor-liquid coexistence curve, we employed correlations for the temperature depend hard sphere diameter σ_{eff} by Tippelskirch et al. [128], whereas the correction factors χ, F_D, F_η were taken from [2, 28].

Figure 6.14 shows the description of viscosity and self-diffusion coefficient by the modified Enskog theory in comparison with our MD simulation results derived by employing the two-body ab initio potential by Schwerdtfeger et al. [112] in combination with our empirical temperature-dependent multi-body contribution. It illustrates that our simulation results for the self-diffusion coefficient describe the slope of the saturation line in reasonable agreement and consistency with the theoretical

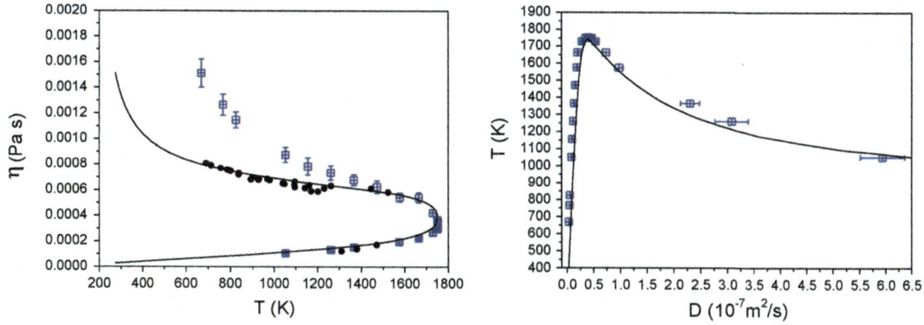

Fig. 6.14 Molecular simulation results for the viscosity and self diffusion coefficient of mercury (*blue squares*) [105] along the vapor-liquid coexistence curve using the semi-empirical effective potential. Shown as *line* are calculations based on the modified Enskog theory. The *black dots* represent experimental viscosities by Tippelskirch et al. [128]

predictions by the modified Enskog theory. Our effective potential also yields reasonable predictions for the shear viscosity of mercury at low and moderate densities, but results in significant deviations at high densities. So at liquid densities between 9.000 and 11.000 kg/m^3 metal-nonmetal transitions occur, which significantly influence the properties of mercury [52]. This can not be described by our simple temperature-dependent effective term, as it would require a density-dependent modeling of the multi-body contributions [52, 128].

6.4.2 *The Influence of Intramolecular Degrees of Freedom, SPC Water Potentials as Case Study*

Both the bond stretching and the angle bending are high frequency modes, and therefore it is commonly believed that these molecular vibrations have a minor effect on thermodynamic properties. Thus, the bond and angle terms in the force field are often omitted, i.e. these modes are frozen to enable larger time steps in the integration of the equation of motion in MD simulations. It therefore allows either saving of computer time or simulations on larger time scales.

This approach is surely appropriate for large and complex compounds for which the structural change in the molecule is dominated by the rotations around the dihedrals. However, we have shown [102–104] that for small-sized polar compounds (such as water) that are often treated as rigid molecules, introducing bond stretching and angle bending terms has an observable effect on both equilibrium properties and dynamic properties. Thus, this section covers our simulation results for a flexible water model as case study to illustrate the effect of introducing flexibility into a fixed point charge water potential. Thereby, simulation results of the widely used rigid SPC model [9] are compared with those of the flexible SPC/Fw model by Wu et al. [137]. In the rigid SPC model [9], the oxygen atom is represented as a partially charged

Lennard Jones bead, whereas the hydrogen atoms are simply represented by partial charges without any Lennard-Jones interactions. The intermolecular interactions are calculated from

$$U_{inter} = \sum_{i} \sum_{j>i} \left\{ 4\varepsilon_{ij} \left[\left(\frac{\sigma_{ij}}{r_{ij}} \right)^{12} - \left(\frac{\sigma_{ij}}{r_{ij}} \right)^{6} \right] + \frac{q_i q_j}{r_{ij}} \right\}. \tag{6.53}$$

In the SPC/Fw model molecular flexibility is added to the SPC model by accounting for bond stretching and angle bending terms via harmonic potentials

$$U_{intra} = \sum k_{r,OH}(r_{OH} - r_{0,OH})^2 + \sum k_{\theta,HOH}(\theta_{HOH} - \theta_{0,HOH})^2. \tag{6.54}$$

The Lennard Jones parameters and partial charges in the flexible models remain identical to those used in the SPC model. By comparing simulation results from the flexible SPC/Fw potential and the corresponding rigid SPC model, we were able to investigate the effect of incorporating intramolecular degrees of freedom on the prediction of the vapor-liquid equilibrium, dielectric constants and transport properties of water (Fig. 6.15).

We also performed simulations for the rigid SPC/E potential [8] that uses the same geometry and LJ parameters as the SPC potential, with the addition of a self-polarization energy correction that slightly increases the partial charges. The force field parameters used in this study are summarized in Table 6.3. Our GEMC simulation studies on the VLE properties of water in [102] show that the introduction of intramolecular degrees of freedom in the SPC/Fw model has a similar effect as the increasing of the partial charges sites in the rigid SPC/E model, i.e. to yield higher liquid densities and lower vapor pressures, as illustrated in Fig. 6.15. We attributed this to the fact that the bond flexibility allows the geometry, and with this the effective dipole moment to vary along the coexistence curve. For all state points studied here, the radius of gyration of the flexible model is higher that the constant value of the *SPC* model, resulting in a higher effective dipole model. Similar observation were made by Engin et al. [29] in their studies on flexible and rigid models for ammonia.

The enhanced dipole moment of the SPC/Fw model in comparison with the rigid SPC model gives reason for its liquid densities being quite similar to those of the SPC/E model. However, the results of the SPC/Fw and SPC/E models differ with respect to the prediction of the vapor pressure, and at elevated temperatures the

Table 6.3 Force field parameters of the SPC, SPC/Fw and SPC/E model

Model	ε_O ($\frac{\text{kJ}}{\text{mol}}$)	σ_O (Å)	q_O (e)	q_H (e)	$k_{r,OH}$ ($\frac{\text{kJ}}{\text{mol Å}^2}$)	$r_{0,OH}$ (Å)	$k_{\theta,HOH}$ ($\frac{\text{kJ}}{\text{mol rad}^2}$)	$\theta_{0,HOH}$ (rad)
SPC	0.65	3.166	−0.82	0.41	Rigid	1.0	Rigid	109.47
SPC/Fw	0.65	3.166	−0.82	0.41	2217.25	1.012	158.89	113.24
SPC/E	0.65	3.166	−0.8476	0.4238	Rigid	1.0	Rigid	109.47

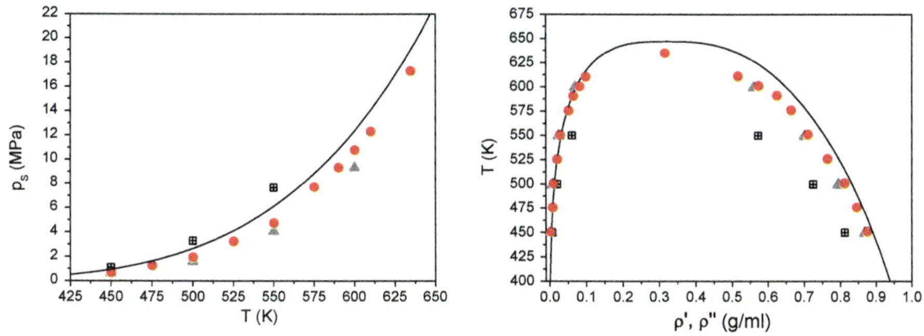

Fig. 6.15 GEMC simulation results [102] for the vapor pressure and vapor liquid coexistence curve of water for the SPC/Fw (*red dots*), SPC/E (*grey triangle*) and the SPC (*crossed squares*) model in comparison with calculations by REFPROP [67] (colour online)

flexible model yields better vapor pressures. This suggests that the introduction of intramolecular degrees of freedom has an additional effect.

This is well illustrated by our MD simulation studies [103] on the ability of SPC/Fw to accurately represent the high dielectric constant ε_r of water over a wide range of state points (for calculation of ε_r see Sect. 7.7). The simulation results of the SPC, SPC/E and SPC/Fw model for the dielectric constant of water at 298.15 and 673.15 K and elevated pressures are given in Fig. 6.16 in comparisons with experimental data [26]. We have analyzed our simulation results for ε_r in terms of equilibrium bond lengths and bond angles, and resulting dipole moments and quadrupole moments. Therein, we determined the quadrupole moment Q_T of SPC/Fw from the simulation results for the dipole moment μ, the equilibrium bond lengths r_{OH} and angles θ_{HOH} by [17]

$$Q_T = \frac{3\,(r_{OH}\sin\theta^*)^2}{4\,r_{OH}\cos\theta^*}\mu, \qquad 2\theta* = \theta_{HOH}. \tag{6.55}$$

The values for μ, r_{OH}, θ_{HOH} and the resulting Q_T at the different state points are given in Table 6.4, together with the corresponding (constant) values of the SPC and SPC/E model.

The SPC/Fw model shows elongated equilibrium bond lengths r_{OH} compared to both the model parameter ($r_{0,OH} = 1.012$ Å) and the rigid SPC and SPC/E model with $r_{OH} = 1$ Å. However, the $O-H$-bond of SPC/Fw slightly contracts with decreasing densities, i.e. at higher temperatures (state points II to IV) or lower pressures (II to I) The equilibrium $H-O-H$ bond angles of SPC/Fw however are remarkably smaller than the bond angle of the rigid models. With decreasing density, the equilibrium $H-O-H$ bond angle increases and approaches the value of the SPC and SPC/E model. These changes in the molecular geometry of the SPC/Fw model allow both the dipole and the quadrupole moment to vary despite fixed charges on the atomic sites. The elongated bond lengths in general result in both higher dipole and quadrupole

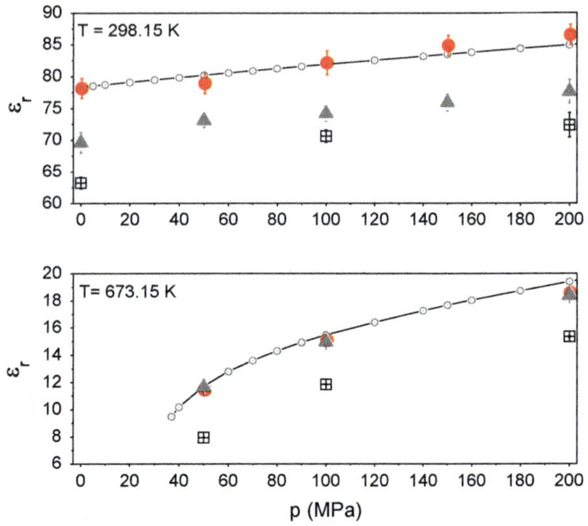

Fig. 6.16 MD simulation results [103] for the dielectric constant ε_r of water in the liquid phase at 298 K, and in the supercritical state at 673 K for the SPC/Fw (*red dots*), SPC/E (*grey triangle*) and the SPC (*crossed squares*) model in comparison with experimental data from Deul ([26], *white circles*)

Table 6.4 Equilibrium bond lengths r_{OH}, bond angles θ_{HOH}, dipole moment μ, and resulting quadrupole moments Q_T of the SPC/Fw model at different state points in comparison with values from the SPC/E and the SPC model

Model	State point	T (K)	p (MPa)	r_{OH} (Å)	θ_{HOH} (°)	Q_T (DÅ)	μ (D)
SPC/Fw	I	298.15	0.1	1.0310	107.70	2.046	2.394
	II	298.15	50	1.0312	107.68	2.046	2.395
	III	473.15	50	1.0271	108.21	2.043	2.336
	IV	673.15	50	1.0228	109.30	2.062	2.327
SPC/E	All	All	All	1.0	109.47	2.037	2.352
SPC	All	All	All	1.0	109.47	1.969	2.274

moments, whereas the reduced bond angle though also yields higher dipole moments but lower quadrupole moments. The dielectric constant in turn increases with increasing dipole moments but decreases with increasing quadrupole moments. Thus, the overall effect of changing r_{OH} in SPC/Fw on ε_r is moderate, as higher bond lengths increase both the dipole moment and the quadrupolar moment which have opposed effect on the dielectric constant. Though the dielectric constant of the flexible water model becomes strongly sensitive to bond angle changes, as stated by Wu et al. [137]. This is due to the fact that a decrease in the equilibrium bond angle gives higher dipole moments and lower quadrupole moments, with both resulting in higher dielectric constants. These effects of bond flexibility on the dielectric constant is well reflected in our simulations. Increasing temperatures yield contracted equilibrium bond lengths and expanded bond angles, both decreasing the molecular dipole moment of the SPC/Fw model. However, these changes in the molecular geometry have opposing effects on the quadrupole moment, and nearly balance out in the liquid phase (states

I, II and III), so that there is little change to the quadrupole moments of the SPC/Fw potential. The lower dielectric constants from the SPC/Fw potential with increasing temperatures are then mainly caused by its decreasing molecular dipole moments. In the supercritical phase at 673 K however, the increase of the quadrupole moment due to the expansion of the bond angle is more pronounced than its decrease with the contracted bond length r_{OH}. Thus, now the dielectric constant is lowered by the superposition of both effects, the smaller dipole and the higher quadrupole moment. Therefore the dielectric constants of SPC/Fw, which are remarkably above the predictions of the SPC/E model at 298 K, now approximate the results of the rigid model at 673 K, as illustrated in Fig. 6.16. The interplay of changes in the molecular geometry and the dipole moment also affects the dynamics of the systems as illustrated by our studies on the viscosity and diffusion coefficient of SPC/Fw [104]. Figure 6.17 shows the MD simulation results for the viscosity η (from Eq. 7.27) of water in the liquid phase at 298 K, and in the supercritical state at 673 K for the SPC/Fw in comparison with the SPC/E and the SPC model and experimental data [66, 125, 136]. The rigid SPC model considerably underestimates the shear viscosity, whereas the introduction of bond flexibility in the SPC/Fw model yields significantly higher viscosities, and with this a good reproduction of the experimental data at 298 K. The effect of bond flexibility is more pronounced than the effect of increased partial charges in the SPC/E model. With increasing temperatures, the bond length of the SPC/Fw model contracts and the dipole moment decreases as discussed above, which both should result in reduced viscosities. The fact that the SPC/Fw model largely overestimates the experimental data at 673 K indicates that the prediction of the viscosity is very sensitive to the expansion of the bond angles with increasing temperatures.

In the Fig. 6.18 the corresponding MD simulation results [104] for the diffusion coefficients D of the different water models are depicted (For the calculation of D

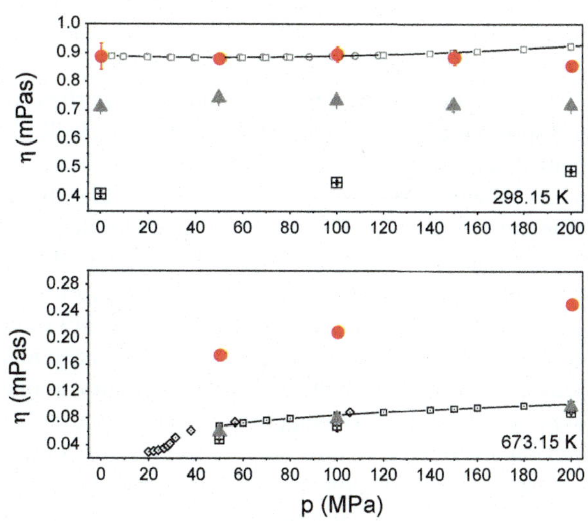

Fig. 6.17 MD simulation results [104] for the viscosity η of water in the liquid phase at 298 K in the supercritical state at 673 K for the SPC/Fw (*red circle*), SPC/E (*grey triangle*) and the SPC model (*crossed squares*) in comparison with experimental data (*rhomb* [66], *circle* [136], *square* [125]). The line through the experimental points is only for guidance

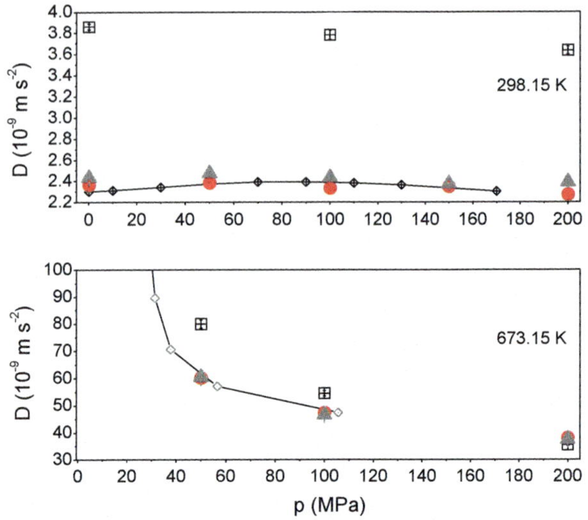

Fig. 6.18 MD simulation results [104] for the diffusion coefficient D of water in the liquid phase at 298 K in the supercritical state at 673 K for the SPC/Fw (*red circle*), SPC/E (*grey triangle*) and the SPC model (*crossed squares*) in comparison with experimental data (*rhomb* [66], *squares* [65]). The line through the experimental points is only for guidance

see Eq. 7.42). At 298 K, the increased dipole moments for both the SPC/Fw and the SPC/E model compared with SPC result in their significant lower diffusion coefficients, which are in good agreement with the experimental data. The elongated bond length in the SPC/Fw model is expected to further slow down its dynamics compared to the SPC/E model, though this effect seems to be partially compensated by its smaller bond angle. Thus, both models yield quite similar results for the diffusion coefficient at 298 K. With increasing temperature, the r_{OH} bond length in SPC/Fw contracts slightly, which results in decreasing dipole moments and a higher diffusivity. Again, the opposing effect of the expanded bond angle with increasing temperature compensates the effect of the reduced bond length, so that also at 673 K the SPC/Fw simulation results for the diffusion coefficient remain similar to those of the SPC/E model.

The results for the phase equilibria properties, the dielectric constant and the transport properties of water discussed above well illustrate that the introduction of bond flexibility allows the geometry to change in response to the thermodynamic state point, which in turn enables both the molecular dipole moment and the quadrupole moment to vary. These changes of the dipole and quadrupole moment with the thermodynamic state point mimics the change of the intermolecular interactions in response to the local environment, and can therefore be regarded as computationally less expensive way to incorporate polarizability effects in water simulations. This agrees with Yu and van Gunsteren [140] who have identified geometric polarization, caused by the change in the molecular geometry, as one of three possible mechanisms of polarization. It is also supported by other studies [72, 75] that have concluded that introducing flexibility transforms a rigid water potential into a polarizable potential. Our studies also demonstrate that the reproduction of the various thermophysical properties are affected in a very complex way and also quite differently by the changes of the equilibrium bond length and bond angle. Recent studies [39, 115] on a flexible version of the widely used TIP4P/2005 water model also suggest that the effect of bond flexibility also depends on the underlying molecular model and the way, the flexibility is introduced into it. The original TIP4P/2005 water model by Abascal and Vega [1] is a four site model with an additional dummy atom M, on which the negative charges is placed that compensates the positive charges located on the hydrogen atoms. In the flexible TIP4P/2005f model [39], flexibility is introduced to the $O-H$ bonds and the $H-O-H$ angle, whereas the $O-M$ distance is not fully flexible but defined by the positions of the hydrogen atoms, i.e. the $O-H$ bond length. For this model no substantial improvement could be observed compared to the performance of the rigid model [39, 115].

References

1. Abascal JLF, Vega C (2005) A general purpose model for the condensed phase of water: TIP4P/2005. J Chem Phys 123:234505
2. Alder BJ, Gass DM, Wainwright TE (1970) Studies in molecular dynamics. VIII. The transport coefficients for a hard-sphere fluid. J Chem Phys 53:3813–3826
3. Allinger NL (1977) Conformational analysis. 130. MM2, a hydrocarbon force field utilizing V_1 and V_2 torsional terms. J Am Chem Soc 99:8127–8134
4. Axilrod BM, Teller E (1943) Interaction of the van der Waals type between three atoms. J Chem Phys 11:299
5. Baskes MI (1992) Modified embedded-atom potentials for cubic materials and impurities. Phys Rev B 46(5):2727–2742
6. Baskes MI, Lee BJ, Kim HI, Cho YK (2001) Second nearest neighbor modified embedded atom method potentials for bcc transition metals. Phys Rev B 64:184102
7. Bedrov D, Borodin O, Li Z, Smith GD (2010) Influence of polarization on structural, thermodynamic, and dynamic properties of ionic liquids obtained from molecular dynamics simulations. J Phys Chem B 114(15):4984–4997
8. Berendsen HJC, Grigera JR, Straatsma TP (1987) The missing term in effective pair potentials. J Phys Chem 91(24):6269–6271
9. Berendsen HJC, Postma JPM, vanGunsteren WF, Hermans J (1981) In: Pullmann B (ed) Intermolecular forces. Reidel, Dordrecht
10. Bhargava BL, Balasubramanian S, Klein ML (2008) Modelling room temperature ionic liquids. Chem Commun 3339–3351
11. Breneman CM, Wiberg KB (1989) Determining atom-centered monopoles from molecular electrostatic potentials. The need for high sampling density in formamide conformational analysis. J Comp Chem 11:361–373
12. Brewer L, Edwards RK (1954) The stability of SiO solid and gas. J Phys Chem 58(4):351–358
13. Bukowski R, Szalewicz K (2001) Complete ab initio three-body nonadditive potential in Monte Carlo simulations of vapor-liquid equilibria and pure phase of argon. J Chem Phys 114(21):9518–9531
14. Canongia Lopes JN, Pádua A (2006) Molecular force field for ionic liquids III: imidazolium, pyridium, and phosphonium cations; chloride, bromide and dycanamide anions. J Phys Chem B 110:19586–19592
15. Cencek W, Patkowski K, Szalewicz K (2009) Full configuration-interaction calculation of the three-body nonadditive contribution to helium interaction potential. J Chem Phys 131:064105
16. Chapmann S, Cowling TG (1970) The mathematical theory of nonuniform gases. Chambridge University Press, London

17. Chatterjee PG, Debenedetti PG, Stillinger FH, Lynden-Bell RM (2008) A computational investigation of thermodynamics, structure, dynamics and solvation behavior in modified water models. J Chem Phys 128:124511
18. Chen B, Siepmann JI (1999) Transferable potentials for phase equilibria. 3. Explicit-hydrogen description of normal alkanes. J Phys Chem B 103:5370–5379
19. Chen IJ, Yin D, MacKerell AD (2002) Combined ab initio/empirical optimization of Lennard-Jones paramaters for polar neutral compounds. J Comp Chem 23:199–213
20. Childs BG, Weintraub S (1949) The measurement of the thermal expansion of single crystals of tin by an interferometric method. Proc Phys Soc B 63:267–277
21. Chirlain LE, Francl MM (1987) Charges fit to electrostatic potentials. J Comp Chem 8:894–904
22. Cornell WD, Cieplak P, Bayly CI, Gould IR, Merz KM Jr, Ferguson DM, Spellmeyer DC, Fox T, Caldwell JW, Kollmann PA (1995) A second generation force field for the simulation of proteins, nucleic acids, and organic molecules. J Am Chem Soc 117:5179–5197
23. Cornell WD, Cieplak P, Bayly CI, Kollmann PA (1993) Application of RESP charges to calculate conformational energies, hydrogen bond energies, and free energies of solvation. J Am Chem Soc 115:9620–9631
24. Daubert TE, Jalowka JW, Goren V (1987) Vapor pressure of 22 pure industrial chemicals. AIChE Symp Series 83:128–156
25. Deshpande VT, Sirdeshmukh DB (1961) Thermal expansion of tetragonal tin. Acta Crystallogr 14:355–356
26. Deul R (1984) Dielektrizitätskonstante und Dichte von Wasser-Benzol-Mischungen bis 400°C und 3000 bar. PhD Thesis Karlsruhe, Germany
27. Dommert F, Wendler K, Qiao B, Delle Site L, Holm C (2014) Generic force fields for ionic liquids. J Mol Liq 192:32–37
28. Dymond JH, Alder BJ (1966) Van der waals theory of transport in dense fluids. J Chem Phys 45:2061–2068
29. Engin C, Merker T, Vrabec J, Hasse H (2011) Flexible or rigid molecular models? A study on vapour-liquid equilibrium properties of ammonia. Mol Phys 109:619–624
30. Ermakova E, Solca J, Steinbrunner G, Huber HH (1998) Ab initio calculation of the three-body potential to be applied in simulations of fluid neon. Chem Eur J 4:377
31. Errington JR, Panagiotopoulos AZ (1999) A new intermolecular potential model for the n-alkanes homologous series. J Phys Chem B 103:6314–6322
32. Eskandari Nasrabad AE (2008) Theory and atomistic simulation of krypton fluid. J Chem Phys 129:244503
33. Finnes MW, Sinclair JE (1984) Philos Mag A 50:45
34. Foiles SM, Baskes MI, Daw MS (1986) Embedded-atom-method functions for the fcc metals Cu, Ag, Au, Ni, Pd, Pt, and their alloys. Phys Rev B 33(12):7983–7991
35. Foloppe N, MacKerell AD Jr, Stote R, Straub J, Watanabe M, Wiorkiewicz-Kuczera J, Yin D, Karplus M (2000) All-atom empirical force field for nuclei acids I. Parameter optimization based on small molecule and condensed phase macromolecular target data. J Comp Chem 21:86–104
36. Fox T, Kollman PA (1998) Application of the RESP mythology in the parameterization of organic solvents. J Phys Chem B 102:8070–8079
37. Frisch MJ, Trucks GW, Schlegel HB, Scuseria GE, Robb MA, Cheeseman JR, Zakrzewski VG, Montgomery JA Jr, Stratmann RE, Burant JC, Dapprich S, Millam JM, Daniels AD, Kudin KN, Strain MC, Farkas O, Tomasi J, Barone V, Cossi M, Cammi R, Mennucci B, Pomelli C, Adamo C, Clifford S, Ochterski J, Petersson GA, Ayala PY, Cui Q, Morokuma K, Malick DK, Rabuck AD, Raghavachari K, Foresman JB, Cioslowski J, Ortiz JV, Baboul AG, Stefanov BB, Liu G, Liashenko A, Piskorz P, Komaromi OI, Gomperts R, Martin RL, Fox DJ, Keith T, Al-Laham MA, Peng CY, Nanayakkara A, Gonzalez C, Challacombe M, Gill PMW, Johnson B, Chen W, Wong MW, Andres JL, Gonzalez C, Head-Gordon M, Replogle ES, Pople JA (1998) Gaussian 98 Revision A.7. Gaussian, Inc., Pittsburgh, PA,

38. Geerke DP, vanGunsteren WF (2007) On the calculation of atomic forces in classical simulation using the charge-on-spring methods to explicity treat electronic polarization. J Chem Theor Comput 3:2128–2137
39. Gonzalez MA, Abascal JLF (2011) A flexible model for water based on TIP4P/2005. J Chem Phys 135:224516
40. Groom CR, Bruno IJ, Lightfoot MP, Ward SC (2016) The cambridge structure database. Acta Crystallogr B72:171–179
41. Götzlaff W (1988) Zustandsgleichung und elektrischer Transport am kritschen Punkt des fluiden Quecksilbers. PhD Thesis Marburg, Germany
42. Haley P, Cybulski SM (2003) Ground state potential energy curves for He-Kr, Ne-Kr, and Kr2: coupled-cluster calculations and comparison with experiment. J Chem Phys 119:5487
43. Halgren TA (1996) Merck molecular force field. V. Extension of MMFF94 using experimental data, additional computational data, and empirical rules. J Comp Chem 15(5+6):616–641
44. Halgren TA (1998) Merck molecular force field. I. Basis, form, scope, parameterization, and performance of MMFF94. J Comp Chem 17(5–6):490–519
45. Halgren TA, Damm W (2001) Polarizable force fields. Current Opin Struct Biol 11:236–242
46. Hellmann R (2013) Ab initio potential energy surface for the nitrogen molecule pair and thermophysical properties of nitrogen gas. Mol Phys 111:387
47. Hellmann R (2014) Ab initio potential energy surface for the carbon dioxide molecule pair and thermophysical properties of dilute carbon dioxide gas. Chem Phys Lett 613:133
48. Hellmann R, Bich E, Vogel E (2007) Ab initio potential energy curve for the helium atom pair and thermophysical properties of the dilute helium gas. I. Helium-helium interatomic potential. Mol Phys 105:3013
49. Hellmann R, Bich E, Vogel E (2008) Ab initio intermolecular potential energy surface and second pressure virial coefficient of methane. J Chem Phys 128:214303
50. Hellmann R, Bich E, Vogel E (2008) Ab initio potential energy curve for the neon atom pair and thermophysical properties of the dilute neon gas. I. Neon-neon interatomic potential and rovibrational spectra. Mol Phys 106:133
51. Hellmann R, Bich E, Vogel E, Vesovic V (2011) Ab initio intermolecular potential energy surface and thermophysical properties of hydrogen sulfide. Phys Chem Chem Phys 13:13749
52. Hensel F, Warren JWW (1999) Fluid metals: the liquid-vapor transition of metals. Princeton University Press, Princeton, NJ
53. Hinchliffe A (2008) Molecular modelling for beginners, 2nd edn. Wiley
54. Hirschfelder JO, Curtiss CF, Bird RB (1954) Molecular theory of gases and liquids. Wiley, New York
55. Hulse R, Singh RR, Pham H (2009) Physical properties of HFO-1234yf. In: 3rd IIR conference on thermophysical properties and transfer processes of refrigerants, Boulder, USA
56. Hulse RJ, Rajat SB, Singh RR, Thomas RHP (2012) Physical properties of HCFO-1233zd(E). J Chem Eng Data 57:3581–3586
57. Jäger B, Hellmann R, Bich E, Vogel E (2009) Ab initio potential energy curve for the argon atom pair and thermophysical properties of the dilute argon gas. I. Argon-argon interatomic potential and rovibrational spectra. Mol Phys 107:2181
58. Jäger B, Hellmann R, Bich E, Vogel E (2011) Ab initio virial equation of state for argon using a new nonadditive three-body potential. J Chem Phys 135:084308
59. Jorgensen WL, Chandrasekhar J, Madure JD, Impey RW, Klein ML (1983) Comparison of simple potential functions for simulating liquid water. J Chem Phys 79:926–935
60. Jorgensen WL, Maxwell DS, Tirado-Rives J (1996) Development and testing of the OPLS all-atom force field on conformational energetics and properties of organic liquids. J Am Chem Soc 118:11225–11236
61. Jorgensen WL, Maxwell DS, Tirado-Rives J (1997) OPLS all-atom force field for carbohydrates. J Comp Chem 18(16):1955–1970
62. Kaminski GA, Friesner RA, Tirado-Rives J, Jorgensen WL (2001) Evaluation and reparametrization of the OPLS-AA force field for proteins via comparison with accurate quantum chemical calculations on peptides. J Phys Chem B 105:6474–6487

63. Klähn M, Seduraman A, Wu P (2008) A force field for guanidinium-based ionic liquids that utilizes the electron charge distribution of the actual liquid: a molecular simulation study. J Phys Chem B 112:10989–11004
64. Kohagen M, Brehm M, Thar J, Zhao W, Müller-Plathe F, Kirchner B (2011) Performance of quantum chemically derived charges and persistence of ion cages in ionic liquids. J Phys Chem B 115:693–702
65. Krynicki K, Green CD, Sawyer DW (1978) Pressure and temperature dependence of self-diffusion in water. Faraday Discuss Chem. Soc. 66:199
66. Lamb WJ, Hoffman GA, Jonas J (1981) Self-diffusion in compressed supercritical water. J Chem Phys 74:6875–6880
67. Lemmon EW, Huber ML, McLinden MO (2013) NIST Standard reference database 23: reference fluid thermodynamic and transport properties—REFPROP, Version 9.1, Standard Reference Data Program. National Institute of Standards and Technology: Gaithersburg, MD
68. Leonhard K, Deiters UK (2002) Monte Carlo simulations of nitrogen using an ab initio potential. Mol Phys 100(15):2571–2585
69. Li S, Zhou Z, Sadus RJ (2007) Role of nonadditive forces on the structure and properties of liquid water. J Chem Phys 127:154509
70. Lide DR, Frederikse HPR (1998) Handbook of chemistry & physics. CRC Press, New York
71. Liessmann G, Schmidt W, Reiffarth S (1995) Data compilation of the Sächsische Olefinwerke Böhlen
72. Liew CC, H I, Arai K (1998) Flexible molecular models for molecular dynamics study of near and supercritical water. Fluid Phase Equilib 144:287–298
73. Liu L, Liu Y, Zybin SV, Sun H, Goddard WA III (2011) ReaxFF-lg: corrections of the ReaxFF reactive force field for london dispersion, with application to the equation of state for energetic materials. J Phys Chem A 115(40):11016–11022
74. Logothetit GE, Ramos GE, Economou IG (2009) Molecular modeling of imidazolium-based [Tf2N-] ionic liquids: microscopic structure, thermodynamic and dynamic properties, and segmental dynamics. J Phys Chem B 113(20):7211–7224
75. Lópes-Lemus J, Chapela GA, Alejandre J (2008) Effect of flexibility on surface tension and coexisting densities of water. J Chem Phys 128:174703
76. Lucas TR, Bauer BA, Patel S (2012) Charge equilibration force fields for molecular dynamics simulations of lipids, bilayers, and integral membrane protein systems. Biochemica et Biophysica Acta (BBA)—Biomembranes 1818:318–329
77. MacKerell JAD (2004) Empirical force fields for biological macromolecules: overview and issues. J Comp Chem 25(13):1584–1604
78. MacKerell AD Jr, Bashford D, Bellott M, Dunbrack RL Jr, Evanseck JD, Field MJ, Fischer S, Gao J, Ha S, Joseph-McCarthy D, Kuchnir J, Kuczera K, Lau FTK, Mattos C, Michnick S, Ngo T, Nguyen DT, Prodhom B, Reiher WE III, Roux B, M S, Smith JC, Stote R, Straub J, Watanabe M, Wiorkiewicz-Kuczera J, Yin D, Karplus M, (1998) All-atom empirical potential for molecular modeling and dynamics studies of proteins. J Phys Chem B 102:3586–3616
79. MacKerrel Jr A (2007) Empirical force fields. In: Xu Y, Xu D, Liang J (eds) Computational methods for protein structure prediction and modeling. Springer, New York
80. Maginn EJ (2007) Atomistic simulation of the thermodynamic and transport properties of ionic liquids. Acc Chem Res 40:1200–1207
81. Marcelli G, Sadus RJ (1999) Molecular simulation of the phase behaviour of noble gases using accurate two-body and three-body intermolecular potentials. J Chem Phys 111(4):1533–1540
82. Marcelli G, Sadus RJ (2000) A link between the two-body and three-body interaction energies of fluids from molecular simulation. J Chem Phys 112:6382–6385
83. Martin MG, Siepmann JI (1998) Transferable potentials for phase equilibria. 1. United-atom description of n-alkanes. J Phys Chem 102:2569–2577
84. Mie G (1903) Zur kinetischen Theorie der einatomigen Körper. Annalen der Physik 11:657–697
85. Morrow TI, Maginn EJ (2002) Molecular dynamics study of the ionic liquid 1-n-butyl-3-methylimidazolium hexafluorophosphate. J Phys Chem B 106(20):12807–12813

86. Nath SK, A EF, dePablo JJ, (1998) On the simulation of vapor-liquid equilibria for alkanes. J Chem Phys 108:9905–9911
87. ofStandard NI, Technology (2006) Computational chemistry comparison and benchmark database, Release 14. http://cccbdb.nist.gov
88. Paricaud P, Předota, Chialvo AA, Cummings PT (2005) From dimer to condensed phases at extreme conditions: accurate predictions of the properties of water by a Gaussian charge polarizable model. J Chem Phys 122:6244511
89. Patel S, Brooks CL III (2003) CHARMM fluctuating charge force field for proteins: I Parameterization and application to bulk organic liquid simulations. J Comput Chem 25(1):1–15
90. Plimpton S (1995) Fast parallel algorithms for short-range molecular dynamics. J Comp Phys 117:1–19
91. Ponder JW, Case DA (2003) Force fields for protein simulations. Adv Prot Chem 66:27–85
92. Ponder JW, Wu C, Ren P, Pande VS, Chodera JD, Mobley DL, Schnieders MJ, Haque DS, ILambrecht, DiStasio JRA, Head-Gordon M, Clark GNI, Johnson ME, Head-Gordon T (2010) Current status of the AMOEBA polarizable force field. J Phys Chem B6 114:2549–2564
93. Potter SC, Tildesley DJ, Burgess AN, Rogers SC (1997) A transferable potential model for the liquid-vapor equilibria of fluoromethanes. Mol Phys 92(5):825–833
94. Raabe G (2010) Development of force field models for alternative refrigerants based on fluoropropenes, including HFO-1234yf. Final report, DFG-Fellowship (Forschungsstipendium) RA 946/2-1
95. Raabe G (2012) Molecular modeling of fluoropropene refrigerants. J Phys Chem B 116:5744–5751
96. Raabe G (2013) Molecular simulation studies on the vapor-liquid-equilibria of binary mixtures of R-1234yf and R-1234ze(E) with R-32 and CO_2. J Chem Eng Data 58:1867–1873
97. Raabe G (2015) Molecular simulation studies on the vapor-liquid-equilibria of the cis- and trans-HCFO-1233zd and the cis- and trans-HFO-1336mzz. J Chem Eng Data 60:2412–2419
98. Raabe G, Maginn EJ (2010) A force field for 3,3,3-fluoro-1-propenes, including HFO-1234yf. J Phys Chem B 114:10133–10142
99. Raabe G, Maginn EJ (2010) Molecular modeling of the vapor-liquid equilibrium properties of the alternative refrigerant 2,3,3,3-tetrafluoro-1-propene (HFO-1234yf). J Phys Chem Lett 1:93–96
100. Raabe G, Mai JP, Köhler J (2013) Herstellung von Rohsilizium (mg-Si) im Mikrowellenofen. Deutsche Bundesstiftung Umwelt AZ-28408
101. Raabe G, Sadus RJ (2003) Monte Carlo simulation of the vapor-liquid coexistence of mercury. J Chem Phys 119(13):6691–6697
102. Raabe G, Sadus RJ (2007) Influence of bond flexibility on the vapor-liquid phase equilibria of water. J Chem Phys 126:044701
103. Raabe G, Sadus RJ (2011) Molecular dynamics simulation of the dielectric constant of water: the effect of bond flexibility. J Chem Phys 134:234501
104. Raabe G, Sadus RJ (2012) Molecular dynamics simulation of the effect of bond flexibility on the transport properties of water. J Chem Phys 127:104701
105. Raabe G, Todd BD, Sadus RJ (2005) Molecular simulation of the shear viscosity and the self-diffusion coefficient of mercury along the vapor-liquid coexistence curve. J Chem Phys 123:034511
106. Rappe AK, Casewit CJ, Colwell KS, Goddard WA III, Skiff W (1992) UFF, a full periodic table force field for molecular mechanics and molecular dynamics simulations. J Am Chem Soc 114:10024–10035
107. Ravelo R, Baskes M (1997) Equilibrium and thermodynamic properties of grey, white, and liquid tin. Phys Rev Lett 70(3):2482–2485
108. Ren P, Ponder JW (2003) Polarizable atomic multipole water model for molecular mechanics simulation. J Phys Chem B 107(24):5933–5947
109. Rick SW, Stuart SJ, Bader JS, Berne BJ (1995) Fluctuating charge force field for aqueous solutions. J Mol Liquids 65(66):31–40

110. Russo JMF, Van Duin ACT (2010) Atomic-scale simulations of chemical reactions: binding from quantum chemistry to engineering. Nucl Instrum Meth B 269:1549–1554
111. Schei A (1967) On the chemistry of ferrosilicon production. Tidsskr Kjemi Bergv 17:152–158
112. Schwerdtfeger P, Wesendrup R, Moyano GE, Sadlej AJ, J G, Hensel F (2001) The potential energy curve and dipole polarizability tensor of mercury dimer. J Chem Phys 115:7401
113. Scott WRP, Hunenberger PH, Tironi IG, Mark AE, Billeter SR, Fennen J, Torda AE, Huber T, Kruger P, vanGunsteren WF (1999) The GROMOS biomolecular simulation program package. J Phys Chem A 103(1):3596–3607
114. Seffert N, Wipff G (2006) The [BMI][Tf_2N] ionic liquid/water binary system: a molecular dynamics study of phase separation and of the liquid-liquid-interface. J Phys Chem B 110(26):13076–13085
115. Shvab I, Sadus RJ (2015) Thermophysical properties of supercritical water and bond flexibility. Phys Rev E 92:012124
116. Singh R (2009) Personal communication
117. Singh UC, Kollmann P (1984) An approach to computing electrostatic charges for molecules. J Comp Chem 5:129
118. Sivera FI, Goldman VV (1978) J Chem Phys 69:4209
119. Stoll J, Vrabec J, Hasse H (2003) A set of molecular models for carbon monoxide and halogenated hydrocarbons. J Chem Phys 119:11396–11407
120. Sugawara S, Sato T, Minamiyama T (1962) Thermodynamic and electrical properties of fluid Hg up to the liquid-vapor critical point. Bull Jap Soc Mech 5:711
121. Sum AK, Biddy MJ, dePablo JJ, Tupy MJ (2003) Predictive molecular model for the thermodynamic and transport properties of triacylglycerols. J Phys Chem B 107:14443–14451
122. Sutton AP, Chen J (1990) Long-range Finnis-Sinclair potentials. Philos Mag Lett 61:139
123. Tanaka K (2014) Personal Communication
124. Tanaka K, Higashi Y (2009) 3rd IIR conference on thermophysical properties and transfer processes of refrigerants. Boulder, USA
125. Tanaka Y, Matsuda Y, Fujiwara H, Kuboty H, Makita T (1987) Viscosity of (water+alcohol) mixtures under high pressure. Int J Thermophys 147-163
126. Thole BT (1981) Molecular polarizabilities calculated with a modified dipole interaction. Chem Phys 59:341–350
127. Hv Tippelskirch (1977) Transport coefficients of expanded fluid metals. Z Naturforsch 32:1146–1151
128. Hv Tippelskirch, Frank EU, Hensel F, Kestin J (1975) Viscosity of fluid mercury to 1529 K and 100 bar. Ber Bunsenges Phys Chem 79:889–897
129. Valadez-Huerta G (2012) Molekulardynamik-Simulationen zu Eigenschaften von Lotmaterialien. Diplomarbeit, Institut für Thermodynamik, TU Braunschweig
130. Van Duin ACT, Dasgupta S, Lorant F, Goddard WA III (2001) ReaxFF: a reactive force field for hydrocarbons. J Phys Chem A 105(41):9396–9409
131. Van Duin ACT, Strachan A, Stewman S, Zhang Q, Xu X, Goddard WA III (2003) ReaxFFSio reactive force field for silicon and silicon oxide systems. J Phys Chem A 107:3803–3811
132. Vanommeslaeghe K, MacKerrel Jr AD (2015) CHARMM additive and polarizable force fields for biophysics and computer-aided drug design. Biochemica et Biophysica Acta (BBA)—General Subjects 1850:861–871
133. Vargaftik NB, Vinogradov YK, Yargin VS (1996) Handbook of physical properties of liquid and gases, 3rd edn. Begell House, New York
134. Wang J, Kollman PA (2001) Automatic parametrization of force field by systematic search and genetic algorithms. J Comp Chem 22(12):1219–1228
135. Wang J, Wolf RM, Caldwell JW, Kollman PA, Case DA (2004) Development and testing of a general AMBER force field. J Comp Chem 25(9):1157–1174
136. Woolf LA (1975) Tracer diffusion of tritiated water (THO) in ordinary water (H_2O) under pressure. J Chem SoC Faraday Trans 71:784–796
137. Wu Y, Tepper HL, Voth GA (2006) Flexible simple point-charge water model with improved liquid-state properties. J Chem Phys 124:024503

138. Yang J, Tian A, Sun H (2000) COMPASS force field for 14 inorganic molecules, He, Ne, Ar, Kr, Xe, H_2, O_2, N_2, NO, CO, CO_2, NO_2, CS_2, and SO_2, in liquid phases. J Phys Chem B 104:4951–4957
139. Youngs TG, Hardacre C (2008) Application of static charge transfer with an ionic-liquid force field and its effect on structure and dynamics. ChemPhysChem 9(11):1548–1558
140. Yu H, vanGunsteren WF (2005) Accounting for polarization in molecular simulation. Comput Phys Commun 172:69–85
141. Zernov VS, Kogan VB, Lyubetskii SG, Duntov FI (1971) J Appl Chem USSR 44:693–696
142. Zhang Y, Maginn EJ (2012) A simple AIMD approach to derive atomic charges for condensed phase simulation of ionic liquids. J Phys Chem B 116(33):10036–10048

Chapter 7
Thermophysical and Structural Properties from Molecular Simulation

This chapter shall provide an overview how several thermophysical and structural properties of pure compounds and mixtures can be determined from analysis of the system's trajectory derived from molecular simulation studies. Once a reliable force field for the system of interest has been established, molecular simulation studies provide a way to predict all its relevant thermodynamic and structural properties within the framework of statistical mechanics. Molecular simulation techniques are therefore increasingly used in chemical engineering as complement to experimental studies, as they

- allow studies at extreme conditions such as very low or high temperatures, high pressures
- enable exploration on substances difficult to handle in experiments (i.e. toxic, hazardous, corrosive compounds)
- provide valuable fundamental insights into the behavior of the compounds on the molecular level, i.e. in which way thermophysical properties are related to the molecular structure.

Compared to the usually employed engineering models on thermophysical properties, molecular simulations offer several advantages [83] as they

- are able to provide predictions on poorly known systems
- enable prediction of mixture properties with no or only minor calibration on binary mixtures
- are capable of predicting differences in isomer properties
- allow to extrapolate predictions

Our molecular simulation study on the vapor-liquid equilibrium of mercury at temperatures up to 1575 K, discussed in the Sect. 6.4.1, demonstrates the valuable role that molecular simulation has in supplementing experimental data for situations in which laboratory measurements are either difficult to perform due to extreme conditions in temperature or pressure, or involve dangerous/toxic substances.

G. Raabe, *Molecular Simulation Studies on Thermophysical Properties*,
Molecular Modeling and Simulation, DOI 10.1007/978-981-10-3545-6_7

The exploration of ionic liquids is a recent example that has proven the value of molecular simulation studies to provide predictions on poorly known compounds. Based on force fields that were optimized to reproduce the few available experimental data, molecular simulation studies have provided predictions of many fundamental properties of ionic liquids and their mixtures (reviews amongst others from [8, 18, 40, 45]). Simulation studies on ionic liquids have also helped to gain a molecular level understanding of their properties [89] and in which way the microstructural characteristics are related to the solubility of the ionic liquids for different solutes [11]. Our own studies on microstructural characteristics of ionic liquid [67] and how they change upon mixing, [66] will serve as an example in this chapter to illustrate how local structures can be analyzed by radial and spherical distribution functions, and how the structural analysis can provide a molecular interpretation of solvation process.

Another example of the use of molecular simulations to complement rare experimental data on newly developed compounds is our study on alternative working fluids based on hydrofluoroolefines. In Sect. 6.1.1.5 we have introduced a transferable force field for hydrofluoroolefines (HFO) and hydrochlorofluoroolefines (HCFO) [63, 65, 68, 69] that enables molecular simulation studies on thermodynamic and transport properties of these compounds and their mixtures. Applications of the molecular modeling for this new class of working fluids will be discussed in detail in Chap. 8. In this chapter, some of our simulation results for ionic liquid, HFO or other compounds will exemplify for the analysis of molecular simulations to derive thermophysical and structural properties.

7.1 *PVT* Properties in a Single Phase

When performing Monte Carlo or MD simulations in the *NVT* ensemble, the resulting pressure in the system for the given temperature and density can be derived according to Eq. 2.62

$$p = k_B T \left(\frac{\partial \ln Q_{NVT}}{\partial V} \right)_{T,V}. \tag{7.1}$$

By introducing the reduced variable $s = r/L$ (see Sect 3.3, Eq. 3.23), the partition function of the *NVT* ensemble can be written by

$$\begin{aligned} Q_{NVT} &= \underbrace{\frac{V^N}{N!\Lambda^{3N}}} \underbrace{\int e^{-U_{conf}(s \cdot L)/k_B T} ds^{3N}} \\ &= Q_{NVT}^{id} \quad \cdot \quad Q_{NVT}^{res}. \end{aligned} \tag{7.2}$$

Accordingly, the pressure can be divided into an ideal gas and a residual part, with

$$p^{id} = k_B T \left(\frac{\partial \ln Q^{id}_{NVT}}{\partial V} \right)_{T,V} = k_B T \frac{N}{V}. \tag{7.3}$$

The residual pressure due to the interactions in a real fluid is then defined by

$$p^{res} = k_B T \left(\frac{\partial \ln Q^{res}_{NVT}}{\partial V} \right)_{T,V} \tag{7.4}$$

$$= k_B T \frac{\int \frac{-\partial(U_{conf}/k_B T)}{\partial V} e^{-U_{conf}/k_B T} ds^{3N}}{\int e^{-U_{conf}(s \cdot L)/k_B T} ds^{3N}}. \tag{7.5}$$

$$= k_B T \left\langle \frac{-\partial(U_{conf}/k_B T)}{\partial V} \right\rangle$$

as the fraction term in Eq. 7.5 represent the definition of the ensemble average of the term $U_{conf}/k_B T$.

The partial derivative of the configurational energy with respect to the volume though is given by

$$\frac{\partial U_{conf}}{\partial V} = \frac{\partial U_{conf}}{\partial r} \cdot \frac{\partial r}{\partial V} = \frac{\partial U_{conf}}{\partial r} \cdot \frac{r}{3V}. \tag{7.6}$$

With this, the residual pressure can in general be determined from

$$p^{res} = -\frac{1}{3V} \left\langle r \frac{\partial U_{conf}}{\partial r} \right\rangle. \tag{7.7}$$

In a system in which the total configurational energy is given by the sum of pairwise interactions energies u_{ij}, the partial derivative of U_{conf} with respect to the interparticle distances r_{ij} becomes

$$r \frac{\partial U_{conf}}{\partial r} = \sum_{i=1}^{N} \sum_{j>i}^{N} r_{ij} \frac{\partial u_{ij}}{\partial r_{ij}} = -\sum_{i=1}^{N} \sum_{j>i}^{N} r_{ij} \cdot f_{ij}. \tag{7.8}$$

The product $r_{ij} \cdot f_{ij}$ is the atomic virial W_{ij}, the sum over all virial contributions then gives the total (residual) virial W. With this, the calculation of the pressure can be expressed by

$$p = p^{id} + p^{res} = k_B T \frac{N}{V} + \frac{\langle W \rangle}{3V}. \tag{7.9}$$

In MD simulations, the ideal gas part of the pressure can be related to the kinetic energy of the system E_{kin}

$$p^{id} = k_B T \frac{N}{V} = \frac{2E_{kin}}{3V} = \frac{1}{3V}\sum_{i=1}^{N} m_i \mathbf{v}_i^2 = \frac{1}{3V}\sum_{i=1}^{N} \frac{\mathbf{p}_i^2}{m_i}. \tag{7.10}$$

Thus, the instantaneous ideal gas pressure can be determined during the simulation from the velocities $\mathbf{v}_i$ or momenta $\mathbf{p}_i$ of the particles. $\mathbf{f}_{ij}$, $\mathbf{r}_{ij}$, $\mathbf{v}_i$ and $\mathbf{p}_i$ though are vector quantities, and the pressure is a tensor

$$p = \begin{pmatrix} p_{xx} & p_{xy} & p_{xz} \\ p_{yx} & p_{yy} & p_{yz} \\ p_{zx} & p_{zy} & p_{zz} \end{pmatrix}. \tag{7.11}$$

The pressure tensor is also referred to as stress tensor σ, with the diagonal element representing the tensile stress, and the off-diagonal elements the shear stress. An element $p_{\alpha\beta}$ of the pressure tensor (or $\sigma_{\alpha\beta}$ of the stress tensor) is calculated from

$$p_{\alpha\beta} = \left\langle \frac{1}{V} \left(\sum_{i=1}^{N} \frac{\mathrm{p}_{i,\alpha}\mathrm{p}_{i,\beta}}{m_i} + \sum_{i=1}^{N}\sum_{j>i}^{N} r_{ij,\alpha} f_{ij,\beta} \right) \right\rangle. \tag{7.12}$$

The total system pressure is then given by the trace of the pressure tensor

$$p = \frac{1}{3}\left(p_{xx} + p_{yy} + p_{zz}\right). \tag{7.13}$$

The long-range correction (LRC) (see Sect. 5.2.1) for contributions beyond the cut-off radius r_{cut} can be derived from [4]

$$p^{LRC} = -\frac{2}{3}\pi N \varrho \int_{r_{cut}}^{\infty} r_{ij}^2 W(r_{ij}) dr_{ij}. \tag{7.14}$$

In the *NpT* ensemble, the pressure is imposed, and the volume of the system will fluctuate during the course of the simulation. Thus, the molar density of the system at given pressure and temperature is derived from the ensemble average of the volume and the included number of particles N by

$$\varrho_m = \left\langle \frac{N}{V} \right\rangle. \tag{7.15}$$

Densities are though often given as mass specific property, for instance in the unit $(kg\ m^{-3})$, which requires the conversion of the molecular simulation result with the molar mass M of the compound. When the volume of the simulation box is given in (m^3), and the molar mass in $(g\ mol^{-1})$, the mass density can be computed from

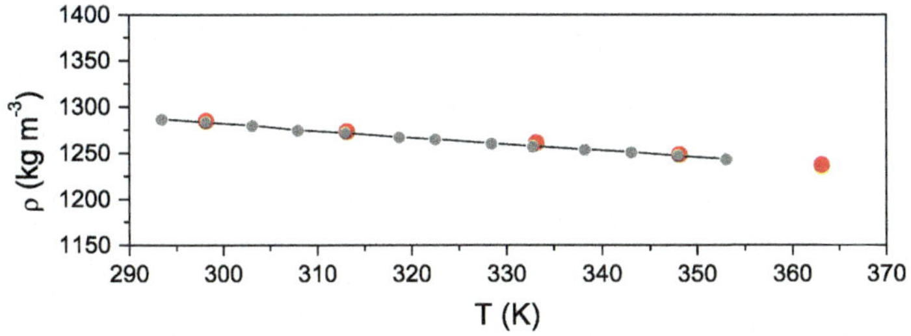

Fig. 7.1 MD simulation results for the liquid density of the ionic liquid $[emim][BF_4]$ (*red circles*) in comparison with experimental data [31] (*grey circles*) (colour online)

$$\varrho\,(kg\,m^{-1}) = \left\langle \frac{N}{V} \right\rangle \frac{M}{1000.0 \cdot N_a}. \tag{7.16}$$

Figure 7.1 shows exemplarily the simulation results for the density of the ionic liquid $[emim][BF_4]$ with the molar mass ($M = 197.97 \frac{g}{mol}$) in the temperature range from $T = (298 - 363)$ K in comparison with experimental data.

7.2 Dynamic and Transport Properties

Monte Carlo simulations sample states according to the ensemble probability distribution. They are not deterministic and therefore do not allow studies on the dynamics behavior. The trajectories from MD simulation though follow the time evolution of the systems, i.e. they provide the positions and velocities of the molecules at different time steps. Thus, the analysis of these trajectories gain insight into the dynamic behavior of the systems and enables the computation of transport coefficients. In the following subsections, we will present common approaches to study the dynamic behavior by equilibrium MD simulations (EMD). We will thereby discuss correlation functions, and the Einstein and Green-Kubo formulas to derive transport coefficients, such as the diffusion coefficient, shear viscosity and thermal conductivity. Non-equilibrium molecular dynamics (NEMD) approaches to compute transport properties are not covered in this section.

7.2.1 Autocorrelation Functions

The dynamic behavior of a molecular system is often studied by autocorrelation functions. A correlation function in general measures the correlation between any two quantities A and B

$$c_{AB} = \frac{\langle A \cdot B \rangle}{\sqrt{\langle A^2 \rangle \langle B^2 \rangle}}. \tag{7.17}$$

In this reduced form, the correlation coefficient c_{AB} adopts values between 0 and 1, with $c_{AB} \approx 1$ indicating a high degree of correlation. Usually, the quantities A and B fluctuate around their ensemble averages $\langle A \rangle$ and $\langle B \rangle$, and the correlation function measures the correlation of these fluctuations or noises. The correlation coefficient than becomes [4]

$$c_{AB} = \frac{\langle (A - \langle A \rangle) \cdot (B - \langle B \rangle) \rangle}{\sqrt{\langle (A - \langle A \rangle)^2 \rangle \langle (B - \langle B \rangle)^2 \rangle}} = \frac{\langle \delta A \cdot \delta B \rangle}{\sigma(A)\sigma(B)}. \tag{7.18}$$

where σ^2 is the variance of the quantity. In a time (cross-) correlation function, the correlation between the quantities A and B at two different times is measured. If the correlation between two values of the same quantity A is evaluated at two different times, $c_{AA}(t)$ becomes an autocorrelation function

$$c_{AA}(t) = \frac{\langle \delta A(t) \cdot \delta A(0) \rangle}{\sigma^2(A)}. \tag{7.19}$$

For many quantities, $c(t)$ decays to zero at long times, and the time integral

$$\tau = \int_0^\infty c(t) dt \tag{7.20}$$

defines the correlation or relaxation time, which can often be related to transport coefficients. The fourier transforms of correlation functions may relate to experimental determined spectra [94]. The dynamics in a fluid are often studied by the velocity autocorrelation function (VACF), defined by

$$c_v(t) = \frac{\langle \mathbf{v}(t) \cdot \mathbf{v}(0) \rangle}{\langle \mathbf{v}(0) \cdot \mathbf{v}(0) \rangle}. \tag{7.21}$$

The VACF can be derived for each molecule in the system individually. However, it is usually averaged over all molecules to improve the statistical precision. For a molecular system, the center-of-mass (COM) velocity is commonly regarded in the VACF. Figure 7.2 shows exemplarily the (reduced) autocorrelation functions for the COM-velocities of the anion and cation of the 1-hexyl-3-methyl-imidazolium chloride $[hmim][Cl]$ ionic liquid. VACFs do not necessarily have monotonic behavior but often exhibit zero-crossings, as shown in Fig. 7.2. The zero-crossing at finite time represent a measure of the mean collision time. The following negative VACFs indicate 'cage' effects [32]. At longer times, the VACF approaches zero, which means that the velocities become uncorrelated.

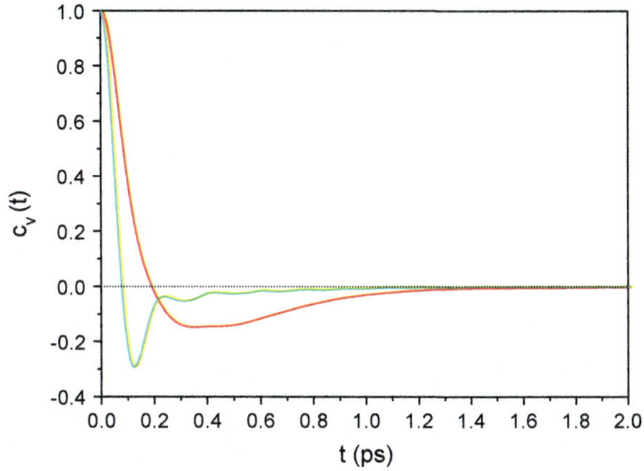

Fig. 7.2 Reduced COM-VACFs of the $[Cl^-]$ anion (*green*) and the $[hmim^+]$ cation (*red*) of the $[hmim][Cl]$ ionic liquid at 298.15 K

The COM-velocity autocorrelation function though only gives a picture of the translational motions in the system. Thus, rotational correlation functions are considered to gain insight into reorientational motions of the molecules. To characterize the rotation along a particular axis, a rotational correlation function can be defined by

$$C^{rot}(t) = \langle \mathbf{r}_i(t) \cdot \mathbf{r}_i(0) \rangle, \tag{7.22}$$

where the vector $\mathbf{r}_i$ describes the orientation of the molecule [7]. The rotational correlation function can also be formulated in terms of

$$C^{rot}(t) = \left\langle \frac{1}{2} \left[3\cos^2 \theta_i(t) - 1 \right] \right\rangle, \tag{7.23}$$

where $\theta_i(t)$ represents the orientational angle of a characteristic axis within the molecule as a function of time [10].

7.2.2 *Transport Coefficients from EMD Simulations*

In phenomenological transport equations, transport coefficients such as the shear viscosity, the diffusion coefficient or the thermal conductivity define the response of a system to an external perturbation that imposes a gradient $\nabla\varphi$ as driving force. This gradient initializes an equalizing flux $\dot{A}$ that is proportional to the driving force with the transport coefficient k as material property being the proportionality factor

$$\dot{A} = -k\nabla\varphi. \tag{7.24}$$

Table 7.1 Principles and analogies among mass, heat and momentum transfer, describing transport laws

	Diffusion	Heat transport	Momentum transport
$\dot{A}$	Mass flow j^D	Heat flux j^q	Momentum flux p
φ	Density ϱ	Temperature T	Velocity v
k	Diffusion coefficient D	Thermal conductivity λ	Viscosity η
Law	Fick	Fourier	Newton

In Table 7.1 we have summarized the application of Eq. 7.24 to the mass, heat and momentum transfer, and the laws that describe the mechanism of these transport processes.

Transport processes due to gradients in density, temperature or velocity imply a non-equilibrium system, and Eq. 7.24 represents the macroscopic law of its relaxation. However, also in an equilibrium system, microscopic thermal fluctuations occur so that the instantaneous value of a property A differs from its equilibrium ensemble average $\langle A \rangle$. The *Onsager regression hypothesis* [53, 54] then states that the regression of the property A to its equilibrium is governed by the same transport laws and transport coefficients as the macroscopic relaxation due to an external perturbation. With this, the Onsager theory allows the computation of transport coefficients from molecular dynamics studies of equilibrium systems. Though, the Onsager hypothesis is only valid for states in the neighborhood of equilibrium for which it can't be distinguished whether the systems is there due to thermal fluctuation or on its final stage of a relaxation towards equilibrium [19]. In the following section we introduce the Green-Kubo and the Einstein formulas to derive transport properties from EMD simulations. In general, both formulas are equivalent and yield theoretically identical results for the transport coefficients. However, due to finite-time simulations in finite-size systems, the different methods are subject to different uncertainties and statistical errors.

7.2.2.1 Green-Kubo Formulas

The Green-Kubo formulas relate the transport coefficients to the integral over the time correlation function of the flux [4]

$$k \propto \int_0^\infty \langle \dot{A}(t)\dot{A}(0) \rangle dt. \tag{7.25}$$

This can be illustrated by the fact that the decay of correlation between instantaneous fluctuations represents the regression of the microscopic fluctuation according to the Onsager theory. More details on the derivation of the Green-Kubo formulas are for instance provided by [19, 25, 98]. The shear viscosity is then given by

$$\eta = \frac{V}{k_B T} \int_0^\infty \langle p_{\alpha\beta}(t) p_{\alpha\beta}(0) \rangle dt, \qquad \alpha \neq \beta \tag{7.26}$$

where $p_{\alpha\beta}$ is an off-diagonal element of the pressure tensor that is determined by Eq. 7.12. To improve the statistics, the autocorrelation function can be averaged over all independent off-diagonal tensor elements, resulting in

$$\eta = \frac{V}{3k_B T} \int_0^\infty \left[\langle p_{xy}(t) p_{xy}(0) \rangle + \langle p_{xz}(t) p_{xz}(0) \rangle + \langle p_{yz}(t) p_{yz}(0) \rangle \right] dt. \tag{7.27}$$

According to Daivis and Evans [16], a modified Green-Kubo correlation can be derived for the shear viscosity, which uses all elements of the pressure tensor, including the diagonal elements

$$\eta = \frac{V}{10 k_B T} \int_0^\infty \left\langle \sum_{\alpha\beta} p_{\alpha\beta}(t) p_{\alpha\beta}(0) \right\rangle dt, \tag{7.28}$$

with

$$p_{\alpha\beta} = \frac{(\sigma_{\alpha\beta} + \sigma_{\beta\alpha})}{2} - \delta_{\alpha\beta} \frac{1}{3} \sum_{\gamma} \sigma_{\gamma\gamma}, \tag{7.29}$$

$$\delta_{\alpha\beta} = 1 \quad \text{for} \quad \alpha = \beta$$
$$\delta_{\alpha\beta} = 0 \quad \text{for} \quad \alpha \neq \beta$$

Therein, the $\sigma_{\alpha\beta}$ represent the elements of the pressure or rather stress tensor (see Eq. 7.11). The factor 10 (instead of 9 for the nine tensor elements considered) is used in the denominator of Eq. 7.28 to compensate that the diagonal elements yield a too high viscosities with [16]

$$\eta_{xx} = \eta_{yy} = \eta_{zz} = \frac{4}{3} \eta. \tag{7.30}$$

As the shear viscosity is a collective dynamic property, it is subject to relative high statistical uncertainties. This is especially true for systems with slowly decaying autocorrelations functions such as ionic liquids or polar systems, for which the autocorrelation functions also tend to fluctuate around zero even at long time scales. When the molecules are modeled by force fields that account for bond stretching and angle bending, the stress autocorrelation function additionally exhibits rapid oscillations due to the high-frequency intramolecular vibration modes [7]. Thus, frequent sampling of the pressure tensor is required to cover the course of the stress relaxation correctly.

In the Green-Kubo formula, the diffusion coefficient is computed from the integral of the velocity autocorrelation (VACF) function

$$D = \frac{1}{3} \int_0^\infty \langle \mathbf{v}_i(t)\mathbf{v}_i(0)\rangle dt. \tag{7.31}$$

As described above, the VACF and with this the diffusion coefficient are single molecule properties, which are however usually averaged over all molecules in the systems to improve the statistics.

The Green-Kubo expression for the thermal conductivity becomes

$$\lambda_T = \frac{V}{k_B T^2} \int_0^\infty \langle \dot{\mathbf{j}}^q(t)\dot{\mathbf{j}}^q(0)\rangle dt, \tag{7.32}$$

wherein the energy flux or heat current vector is given by [22, 59]

$$\dot{\mathbf{j}}^q = \frac{1}{V}\frac{d}{dt}\left(\sum_i \mathbf{r}_i e_i\right) = \frac{1}{V}\left[\sum_i \mathbf{v}_i e_i + \frac{1}{2}\sum_i\sum_j (\mathbf{f}_{ij}\mathbf{v}_i)\mathbf{r}_{ij}\right]. \tag{7.33}$$

The quantity e_i is the total energy of the atom or site and consists of both its kinetic and potential energy. The potential energy due to interactions between two atoms is thereby equally divided between them [4]. Thus, the first term represents the diffusive contribution to the heat flux due to the atomic motion. The second term then yields the contribution of intermolecular interactions to the heat transfer.

In a similar way, the electrical conductivity can be derived from the electrical current autocorrelation function [23]

$$\sigma^{elec} = \frac{1}{3k_B T V} \int_0^\infty \langle \dot{J}^{elec}(t)\dot{J}^{elec}(0)\rangle dt, \tag{7.34}$$

where the total electric current is given by

$$\dot{J}^{elec}(t) = \sum_i \mathbf{v}_i(t) q_i. \tag{7.35}$$

For an ionic system $\dot{J}^{elec}(t)$ is often expressed by

$$\dot{J}^{elec}(t) = \sum_i \mathbf{v}_{i+}(t) - \sum_j \mathbf{v}_{j-}(t). \tag{7.36}$$

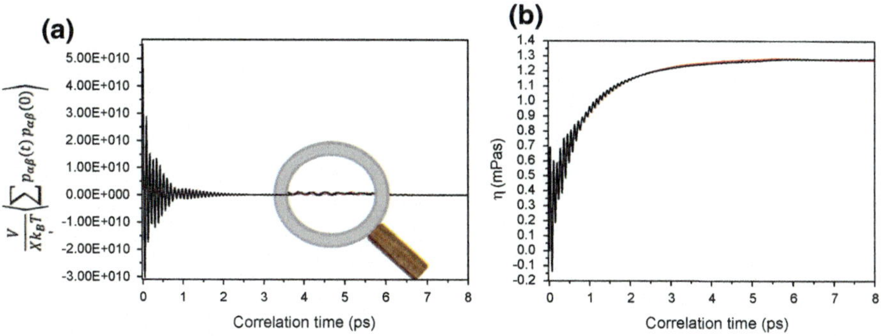

Fig. 7.3 **a** Ensemble average of the pressure correlation functions according to Eq. 7.27 (*black line*) and Eq. 7.28 (*red line*). The ensemble average of the pressure correlation functions are multiplied with the corresponding prefactors. **b** Resulting instantaneous values of the viscosity as function of the correlation time

Example: Viscosity of SPC/Fw Water

In this example, we illustrate the calculation of the viscosity of the flexible water model SPC/Fw [95] (see discussion in Sect. 6.4.2) by the Green-Kubo formula. We first performed MD simulations in the *NpT*-ensemble to determine the systems average densities for given values for temperature and pressure. To derive the viscosity, we performed additional simulations in the *NVT*-ensemble at the averaged densities from the *NpT*-simulations (see [70] for more details). We here present simulation results at $T = 277.15$ K and $p = 0.1$ MPa for which we performed production runs of 15 ns in the *NVT*- ensemble with $\delta t = 0.001$ ps. The pressure tensors were saved every 0.01 ps for further analysis by Eqs. 7.27 and 7.28. Figure 7.3a shows the ensemble average of the pressure correlation functions, employing the three independent off-diagonal tensor elements (p_{xy}, p_{yz}, p_{xz}) according to Eq. 7.27, and using all elements following Eq. 7.28 (The ensemble average of the pressure correlation functions are multiplied by the corresponding prefactors). As discussed above, the autocorrelation functions exhibit rapid oscillations due to the intramolecular vibration modes from bond stretching and angle bending in the flexible water model SPC/Fw. The enlarged detail illustrates the still present albeit slight fluctuation of the autocorrelation functions around zero at longer correlation times. Figure 7.3b shows the resulting instantaneous values of the viscosity from integrating both functions shown in (a). As expected, both Green-Kubo formulae converge to the same value of the viscosity within their statistical errors.

7.2.2.2 Einstein Formulas

Whereas the Green-Kubo formulas describe transport coefficients k in terms of time correlation function of the flux $\dot{A}$, the Einstein formulas relate k to the mean square displacement (msd) of the quantity $A(t)$ from its value $A(0)$ at a selected initial time [4]

$$2tk = \langle [A(t) - A(0)]^2 \rangle . \tag{7.37}$$

The Einstein relation for the shear viscosity is given by

$$2t\eta = \frac{V}{k_B T} \langle [L_{\alpha\beta}(t) - L_{\alpha\beta}(0)]^2 \rangle \tag{7.38}$$

with

$$L_{\alpha\beta} = \frac{1}{V} \sum_i r_{i\alpha} p_{i\beta} . \tag{7.39}$$

The corresponding Einstein relation for the thermal conductivity is

$$2t\lambda_T = \frac{V}{k_B T^2} \langle [\delta u_i(t) - \delta u_i(0)]^2 \rangle , \tag{7.40}$$

wherein δu_i is derived from

$$\delta u_i = \frac{1}{V} \sum_i r_i (u_i - \langle u_i \rangle) . \tag{7.41}$$

The most familiar Einstein relation though is the expression for the diffusion coefficient

$$2tD = \frac{1}{3} \langle [r(t) - r(0)]^2 \rangle = \frac{\langle msd \rangle}{3} . \tag{7.42}$$

Normal diffusive behavior is associated with a linear increase of the msd with time, from which the diffusion coefficient is derived from the gradient. This is exemplarily shown in Fig. 7.4 for simulation results from MD studies on the refrigerant R-1234yf. It has to be pointed out that the true displacements of the molecules, i.e. without the application of periodic boundary conditions, need to be accumulated in the Einstein formula.

Ionic liquids or glass formers in the supercooled region though exhibit subdiffusive dynamics with

$$\langle |\delta r(t)|^2 \rangle \approx t^{\beta}, \qquad \beta < 1 \tag{7.43}$$

and the diffusive regime is only reached at longer time scales [17]. These subdiffusive dynamics are ascribed to the formation of domains of enhanced or reduced mobility of the molecules due to caging effects, formation of cluster and the like.

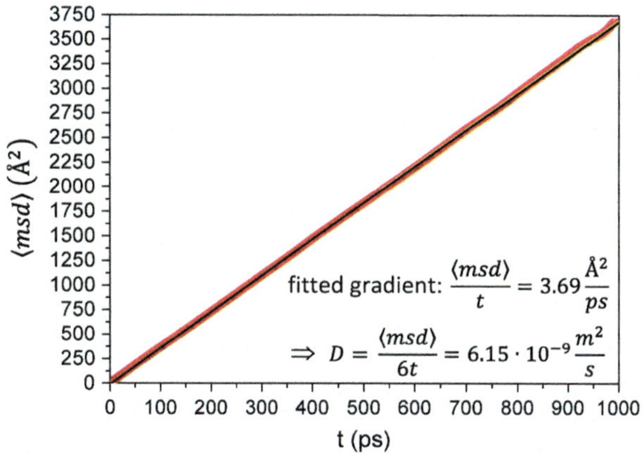

Fig. 7.4 Ensemble average of $\langle msd \rangle(t)$ from MD studies on HFO-1234*yf* at 298 K and 2 MPa (*red squares*). A linear fit (*black line*) yields the gradient of the msd versus time, from which the diffusion coefficient can be derived by Eq. 7.42

These heterogeneous dynamics can be characterized by the selfpart of the van Hove correlation function [17, 84]

$$G_s(r,t) = \frac{1}{N}\left\langle \sum_{i=1}^{N} \delta(\mathbf{r}_i(t) - \mathbf{r}_i(0) - \mathbf{r}) \right\rangle. \tag{7.44}$$

It represents a correlation between the positions of a molecule at different times, and gives the probability of finding the molecule at the time t at position $\mathbf{r}(t)$ when it was at $\mathbf{r}(0)$ at $t = 0$ [17]. For freely moving particles, the van Hove self-correlation function $G_s(r,t)$ shows a Gaussian form, whereas heterogeneous dynamic behavior results in deviations from it. The degree of heterogeneity of the molecular motions can therefore be measured by a non-Gaussian parameter [71]

$$\alpha(t) = \frac{3}{5}\frac{\langle |\Delta r(t)|^4 \rangle}{\langle |\Delta r(t)|^2 \rangle} - 1 \tag{7.45}$$

that becomes zero if the system exhibits diffusive behavior with a Gaussian distribution of positions $\mathbf{r}_i$.

7.3 Second-Derivative Properties

Thermophysical properties such as heat capacities, the isothermal compressibility, or the volumic expansivity can be derived as second derivative of the corresponding partition function of the systems [35]. It will be shown that these properties are related to fluctuations of first order derivatives, thus they are also referred to as thermophysical properties from fluctuation.

7.3.1 Heat Capacity at Constant Volume C_V

The molar heat capacity at constant volume C_V is defined as change in the internal molar energy due to changing temperatures at constant volume

$$C_V = \left(\frac{\partial U_m}{\partial T}\right)_V \tag{7.46}$$

and is usually computed in the *NVT*-ensemble. As discussed in Sect. 2.2, the internal energy is obtained from the partition function Q_{NVT} of the ensemble by

$$U_m = k_B T^2 \frac{\partial \ln Q_{NVT}}{\partial T} = -\frac{\partial \ln Q_{NVT}}{\partial \beta}. \tag{7.47}$$

Thus, the derivation of the expression for C_V is in the easiest way illustrated by the example of a monoatomic gas for which the partition function is given by (see Sect. 2.3.1, Eq. 2.86)

$$Q_{NVT}(\beta) = \frac{1}{N!}\frac{1}{\Lambda^{3N}} \int e^{-\beta U_{conf}(r)} dr^{3N} \tag{7.48}$$

$$\Lambda(\beta) = \sqrt{\frac{h^2 \beta}{2\pi m}}. \tag{7.49}$$

Thus, $\ln Q_{NVT}$ can be written as

$$\ln Q_{NVT} = \ln\left(\int e^{-\beta U_{conf}} dr^{3N}\right) - \ln N! - 3N \ln \Lambda \tag{7.50}$$

The internal energy as negative first derivation of $\ln Q_{NVT}$ with respect to the reciprocal temperature β is

$$\begin{aligned} U_m &= -\frac{\partial \ln Q_{NVT}}{\partial \beta} \\ &= -\left(\frac{\int (-U_{conf}) e^{-\beta U_{conf}} dr^{3N}}{\int e^{-\beta U_{conf}} dr^{3N}} - 3N \frac{1}{\Lambda}\frac{\partial \Lambda}{\partial \beta}\right) \\ &= \frac{\int U_{conf} e^{-\beta U_{conf}} dr^{3N}}{\int e^{-\beta U_{conf}} dr^{3N}} + \frac{3N}{2\beta} && (7.51) \\ &= \langle U_{conf} \rangle_{NVT} + \frac{3}{2} N k_B T && (7.52) \end{aligned}$$

as the fraction term in Eq. 7.51 represent the definition of an ensemble average in the *NVT*–ensemble. Thus, the internal energy is given by the sum of the ensemble average of the configurational energy $\langle U_{conf} \rangle$ and the kinetic energy of the monatomic gas $\frac{3}{2} N k_B T$ according to the equipartition relation. The heat capacity at constant volume C_V can now be derived as second derivative of the partition function

$$C_V = \left(\frac{\partial U_m}{\partial T}\right)_V = \left(\frac{\partial U_m}{\partial \beta}\right)\left(\frac{\partial \beta}{\partial T}\right) = \frac{1}{k_B T^2}\left(\frac{\partial^2 \ln Q_{NVT}}{\partial \beta^2}\right). \tag{7.53}$$

By using Z_{NVT} as abbreviation for the configurational integral $\int \exp\left(-\beta U_{conf}\right) dr^{3N}$, the second derivative of the $\ln Q_{NVT}$ (see Eq. 7.51) becomes

$$\begin{aligned}
\frac{\partial^2 \ln Q_{NVT}}{\partial \beta^2} \\
&= \frac{\left[\int U_{conf}^2 e^{-\beta U_{conf}} dr^{3N}\right] \cdot Z_{NVT} - [\int (-U_{conf}) e^{-\beta U_{conf}} dr^{3N})]^2}{Z_{NVT}^2} \\
&\quad + \frac{3N}{2\beta^2} \\
&= \frac{\int U_{conf}^2 e^{-\beta U_{conf}} dr^{3N}}{Z_{NVT}} - \left(\frac{\int (-U_{conf}) e^{-\beta U_{conf}} dr^{3N}}{Z_{NVT}}\right)^2 + \frac{3N}{2\beta^2} \\
&= \left\langle U_{conf}^2 \right\rangle_{NVT} - \left\langle U_{conf} \right\rangle_{NVT}^2 + \frac{3N}{2\beta^2}.
\end{aligned} \tag{7.54}$$

Thus, the heat capacity at constant volume of the monatomic gas in the *NVT*–ensemble is

$$C_V = \underbrace{\frac{1}{k_B T^2}\left(\langle U_{conf}^2\rangle_{NVT} - \langle U_{conf}\rangle_{NVT}^2\right)}_{C_V^{res}} + \underbrace{\frac{3}{2} k_B N}_{+\ C_V^{id}}. \tag{7.55}$$

The term C_V^{id} represents the molar isochoric heat capacity of a monatomic gas that only has three degrees of freedom, $N_f = 3$, for the translational motion in the three directions in space. For polyatomic ideal gas, C_V^{id} is given by

$$C_V^{id} = \frac{N_f}{2} k_B N. \tag{7.56}$$

according to the theorem of equipartition of energy. The difference $\langle U_{conf}^2\rangle - \langle U_{conf}\rangle^2$ defines the root mean square (RMS) deviation of the configurational energy $\sigma^2(U_{conf}) = \langle \delta U_{conf}^2\rangle$. In Monte Carlo simulations, the kinetic, ideal part of the energy cannot be addressed. Therefore, only the residual part of the heat capacity is determined in the simulation by analyzing the fluctuation of the configurational energy. The ideal gas contribution of the component is then obtained from experiment, group contribution methods [73] or ab initio calculations [52]. In general, MD simulations can yield both, the ideal, kinetic and the residual part of the heat capacity.

7.3.2 Isothermal Compressibility κ_T

The isothermal compressibility κ_T quantifies the change in the molar volume with the pressure at a given temperature

$$\kappa_T = -\frac{1}{V_m}\left(\frac{\partial V_m}{\partial p}\right)_T, \tag{7.57}$$

and is commonly computed in the NpT-ensemble, for which the partition function Q_{NpT} is given by (see Eq. 3.24)

$$Q_{NpT} = \frac{1}{N!}\frac{1}{\Lambda^{3N}}\frac{1}{V_0}\int e^{-\beta pV}dV\int e^{-\beta U_{conf}(r)}dr^{3N}. \tag{7.58}$$

The molar volume is then derived from the partition function Q_{NpT} by

$$\begin{aligned} V_m &= -\frac{1}{\beta}\left(\frac{\ln Q_{NpT}}{\partial p}\right) \\ &= -\frac{1}{\beta}\left(\frac{\int(-\beta V)e^{-\beta pV}dV\int e^{-\beta U_{conf}}dr^{3N}}{\int e^{-\beta pV}dV\int e^{-\beta U_{conf}}dr^{3N}}\right) \\ &= \langle V\rangle_{NpT}. \end{aligned} \tag{7.59}$$

Thus, the isothermal compressibility κ_T can be calculated from the second derivative of the partition function Q_{NpT} with respect to the pressure

$$\begin{aligned} \kappa_T &= -\frac{1}{\langle V\rangle_{NpT}}\frac{\partial}{\partial p}\left(-\frac{1}{\beta}\left(\frac{\ln Q_{NpT}}{\partial p}\right)\right)_T \\ &= -\frac{1}{\langle V\rangle_{NpT}}\frac{\partial}{\partial p}\left(\frac{\int Ve^{-\beta pV}dV\int e^{-\beta U_{conf}}dr^{3N}}{\int e^{-\beta pV}dV\int e^{-\beta U_{conf}}dr^{3N}}\right). \end{aligned} \tag{7.60}$$

By using Z_{NpT} as abbreviation for the configurational integral $\int e^{-\beta pV}dV \int e^{-\beta U_{conf}}dr^{3N}$, the derivative yields

$$\begin{aligned} \kappa_T &= -\frac{1}{\langle V\rangle_{NpT}}\left(\frac{\int(-\beta V^2)e^{-\beta pV}dV\int e^{-\beta U_{conf}}dr^{3N}}{Z_{NpT}}\right. \\ &\qquad \left. + \frac{\beta\left[\int Ve^{-\beta pV}dV\int e^{-\beta U_{conf}}dr^{3N}\right]^2}{Z_{NpT}^2}\right) \\ &= \frac{\beta\left(\langle V^2\rangle_{NpT} - \langle V\rangle_{NpT}^2\right)}{\langle V\rangle_{NpT}} = \frac{1}{k_BT}\frac{\left(\langle V^2\rangle_{NpT} - \langle V\rangle_{NpT}^2\right)}{\langle V\rangle_{NpT}}. \end{aligned} \tag{7.61}$$

Thus, the isothermal compressibility κ_T is related to the RMS deviation of the volume $\sigma^2(V) = \langle \delta V^2 \rangle = \left(\langle V^2 \rangle - \langle V \rangle^2\right)$, and can be derived from monitoring the fluctuation of the volume during an *NpT*–simulation. However, isothermal compressibility κ_T computed by this fluctuation formula are in general subject to large statistical uncertainties [10]. Thus, κ_T can alternatively be estimated from a series of isothermal simulations in which the molar densities or volumes are computed for varying pressures [87]

$$\kappa_T = -\frac{1}{V_m}\left(\frac{\partial V_m}{\partial p}\right)_T = \frac{1}{\varrho_m}\left(\frac{\partial \varrho_m}{\partial p}\right)_T = \left(\frac{\partial \ln \varrho_m}{\partial p}\right)_T \approx \left(\frac{\ln(\varrho_{m,2}/\varrho_{m,1})}{p_2 - p_1}\right)_T \tag{7.62}$$

7.3.3 Thermal Expansivity α_P

The thermal expansivity α_P, also referred to as cubic expansion coefficient, quantifies the change in the molar volume with the temperature at a given pressure

$$\alpha_P = \frac{1}{V}\left(\frac{\partial V_m}{\partial T}\right)_p . \tag{7.63}$$

This property is again computed from simulations in the *NpT*–ensemble, for which the molar volume is derived as derivative of the partition function Q_{NpT} with respect to the pressure as given by Eq. 7.59. Thus, the thermal expansivity α_P can be determined from the second partial derivative of Q_{NpT}

$$\begin{aligned}
\alpha_P &= \frac{1}{\langle V \rangle_{NpT}}\left(\frac{\partial}{\partial \beta}\left(-\frac{1}{\beta}\frac{\partial \ln Q_{NpT}}{\partial p}\right)_T \frac{\partial \beta}{\partial T}\right) \qquad (7.64)\\
&= -\frac{1}{k_B T^2 \langle V \rangle_{NpT}}\left(\frac{\partial}{\partial \beta}\left(\frac{\int V e^{-\beta p V} dV \int e^{-\beta U_{conf}} dr^{3N}}{\int e^{-\beta p V} dV \int e^{-\beta U_{conf}} dr^{3N}}\right)\right)\\
&= -\frac{1}{k_B T^2 \langle V \rangle_{NpT}}\left(\frac{-\int V e^{-\beta p V} dV \int (U_{conf} + pV) e^{-\beta U_{conf}} dr^{3N}}{Z_{NpT}}\right.\\
&\quad + \frac{\int V e^{-\beta p V} dV \int e^{-\beta U_{conf}} dr^{3N}}{Z_{NpT}}\\
&\quad \left. \cdot \frac{\int e^{-\beta p V} dV \int (U_{conf} + pV) e^{-\beta U_{conf}} dr^{3N}}{Z_{NpT}}\right) .
\end{aligned}$$

The expression $U_{conf} + pV$ is the configurational enthalpy H_{conf} of the system. Thus, the first fraction represent the ensemble average $\langle V\, H_{conf} \rangle_{NpT}$, whereas the second term is the product of the ensemble averages $\langle V \rangle_{NpT} \langle H_{conf} \rangle_{NpT}$. The thermal expansivity α_P can therefore be computed from

$$\alpha_P = \frac{\langle V\, H_{conf}\rangle_{NpT} - \langle V\rangle_{NpT}\langle H_{conf}\rangle_{NpT}}{k_B T^2 \langle V\rangle_{NpT}}. \tag{7.65}$$

Again the computation of α_P from the fluctuation formula might result in large statistical uncertainties. Thus, it can alternatively be derived from a series of isobaric simulations in which the molar densities or volumes are now computed for varying temperature

$$\alpha_P = \frac{1}{V_m}\left(\frac{\partial V_m}{\partial T}\right)_p = -\frac{1}{\varrho_m}\left(\frac{\partial \varrho_m}{\partial T}\right)_p \approx -\left(\frac{\ln(\varrho_{m,2}/\varrho_{m,1})}{T_2 - T_1}\right)_p \tag{7.66}$$

7.3.4 *Heat Capacity at Constant Pressure C_P*

The molar heat capacity at constant pressure C_P is defined as change in the molar enthalpy with temperature at constant pressure

$$C_P = \left(\frac{\partial H_m}{\partial T}\right)_p. \tag{7.67}$$

As described in Sect. 7.3.1 on the computation of C_V, the expression for heat capacities has to be separated in a residual and an ideal part, as the ideal, kinetic contribution can not be considered in Monte Carlo simulations. In their derivation of the expression for the C_P^{res}, Lagache et al. [35] divided the system's enthalpy into an ideal and residual contribution according to

$$H_m = H_m^{id} + H_m^{res} \tag{7.68}$$

$$H_m^{id} = U^{intra} + U^{kin} + Nk_BT \tag{7.69}$$

$$H_m^{res} = U^{inter} + pV - Nk_BT. \tag{7.70}$$

The ideal gas enthalpy comprises the kinetic energy and the contributions from intramolecular degrees of freedom. Both are independent of the system's density and only a function of the temperature. The residual enthalpy includes the contribution from intermolecular interactions. The residual part of the heat capacity C_P^{res} is then computed from

$$C_P^{res} = \left(\frac{\partial H_m^{res}}{\partial T}\right)_p = \left(\frac{\partial U^{inter}}{\partial T}\right)_p + p\left(\frac{\partial V}{\partial T}\right)_p - Nk_B \tag{7.71}$$

The terms for the internal energy U^{inter} and the volume V are again derived as first derivative of the partition function Q_{NpT}. The derivation of the fluctuation formula for the thermal expansivity in Eqs. 7.64 and 7.65 has shown that the derivative of an ensemble average $\langle X\rangle_{NpT}$ with respect to β yields

$$\frac{\partial \langle X \rangle_{NpT}}{\partial \beta} = -\left(\langle X H_{conf} \rangle_{NpT} - \langle X \rangle_{NpT} \langle H_{conf} \rangle_{NpT}\right). \tag{7.72}$$

Thus, the derivative of U^{inter} with respect to the temperature becomes

$$\begin{aligned} \frac{\partial U^{inter}}{\partial T} &= \frac{\partial \langle U^{inter} \rangle_{NpT}}{\partial \beta} \frac{\partial \beta}{\partial T} \\ &= \frac{1}{k_B T^2} \left(\langle U^{inter} H_{conf} \rangle_{NpT} - \langle U^{inter} \rangle_{NpT} \langle H_{conf} \rangle_{NpT}\right). \end{aligned} \tag{7.73}$$

The expression for the derivative of the volume with respect to β can be taken from Eq. 7.65. Thus, the residual heat capacity at constant pressure can be computed from

$$\begin{aligned} C_P^{res} &= \frac{1}{k_B T^2} \left(\langle U^{inter} H_{conf} \rangle_{NpT} - \langle U^{inter} \rangle_{NpT} \langle H_{conf} \rangle_{NpT}\right) \\ &+ \frac{p}{k_B T^2} \left(\langle V H_{conf} \rangle_{NpT} - \langle V \rangle_{NpT} \langle H_{conf} \rangle_{NpT}\right) - N k_B. \end{aligned} \tag{7.74}$$

7.3.5 Related Properties

Once the heat capacities, the thermal expansivity α_P and the isothermal compressibility κ_T of the system are known, it is also possible to derive other related thermodynamic quantities, such as the Joule-Thomson coefficient or the speed of sound. The Joule-Thomson coefficient μ_{JT} describes the rate of temperature change with respect to the pressure at constant values of the enthalpy, i.e. during adiabatic throttling. It is thermodynamically defined by [90]

$$\mu_{JT} = \left(\frac{\partial T}{\partial p}\right)_H = -\frac{1}{C_p}\left(\frac{\partial H_m}{\partial p}\right)_T = -\frac{1}{C_p}\left[V_m - T\left(\frac{\partial V_m}{\partial T}\right)_p\right]. \tag{7.75}$$

Thus, it can be derived from the simulation results for the heat capacity at constant pressure C_P and the thermal expansivity α_P by

$$\mu_{JT} = \frac{V_m}{C_P} \left(T \cdot \alpha_P - 1\right). \tag{7.76}$$

The isentropic speed od sound v_S is defined by

$$v_S = \sqrt{\left(\frac{\partial p}{\partial \varrho}\right)_S} = V_m \sqrt{-\left(\frac{\partial p}{\partial V_m}\right)_S}. \tag{7.77}$$

For a real fluid Eq. 7.77 becomes [5]

$$v_S = V_m \sqrt{-\left(\frac{\partial p}{\partial V_m}\right)_T \frac{C_p}{C_V}}. \tag{7.78}$$

Thus it can be derived from molecular simulation results for the isobaric and isochoric heat capacities and the isothermal compressibility κ_T by

$$v_S = \sqrt{\frac{V_m}{\kappa_T} \frac{C_P}{C_V}}. \tag{7.79}$$

7.3.6 Example: Thermal Expansivity α_P of HFO-1234*yf*

In [68] we determined second-derivative properties such as the thermal expansivity α_P for the refrigerant HFO-1234*yf*. We performed MD simulation studies in the *NpT* ensemble to determine the system densities at different pressures and temperatures. The systems were equilibrated for 2.5 ns. Then a production run of 5 ns followed, which we divided into ten blocks to derive ensemble averages for the liquid densities and their standard deviations. The simulation results at p = 2 MPa and different temperatures are summarized in Table 7.2. Within a limited temperature range, the change of ln $\varrho(T)$ is approximately linear. Thus, we determined α_P from Eq. 7.66 by fitting a straight line to the simulated $ln \varrho$ at three successive temperatures. For each of these temperature intervals, α_P was assumed to be constant and the fitting result was assigned to the middle temperatures. That is, the α_P value at 283.15 K and 2 MPa for instance has been calculated from a linear fit of the simulation results for $ln \varrho$ at 273.15, 283.15 and 298.15 K.

The comparison of these simulation results with REFPROP [38] calculations is shown in Fig. 7.5.

The stated uncertainties for the α_P results mainly account for the uncertainties of the MD simulations for computing the liquid density, i.e. the relation of the standard deviation to the simulation result for the density. It also includes the error

Table 7.2 Simulation results [68] for the liquid density $\varrho(T)$ of HFO-1234*yf* at 2 MPa and T = (273.15 − 310) K, and derived results for the thermal expansivity $\alpha_P(T)$ from Eq. 7.66

T (K)	ϱ(kg m^{-3})	$\alpha_P(10^{-3}\text{K}^{-1})$
273.15	1187.6 ± 23.2	
283.15	1156.8 ± 25.3	2.88 ± 0.14
298.15	1105.4 ± 28.4	3.23 ± 0.29
303.15	1083.1 ± 29.1	3.38 ± 0.28
308.15	1062.1 ± 32.5	3.89 ± 0.38
310.00	1054.7 ± 30.9	

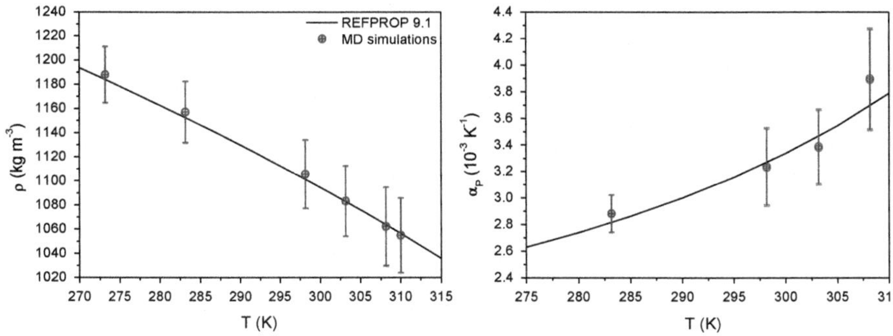

Fig. 7.5 MD Simulation results for the density $\varrho(T)$ of HFO-1234yf at 2 MPa and derived thermal expansivity α_P from Eq. 7.66. Comparison with REFPROP calculations

range of the linear fit. However, the stated uncertainties for α_P do not account for the fact that the change of the $\varrho(T)$ in the regarded temperature interval is in the same order of magnitude as the uncertainty of the simulated density. With this, the α_P results are in fact subject of higher uncertainties. This is also true for other second-derivative properties: The simulation result for the density of HFO-1234yf at 273.15 K, 1 MPa for instance is $\varrho = (1184.7 \pm 23.6)\,\text{kg}\,\text{m}^{-3}$ [68]; at 273.15 K, 2 MPa it is $\varrho = (1187.6 \pm 23.2)\,\text{kg}\,\text{m}^{-3}$. Thus, when we intend to calculate the isothermal compressibility κ_T from Eq. 7.62, we have to face the problem that the change of the density with pressure is even one order of magnitude smaller than the stated uncertainties for the simulated densities.

7.3.7 Second Derivative Properties from Phase Space Functions

An alternative approach to derive second derivative properties, and also other thermodynamic properties, is the method proposed by Lustig [41], which is shortly introduced in this section. The basic idea is that all thermodynamic properties of the *NVEP* ensemble of a molecular dynamics simulation, and also derivatives to arbitrary order, can be derived from basic phase space functions Ω_{mn}. Thereby, Lustig introduced the phase space functions Ω_{mn} as abbreviation for the derivatives of the phase space volume Ω. All thermodynamic properties can be expressed in terms of these phase space functions, which in turn can be related to ensemble averages of volume derivatives of the configurational energy U_{conf} and of some power of the kinetic energy E_{kin}. These ensemble averages can be sampled in the course of the simulation so that no post-simulation analysis is required to derive the thermodynamic properties.

It was later realized that in classical MD simulations, not only NVE and the total momentum $\mathbf{p}$ are conserved (see Sect. 4.1), but also the quantity $\mathbf{G}$

$$\mathbf{G} = \sum_{i=1}^{N} \mathbf{p}_i t - \sum_{i=1}^{N} m_i \mathbf{r}_i, \tag{7.80}$$

which is determined by the initial position of the center of mass of the system [48]. Using the formalism developed by Lustig, Meier and Kabelac [48] derived a general expression for the phase space functions of the *NVEPG* ensemble, given by

$$\begin{aligned}
\Omega_{mn} = & \left(\frac{N}{V}\right)^n \frac{1}{N^n}(-1)^m \frac{2}{3N-3}\left(-\frac{3N-3}{2}\right)_m (-1)^n \\
& \cdot(-[N-1])_n \left\langle E_{kin}^{-(m-1)} \right\rangle + (1+\delta_{0n}) \sum_{i=1}^{n} \binom{n}{i} (-1)^{n-i} \\
& \cdot(-[N-1])_{n-i} \left(\frac{N}{V}\right)^{n-i} \frac{1}{N^{n-i}} \cdot \frac{2}{3N-3} \sum_{l=1}^{i} (-1)^{m+l} \left(-\frac{3N-3}{2}\right)_{m+l} \\
& \cdot \left\langle E_{kin}^{-(m+l-1)} \left(\sum_{k=1}^{k_{max}(i,l)} c_{ilk} W_{ilk} \right) \right\rangle .
\end{aligned} \tag{7.81}$$

Therein, δ_{ij} is the Kronecker delta that is $\delta_{ij} = 1$ if $i = j$ and $\delta_{ij} = 0$ if not. The subscripts at the brackets stands for $(x)_n = x(x+1)(x+2)\cdots(x+n-1)$, the W_{ilk} terms represent products of volume derivatives of the configurational energy, and the c_{ijk} are their combinatorial number. A detailed derivation and description is provided by Lustig [41] or Meier and Kabelac [48].

The isochoric heat capacity C_V, for instance, can be expressed in terms of phase space functions by

$$C_V = \frac{k_B}{1 - \Omega_{00}\Omega_{20}}. \tag{7.82}$$

The phase space functions Ω_{00} and Ω_{20} of the *NVEPG* ensemble with $\mathbf{p} = 0$ are given by

$$\Omega_{00} = k_B T = \frac{2}{3N-3}\langle E_{kin}\rangle \tag{7.83}$$

$$\Omega_{20} = -\left[1 - \frac{3N-3}{2}\right]\langle E_{kin}^{-1}\rangle. \tag{7.84}$$

Expressions for other second order derivative such as the isothermal compressibility κ_T, or speed of sound v_S as well as for the required phase space functions are also provided by Lustig [41] or Meier and Kabelac [48].

7.4 Free Energy, Chemical Potential and Related Properties

In Sect. 2.2 we have discussed thermodynamic potentials whose minimum defines the equilibrium condition in the respective ensemble. For systems at constant volume and temperature, the Helmholtz free energy F represents the thermodynamic potential, at conditions of constant pressure and temperature, it is the Gibbs free energy G. Thus, the free energies functions F and G are important quantities to define equilibrium and stability at conditions relevant for experiments and processes. The bridge equations relate the free energy functions directly to the partition functions of the respective ensembles

$$F = -k_B T \ln Q_{NVT},$$
$$G = -k_B T \ln Q_{NpT}.$$

This also illustrates the difficulties in determining these functions in standard molecular simulation techniques: The partition function Q is the sum of all available microstates for the given constrained properties of the ensemble. Thus, a summation over all possible state of the system is required to obtain accurate estimates for the free energy function. Standard MC or MD simulations however sample according to the equilibrium distribution i.e. preferentially in the lower energy region. Whereas it is not feasible to sample the full configurational space to obtain accurate absolute values for the free energy, it is often possible to determine the difference in the free energy between a reference state 0 and the system of interest I by evaluating the ratio of its partition functions,

$$\Delta F = F_I - F_0 = -k_B T \ln\left(\frac{Q_{NVT}^{(I)}}{Q_{NVT}^{(0)}}\right), \quad (7.85)$$

$$\Delta G = G_I - G_0 = -k_B T \ln\left(\frac{Q_{NpT}^{(I)}}{Q_{NpT}^{(0)}}\right). \quad (7.86)$$

Calculation of free energy differences also play an important role in drug discovery, for instance to rank drugs by their binding affinity to target proteins or by their dissolution ability. Thus, there has been a surge in research on the methodological improvement in free energy computation in recent years. A comprehensive review on the recent literature in free energy calculation is out of the scope of this chapter, and the interested reader is referred to the textbook 'Free Energy Calculations' by the editors Chipot and Pohorille [14], or reviews, for instance by Shirts et al. [78] or Klimovich et al. [30]. Thus, the following subsections will provide a brief introduction to the equilibrium methods for free energy computation, i.e. the Thermodynamic Integration (TI, [28]) and the Exponential Averaging (EXP, [97])—nonequilibrium methods are also not covered by this chapter. The introduction to the TI and EXP method mainly focus on formalisms to calculate the Helmholtz free energy differences in the *NVT* ensemble, though corresponding expressions can be derived to calculate Gibbs free energy differences in the *NpT* ensemble.

Free energies can also be obtained from the chemical potential μ. The chemical potential is defined as the change of the thermodynamic potential of the respective ensemble with the number of particles, i.e.

$$\mu = \left(\frac{F}{N}\right)_{V,T} \quad \text{in the } NVT\text{-ensemble} \tag{7.87}$$

$$\mu = \left(\frac{G}{N}\right)_{p,T} \quad \text{in the } NpT\text{-ensemble.} \tag{7.88}$$

Thus, for a pure compound, the chemical potential is identical with the molar free energy. For a mixture, the chemical potential of a component μ_i corresponds to its partial molar free energy. In Sects. 7.4.4 and 7.4.5 we will discuss the Widom approach of test particle insertion [91] and the Expanded Ensemble method [44] to evaluate the chemical potential. Section 7.4.6 then provides a general relation between the chemical potential and the Gibbs free energy change calculated by Free Energy methods. In classical thermodynamic calculations, the chemical potential is expressed by means of corrective coefficients, i.e. the fugacity coefficient φ or the activity coefficient γ_i. The fugacity coefficient describes the deviation of chemical potential of a real fluid from that of an ideal gas—both in the pure fluid or in the mixture. The activity coefficient though only exists in mixtures and yields the deviation of the chemical potential of a component in a real mixture from its values in an ideal mixture with no excess properties. The activity coefficient is therefore related to the excess chemical potential. Thus, Sects. 7.4.7 and 7.4.8 describe approaches to derive fugacity and activity coefficients from molecular simulations. Though, the chemical potentials, and accordingly the fugacity and activity coefficients, of the different compounds in a mixture are not independent, but related by the Gibbs-Duhem equation. We will therefore finally also discuss, how we can take advantage of this relation.

7.4.1 Thermodynamic Integration

In the Thermodynamic Integration (TI) method, originally proposed by Kirkwood [28], a coupling parameter $0 \leq \lambda \leq 1$ is introduced that connects the reference and the target state, i.e. a value of $\lambda = 0$ represents the references state 0, and $\lambda = 1$ the new state of interest I. Thus, the Helmholtz free energy difference between the states 0 and I can be calculated by

$$F_I(\lambda = 1) - F_0(\lambda = 0) = \int\limits_{\lambda=0}^{\lambda=1} \left(\frac{\partial F}{\partial \lambda}\right)_{NVT} d\lambda. \tag{7.89}$$

With the relation between Helmholtz free energy and the partition function Q_{NVT} given by Eq. 7.48, the partial derivative of F with respect to the coupling parameter λ can be expressed by

$$\left(\frac{\partial F}{\partial \lambda}\right)_{NVT} = -k_B\, T \left(\frac{\partial \ln Q_{NVT}(\lambda)}{\partial \lambda}\right) \tag{7.90}$$

$$= -k_B T \frac{1}{Q_{NVT}} \left(\frac{\partial Q_{NVT}(\lambda)}{\partial \lambda}\right) \tag{7.91}$$

$$= \frac{\int \left(\frac{\partial U_{conf}(\lambda)}{\partial \lambda}\right) e^{-\beta U_{conf}} dr^{3N}}{\int e^{-\beta U_{conf}} dr^{3N}} \tag{7.92}$$

$$= \left\langle \frac{\partial U_{conf}(\lambda)}{\partial \lambda} \right\rangle . \tag{7.93}$$

With this, the Helmholtz free energy can be obtained from

$$F_I(\lambda = 1) - F_0(\lambda = 0) = \int\limits_{\lambda=0}^{\lambda=1} \left\langle \frac{\partial U_{conf}(\lambda)}{\partial \lambda} \right\rangle d\lambda \tag{7.94}$$

For discrete values of λ, standard molecular simulations, either MC or MD, are performed to derive the ensemble average of the derivative of $\partial(U_{conf}(\lambda))/\partial\lambda$. Therein $U_{conf}(\lambda)$ is often determined by a linear coupling of the configurational energies in the reference state $U^{(0)}_{conf}$ and the final state, $U^{(I)}_{conf}$

$$U_{conf}(\lambda) = (1 - \lambda) \cdot U^{(0)}_{conf} + \lambda \cdot U^{(I)}_{conf} \tag{7.95}$$

Though it is also possible to use other coupling functions $f(\lambda)$, so that $U_{conf}(\lambda)$ is generally determined by

$$U_{conf}(\lambda) = (1 - f(\lambda)) \cdot U^{(0)}_{conf} + f(\lambda) \cdot U^{(I)}_{conf} . \tag{7.96}$$

In the MD simulation program DL_POLY [80] for instance, different options for $f(\lambda)$ are available, such as trigonometric or polynomic coupling functions etc.

The reference state of known free energy could be the ideal gas to derive free energies of fluids, or the Einstein crystal for real solids [20]. It is important that the path defined by λ is reversible, i.e. does not exhibit a hysteresis. As hysteresis are common in first-order phase transitions, the two-phase region needs to be avoided on the path from the ideal gas 0 to a target liquid phase. Though it should be noted that the path described by the change of λ does not have to be physically/thermodynamically meaningful [20].

7.4.2 Exponential Averaging (EXP)

The Exponential Averaging (EXP) by Zwanzig [97] (in older literature called the Free Energy Perturbation (FEP)) is based on the fact, that the free energy difference between a reference system 0 and a target system I can be expressed by the ratio of their partition functions

$$\Delta F_{0\rightarrow I} = F_I - F_0 = -k_B\, T \ln\left(\frac{Q_{NVT}^{(I)}}{Q_{NVT}^{(0)}}\right) \tag{7.97}$$

$$= -k_B\, T \ln\left(\frac{\int e^{-\beta U_{conf}^{(I)}} dr^{3N}}{\int e^{-\beta U_{conf}^{(0)}} dr^{3N}}\right). \tag{7.98}$$

By replacing the configurational energy of the target system $U_{conf}^{(I)}$ by

$$U_{conf}^{(I)} = \Delta U_{conf}^{(0\rightarrow I)} + U_{conf}^{(0)}$$

with

$$\Delta U_{conf}^{(0\rightarrow I)} = U_{conf}^{(I)} - U_{conf}^{(0)}$$

Equation 7.97 becomes

$$\Delta F_{0\rightarrow I} = -k_B\, T \ln\left(\frac{\int e^{-\beta \Delta U_{conf}^{(0\rightarrow I)}} e^{-\beta U_{conf}^{(0)}} dr^{3N}}{\int e^{-\beta U_{conf}^{(0)}} dr^{3N}}\right)$$

$$= -k_B\, T \ln\left\langle e^{-\beta \Delta U_{conf}^{(0\rightarrow I)}}\right\rangle_0 . \tag{7.99}$$

The index 0 indicates that the ensemble average is derived by sampling configurations from the reference system 0. When the EXP method is for instance used to investigate solvation processes, the transition from state 0 to state I corresponds to the creation of a solute molecule in the solvent, the transition from I to state 0 to its annihilation, respectively. Thus, in the reference state, the interaction of the solute with the solvent are turned off, in the target state "I" the solute is fully interacting with the solvent. Thus, either MC or MD simulations in the equilibrium state of the reference systems are performed, and the interactions with the solute are switched on to determine the difference in the configurational energy $\Delta U_{conf}^{(0\rightarrow I)}$. Though the energy difference between the two states 0 and I needs to be small enough, which means that the important regions in the phase spaces of both systems have to overlap sufficiently. Otherwise the conformation of the reference system 0 will not be representative for a conformation of the target system I, and the calculated free energy difference will not be accurate. When the energy difference is $\Delta U_{conf}^{(0\rightarrow I)} \gg k_B T$

[37], the transition from 0 to I is divided into a series of intermediate states, again described by a coupling parameter $0 \leq \lambda \leq 1$, with $\lambda = 0$ in 0, and $\lambda = 1$ in I.

For each transition from an intermediate state λ_i to λ_{i+1}, the free energy difference $\Delta Fi \rightarrow i+1$ is obtained from Eq. 7.99. The total free energy difference $\Delta F_{0\rightarrow I}$ is then calculated by summing up the free energy changes of all intermediate transformations [14]

$$\Delta F_{0\rightarrow I} = -k_B\, T \sum_{i=1}^{N-1} \ln \left\langle e^{-\beta \Delta\lambda_i \Delta U_{conf}^{(\Delta\lambda_i)}} \right\rangle_{\lambda_i} . \tag{7.100}$$

The concept of Exponential Averaging is widely applicable. The coupling parameter λ could as well represent changes in internal degrees of freedom to study free energy differences between conformations, or describe the separation between a ligand and a protein to calculate free energies of binding. More details on the EXP method and its applications, practical aspects and techniques to improve its efficiency are given in the textbook by Chipot and Pohorille [14].

7.4.3 Alchemical Pathway

In both, the Thermal integration (TI) and the Exponential Averaging (EXP), the transition from 0 to I is divided into a series of intermediate states that normally do not correspond to real physical states. They are therefore called alchemical intermediates. The transformation from 0 to I is described by an alchemical pathway that involves transitions between two λ values. The suitable choice of the intermediate states—with regard to both, the number of intermediate states and their spacing—essentially influences the efficiency and accuracy of the free energy calculation. In general, an uneven spacing is required to allow for a closer spacing when the curvature $\langle dU/d\lambda \rangle$ is large [30]. Free energy simulations are usually started with even spaced λ values, and then the spacing is manually adjusted in the λ regions where the phase space did not overlap adequately. For identifying regions with poor overlaps in phase space, the Python tool within the pymbar package [77] for instance can be used that employs an analysis technique based on an overlapping distribution method. Based on this analysis, we have in [47] proposed a simple approach to optimize the λ spacing by imposing equal standard deviations of the partial free energies $\Delta G_{\lambda_i \rightarrow \lambda_{i+1}}$ as optimization criterion.

For studies on solvation processes, the intermediate states represent the gradually turning on/off of the solute-solvent interactions by scaling the potential energy by λ. However, this may result in inaccuracies, numerical instabilities and slow convergence due to singularities in the coulomb and repulsive part of the Lennard-Jones interactions [6, 78, 81]. To eliminate the singularities and instabilities, and to ensure a smooth behavior along the alchemical pathway, Beutler et al. [6] proposed a so called "soft-core potential" by introducing an $\alpha(1-\lambda)^2$ term to change the potential function according to

$$u_{ij}(r_{ij},\lambda) = \lambda^n 4\varepsilon_{ij}\left[\left(\frac{1}{\alpha_{LJ}(1-\lambda)^2+\left(\frac{r_{ij}}{\sigma_{ij}}\right)^s}\right)^{12/s} - \left(\frac{1}{\alpha_{LJ}(1-\lambda)^2+\left(\frac{r_{ij}}{\sigma_{ij}}\right)^s}\right)^{6/s}\right] + \frac{q_i q_j}{4\pi\varepsilon_0[\alpha_c(1-\lambda)^2+r_{ij}^p]^{1/p}}. \tag{7.101}$$

Therein, α, s and p are arbitrary constants. Beutler et al. [6] suggest to chose $\alpha = 0.5$ for the most continuous change in the interaction energy, and $s = 6$, $p = 2$ to yield the computational most advantageous form of the soft-core potential.

Equation 7.101 represents a so called one-step transformation [81], in which the soft-core potential is used for both, the Lennard Jones and the coulombic interactions. Though more often, a two-step transformation is applied, in which the changes in the coulomb and LJ interactions are separated from each other, and the electrostatic interactions are first turned off, before the LJ interactions are altered [30, 78]. The electrostatic interactions are scaled linearly to maintain the optimal computational efficiency, and the soft core approach is only used for the LJ interactions. Pham and Shirts [60, 61] proposed a more generalized form of the LJ soft-core potential, represented by

$$u_{ij}(r_{ij},\lambda) = \lambda^a 4\varepsilon_{ij}\left[\left(\frac{1}{\alpha(1-\lambda)^b+(r_{ij}/\sigma_{ij})^c}\right)^{12/c} - \left(\frac{1}{\alpha(1-\lambda)^b+(r_{ij}/\sigma_{ij})^c}\right)^{6/c}\right]. \tag{7.102}$$

Setting the constants $a = 1, b = 2, c = 6$ ("1-2-6 form") then yields the original Beutler soft-core potential. Pham and Shirts [60, 61] and Naden et al. [49] have evaluated different choices for the constants of the so called "soft-core a-b-c form" with regard to the efficiency and accuracy of the calculated free energy difference. Naden et al. [49] state that the "1-1-6" form with $\alpha = 0.5$ is by 30% more efficient then the original soft-core formulation, though Pham and Shirts [60, 61] suggest that an optimized "1-1-48" form with $\alpha = 0.0025$ results in the most efficient and lowest variance pathway. Different soft-core potentials are for instance also evaluated by Steinbrecher et al. [81]. Finally, it should be noted that the MBAR (Sect. 3.5.3) or the WHAM (Sect. 3.5.2) methods can be applied to make use of the information from all intermediate states in order to optimize the estimate of the free energy difference $\Delta F_{0\rightarrow I}$.

7.4.4 Chemical Potential from Widom Method

The Widom Particle Insertion Method [91] is based on the thermodynamic identity that the chemical potential in the *NVT*-ensemble is defined by the change of the Helmholtz free energy with the number of particles. This can be approximated by the difference in the Helmholtz free energies between a system with $N+1$ particle and an N particle system. This difference in free energies in turn is evaluated by the ratio of the partition functions

$$\mu_{NVT} = \left(\frac{\partial F}{\partial N}\right)_{V,T} \approx F_{N+1} - F_N = -k_B\,T \ln\left(\frac{Q_{NVT,N+1}}{Q_{NVT,N}}\right). \tag{7.103}$$

When expressing the partition functions Q_{NVT} by using the reduced coordinates s (see Eq. 3.23), the chemical potential is derived from

$$\mu_{NVT} = -k_B\,T \ln\left(\frac{V^{N+1}N!\Lambda^{3N}}{(N+1)!\Lambda^{3N+3}V^N}\frac{\int e^{-U_{conf,N+1}/k_BT}ds^{3(N+1)}}{\int e^{-U_{conf,N}/k_BT}ds^{3N}}\right). \tag{7.104}$$

The configurational energy in the $N+1$ system can be separated in the contribution from interactions of the N particles among themselves, $U_{conf,N}$, and the contribution of the interaction of the $(N+1)$th particle with the other N particles $\Delta U_{conf,N+1}$. With this, Eq. 7.104 becomes

$$\begin{aligned}\mu_{NVT} &= -k_B\,T \ln\left(\frac{V}{(N+1)\Lambda^3}\frac{\int e^{-\Delta U_{conf,N+1}/k_BT}e^{-U_{conf,N}/k_BT}ds^{3N}}{\int e^{-U_{conf,N}/k_BT}ds^{3N}}\right)\\ &= -k_B\,T \ln\left(\frac{V}{(N+1)\Lambda^3}\left\langle e^{-\Delta U_{conf,N+1}/k_BT}\right\rangle_{NVT}\right).\end{aligned} \tag{7.105}$$

Thus, the chemical potential can be derived from the ensemble average of the interaction energy that an $N+1$th particle experiences, when it is inserted into an N-particle system. This ensemble average can be determined from conventional Monte Carlo simulations in a N-particle system, in which at frequent intervals, a test particle is inserted at a randomly generated position [20]. In general, the $(N+1)$th particle is not actually inserted, but only represents a 'ghost' particle.

In an ideal gas with no interactions between the particles, $\Delta U_{conf,N+1} = 0$, and Eq. 7.105 reduces to

$$\mu_{NVT}^{id} = -k_B\,T \ln\left(\frac{V}{(N+1)\Lambda^3}\right), \tag{7.106}$$

the expression for the chemical potential of an ideal gas in the *NVT*-ensemble. Thus, the expression for μ_{NVT} in Eq. 7.105 can be separated into the ideal gas contribution and the residual chemical potential that describes the deviation of a real fluid from an ideal gas at the same temperature and volume

$$\mu_{NVT} = \underbrace{-k_B\,T\ln\left(\frac{V}{(N+1)\Lambda^3}\right)}_{} \underbrace{-k_B\,T In\left\langle e^{-\Delta U_{conf,N+1}/k_B T}\right\rangle_{NVT}}_{}$$
$$= \quad \mu^{id}_{NVT} \quad + \quad \mu^{res}_{NVT}. \tag{7.107}$$

The chemical potential of the ideal gas can be evaluated analytically [20], so that only the residual part is determined in molecular simulations. The long-range correction (LRC) to the chemical potential due to interactions beyond r_{cut} is given by

$$\mu^{LRC} = 4\pi\varrho\int\limits_{r_{cut}}^{\infty} r_{ij}^2 u_{ij}(r_{ij})dr_{ij}. \tag{7.108}$$

In a mixtures composed of N_1, N_2, ..., N_i particles of the different compounds, the residual chemical potential $\mu_{i,NVT}$ of a component i is accordingly derived from the ensemble averages of the interaction energy $\Delta U_{conf,N_i+1}$ that one of its molecules experience when it is inserted into the mixture

$$\mu^{res}_{i,NVT} = -k_B\,T In\left\langle e^{-\Delta U_{conf,N_i+1}/k_B T}\right\rangle_{NVT}. \tag{7.109}$$

In the *NpT*-ensemble though, the chemical potential is defined by the change of the Gibbs free energy with the number of particles, and can therefore be derived from the ratio of the partition functions $Q_{NpT,N+1}$ and $Q_{NpT,N}$

$$\mu_{NpT} = \left(\frac{G}{N}\right)_{p,T} \approx G_{N+1} - G_N = -k_B\,T\ln\left(\frac{Q_{NpT,N+1}}{Q_{NpT,N}}\right). \tag{7.110}$$

When using the reduced coordinates s, the partition function of the *NpT*-ensemble can be expressed by (see Eq. 3.24)

$$Q_{NpT}(s^{3N}) = \frac{1}{V_0}\int\frac{V^N}{N!\lambda^{3N}}e^{-pV/k_B T}dV\int e^{-U_{conf}/k_B T}ds^{3N}.$$

With this, the chemical potential is evaluated from

$$\mu_{NpT} = -k_B\,T\ln\left(\frac{\int\frac{V^{N+1}}{(N+1)!\Lambda^{3(N+1)}}e^{-pV/k_B T}dV\int e^{-U_{conf,N+1}/k_B T}ds^{3(N+1)}}{\int\frac{V^N}{N!\lambda^{3N}}e^{-pV/k_B T}dV\int e^{-U_{conf}/k_B T}ds^{3N}}\right). \tag{7.111}$$

When we again divide the interaction energy of the $N+1$ system into the interaction in the N particle system $U_{conf,N}$ and the interaction of the inserted $N+1$th particle $\Delta U_{conf,N+1}$, we derive a similar expression for the chemical potential in the *NpT*-ensemble as before for μ_{NVT}

$$\mu_{NpT} = -k_B T \left\langle \ln \frac{V}{(N+1)\Lambda^3} e^{-\Delta U_{conf,N+1}/k_B T} \right\rangle_{NpT} . \quad (7.112)$$

Though, when we now set the interaction energy $\Delta U_{conf,N+1} = 0$ to deduce the expression of the chemical potential of an ideal gas in the *NpT*-ensemble, we have to take into account, that the volume fluctuates. Thus, $\mu_{i,NpT}^{id}$ is not a constant value that can be evaluated analytically, rather is has to be determined from the ensemble average of the volume

$$\mu_{NpT}^{id} = -k_B T \ln \left\langle \frac{V}{(N+1)\Lambda^3} \right\rangle_{NpT} . \quad (7.113)$$

When we then separate the expression for the chemical potential from Eq. 7.112 into an ideal gas and a residual part, we obtain [76]

$$\begin{aligned} \mu_{NpT} &= \underbrace{-k_B T \ln \left\langle \frac{V}{(N+1)\Lambda^3} \right\rangle_{NpT}} \underbrace{- k_B T \ln \left(\frac{\left\langle V e^{-\Delta U_{conf,N+1}/k_B T} \right\rangle_{NpT}}{\langle V \rangle} \right)} \\ &= \qquad \mu_{NpT,V}^{id} \qquad + \qquad \mu_{NpT,V}^{res} . \end{aligned} \quad (7.114)$$

The residual chemical potential though describes the deviation from the chemical potential of an ideal gas at the same values for temperature and volume, expressed by the subscript "V". For simulations in the *NpT* ensemble it is more common to determine $\mu_{NpT,p}^{res}$ as deviation from the chemical potential of the ideal gas at the same temperature and pressure. The expression for $\mu_{NpT,p}^{res}$ can be derived by extending the term in brackets in Eq. 7.112 by $pk_B T / pk_B T$. By applying the ideal gas law

$$pV = (N+1) k_B T, \quad (7.115)$$

the chemical potential of the ideal gas in the *NpT*-ensemble can be written as

$$\mu_{NpT}^{id} = -k_B T \ln \left(\frac{k_B T}{p\Lambda^3} \right)_{NpT} . \quad (7.116)$$

As both the pressure and the temperature are imposed, no fluctuations have to be considered, and the ideal gas part can again be determined analytically. Separating the modified Eq. 7.112 then into the ideal gas and residual part yields

$$\begin{aligned} \mu_{NpT} &= \underbrace{-k_B T \ln \left(\frac{k_B T}{p\Lambda^3} \right)_{NpT}} \underbrace{- k_B T \ln \left\langle \frac{pV}{k_B T (N+1)} e^{-\Delta U_{conf,N+1}/k_B T} \right\rangle_{NpT}} \\ &= \qquad \mu_{NpT,p}^{id} \qquad + \qquad \mu_{NpT,p}^{res} . \end{aligned} \quad (7.117)$$

The Widom Particle Insertion Method is becoming increasingly inefficient with increasing density of the fluid and/or molecular size of the particles. This is due to the fact that it then becomes more and more difficult to find an appropriate void in the system to insert the particle [37]. In consequence of frequent overlaps with other particles, the simulation samples preferentially in the high energy region, but not in the full configurational space as it would be required to obtain accurate estimates for the chemical potential. Thus, alternative approaches to derive the chemical potential have been developed, such as the Expanded Ensemble Method [44] that will be discussed in the next section.

7.4.5 Chemical Potential from Expanded Ensemble Method

The difficulties encountered by the Widom Method in dense fluids could be reduced or avoided by a gradual insertion of the particle. This could be realized by Thermal Integration (see Sect. 7.4.1) when the coupling parameter λ represents the progressive turning-on of the interaction of the $N+1$th particle with the other N molecules, i.e. for $\lambda = 0$, the particle is in the ideal gas state, for $\lambda = 1$ it is fully interacting with the N particle system. This approach would require a number of independent simulations for different values of the coupling parameter λ. Though the Expanded Ensemble (EE) approach by Lyubartsev et al. [44] allows for the evaluation of the chemical potential in one single simulation by using a sum M of subensembles from which each corresponds to a different value of the coupling parameter. During the simulation, a Metropolis MC random walk over the subensembles is performed [43], i.e. attempts are made to change the coupling or insertion parameter to another (neighboring) value. Additionally, a bias potential or balancing factor η is introduced to ensure sufficient large transition probabilities between the subensembles, and with this an almost uniform sampling among different λ values (see Sect. 3.6.4 on flat histogram methods).

For an arbitrary insertion step, described by the index m with $m = 0$ for $\lambda = 0$ and $m = M$ for $\lambda = 1$, the partition function of the *NVT*-ensemble is given by

$$Q_{NVT,m} = \frac{V^{N+\lambda_m}}{(N+\lambda_m)!\Lambda^{3(N+\lambda_m)}} \int e^{-(U_{conf,N}+\lambda_m \Delta U_{conf,N+1})/k_B T} ds^{3(N+\lambda_m)}. \quad (7.118)$$

For $m = 0$, we obtain the partition function of the N-particle system, for $m = M$ that of the $N+1$ system

$$Q_{NVT,0} = \frac{V^N}{N!\Lambda^{3N}} \int e^{-U_{conf,N}/k_B T} ds^{3N} \quad (7.119)$$

$$Q_{NVT,M} = \frac{V^{N+1}}{(N+1)!\Lambda^{3(N+1)}} \int e^{-U_{conf,N+1}/k_B T} ds^{3N}. \quad (7.120)$$

The chemical potential is derived from the ratio of the $Q_{NVT,M}$ and $Q_{NVT,0}$, and using Z_{NVT} as abbreviation for the configurational integral yields

$$\mu_{NVT} = -k_B\,T \ln\left(\frac{Q_{NVT,M}}{Q_{NVT,0}}\right) = -k_B\,T \ln\left(\frac{V}{(N+1)\Lambda^3}\,\frac{Z_{NVT,M}}{Z_{NVT,0}}\right). \quad (7.121)$$

With the balancing factor taking into account, the Metropolis acceptance criterion for a transition between two subensembles λ_m and λ_{m+1} is given by [1]

$$acc(\lambda_m \rightarrow \lambda_{m+1}) \quad (7.122)$$
$$= min\left[1, exp\left(\frac{-(U_{conf,N+1}(\lambda_{m+1}) - U_{conf,N+1}(\lambda_m))}{k_B\,T} - (\eta_{m+1} - \eta_m)\right)\right].$$

With this, the ratio of the probabilities $\wp$ of the two extreme states $m = 0$ and $m = M$ becomes [43]

$$\frac{\wp_M}{\wp_0} = \frac{Z_{NVT,M}}{Z_{NVT,0}}\,e^{(\eta_M - \eta_0)}. \quad (7.123)$$

When we now solve Eq. 7.123 for the ratio of the configuration integral and insert this expression in Eq. 7.121, we obtain

$$\mu_{NVT} = -k_B\,T \ln\left(\frac{V}{(N+1)\Lambda^3}\right) - k_B\,T \ln\frac{\wp_M}{\wp_0} + k_B\,T\,(\eta_M - \eta_0). \quad (7.124)$$

The first term in Eq. 7.124 again represents the chemical potential of the ideal gas in the *NVT*-ensemble. Thus, in the Expanded Ensemble method, the residual chemical potential is derived from

$$\frac{\mu^{res}_{NVT}}{k_B\,T} = -\ln\frac{\wp_M}{\wp_0} + (\eta_M - \eta_0). \quad (7.125)$$

The efficiency of the Expanded Ensemble method depends on an appropriate choice of the number M of subensembles, on the distribution of the λ_m values in the range [0, 1], and particularly on the optimal choice of the balancing factor η_m. Thus, the inclusion of an automated optimization procedure for η_m can significantly improve the efficiency of the EE method [96]. In the MD simulation program M. Dyna*Mix* by Lyubartsev and Laaksonen [42], an optimization procedure based on the approach proposed by Wang and Landau [86] (see Sect. 3.6.4.2) is implemented. In the different subensembles, the interaction between the inserted particle and the $N-$particle systems needs to be scaled according to the coupling parameter λ. A linear scaling with λ has turned out to be disadvantageous, as the repulsive term of the Lennard-Jones potential only decrease slowly with decreasing λ [26]. Thus, the Lennard-Jones and electrostatic interactions as function of λ can, for instance, be determined by [26]

$$U_{conf,N+1}(\lambda_m) = \lambda_m^4 U_{LJ} + \lambda_m^2 U_{elec}. \tag{7.126}$$

In [57], Palusch et al. have used the 'soft-core' scaling of the Lennard-Jones interactions (see Sect. 7.4.3) to avoid instabilities in the simulation when λ_m approaches zero.

7.4.6 Chemical Potential from Free Energy Methods

In [79], Shirts et al. derived a general relation between the chemical potential of a compound in a liquid mixture, and the Gibbs free energy change calculated by free energy methods in the coupling process from $\lambda = 0$ to $\lambda = 1$ (see Sects. 7.4.1, 7.4.2, 7.4.3). The chemical potential of a component i in a binary mixture with component j can be derived from

$$\begin{aligned}\mu_i(N_i, N_j, p, T) &= G(N_i, N_j, p, T) - G(N_i - 1, N_j, p, T)\\ &= -k_B T \ln\left(\frac{Q_{NpT}(N_i, N_j, p, T)}{Q_{NpT}(N_i - 1, N_j, p, T)}\right).\end{aligned} \tag{7.127}$$

Shirts et al. state that the introduction of the molecule of type i from $N_i - 1$ to N_i by free energy simulation techniques can be separated into sub-processes according to

$$\begin{aligned}\ln\left(\frac{Q_{NpT}(N_i, N_j, p, T)}{Q_{NpT}(N_i - 1, N_j, p, T)}\right) &= \ln\left(\frac{Q_{NpT}(N_i, N_j, p, T)}{Q_{NpT}(N_i, N_j, p, T, \lambda = 1)}\right)\\ &\cdot \ln\left(\frac{Q_{NpT}(N_i, N_j, p, T, \lambda = 1)}{Q_{NpT}(N_i, N_j, p, T, \lambda = 0)}\right)\\ &\cdot \ln\left(\frac{Q_{NpT}(N_i, N_j, p, T, \lambda = 0)}{Q_{NpT}(N_i - 1, N_j, p, T)}\right),\end{aligned} \tag{7.128}$$

which are all associated with a change in free energy. The first term describes the free energy difference resulting from the selection of a single molecule of i for the coupling process. Due to the selection, only $N_i - 1$ molecules of component i remain undistinguishable in the system. This changes the combinatorial term from $\frac{1}{N_i!}$ in $Q_{NpT}(N_i, N_j, p, T)$ to $\frac{1}{(N_i-1)!}$ in $Q_{NpT}(N_i, N_j, p, T, \lambda = 1)$, and with this

$$\frac{Q_{NpT}(N_i, N_j, p, T)}{Q_{NpT}(N_i, N_j, p, T, \lambda = 1)} = \frac{1}{N_i}. \tag{7.129}$$

The second term covers the free energy difference between the uncoupled system at $\lambda = 0$ and the coupled system at $\lambda = 1$, in which the introduced particle of i is fully interacting with all other particles. This is the free energy change calculated by the free energy simulation techniques, when the interaction of the introduced particle is *coupled* with the other particles in the system. Following the derivation by Shirts et al. [79], it will be denoted $\Delta G_{sim,i}$

$$-k_B T \cdot \ln\left(\frac{Q_{NpT}(N_i, N_j, p, T, \lambda = 1)}{Q_{NpT}(N_i, N_j, p, T, \lambda = 0)}\right) = \Delta G_{sim,i}. \quad (7.130)$$

The thirds term in Eq. 7.128 though covers the free energy change in the decoupling process at $\lambda = 0$, in which the system is separated into two independent subsystems: the $(N_i - 1, N_j)$ system, and the system of the isolated molecule of i in the ideal gas phase. Both systems have the same volume V^*, which is the mean volume of the $(N_i - 1, N_j)$ system, i.e. $V^*(N_i - 1, N_j, p, T)$. The canonical partition function at $\lambda = 0$ can therefore be separated into

$$Q_{NVT}(N_i, N_j, V^*, T, \lambda = 0) = Q_{NVT}(N_i - 1, N_j, V^*, T) \cdot Q_{NVT}(N_i = 1, N_j = 0, V^*, T). \quad (7.131)$$

Thereby, the canonical partition function of the single molecule of i in the ideal gas phase is given by

$$Q_{NVT}(N_i = 1, N_j = 0, V^*, T) = Q^{id}_{NVT}(N_i = 1) = \frac{V^*}{\Lambda_i^3}. \quad (7.132)$$

Based on Eq. 7.131 and the relation between the Q_{NVT} and Q_{NpT} given by Eq. 2.51, the partition function $Q_{NpT}(N_i, N_j, p, T, \lambda = 0)$ can be expressed by

$$\begin{aligned} &Q_{NpT}(N_i, N_j, p, T, \lambda = 0) \\ &= \frac{1}{V_0}\int e^{-pV/k_BT} Q_{NVT}(N_i - 1, N_j, V^*, T) \cdot Q_{NVT}(N_i = 1, N_j = 0, V^*, T)dV \\ &= \frac{1}{V_0}\frac{1}{\Lambda_i^3}\int V^* e^{-pV/k_BT} Q_{NVT}(N_i - 1, N_j, V^*, T)dV. \end{aligned} \quad (7.133)$$

Extending Eq. 7.133 by $Q_{NpT}(N_i - 1, N_j, p, T)$ yields

$$Q_{NpT}(N_i, N_j, p, T, \lambda = 0) = \frac{1}{\Lambda_i^3} Q_{NpT}(N_i - 1, N_j, p, T) \cdot \frac{1}{V_0}\frac{\int V^* e^{-pV/k_BT} Q_{NVT}(N_i - 1, N_j, V^*, T)dV}{Q_{NpT}(N_i - 1, N_j, p, T)}.$$

Therein the term in the second line defines the ensemble average $\langle V^* \rangle$. Thus, the ratio of partition function of the $(N_i, N_j, \lambda = 0)$ and the $(N_i - 1, N_j)$ system is given by

$$\frac{Q_{NpT}(N_i, N_j, p, T, \lambda = 0)}{Q_{NpT}(N_i - 1, N_j, p, T)} = \frac{\langle V^*(N_i - 1, N_j, p, T)\rangle}{\Lambda_i^3}. \quad (7.134)$$

With this, the chemical potential of the component i, defined by Eqs. 7.127 and 7.128, can be derived from

$$\mu_i(N_i, N_j, p, T) = -k_B T \ln\left(\frac{1}{N_i}\right) + \Delta G_{sim,i} - k_B T \ln \frac{\langle V^*(N_i-1, N_j, p, T)\rangle}{\Lambda_i^3}$$
$$= \Delta G_{sim,i} - k_B T \ln\left\langle \frac{V^*(N_i-1, N_j, p, T)}{N_i \Lambda_i^3}\right\rangle. \quad (7.135)$$

Adding and subtracting a $k_b T \ln \left(V(N_i, N_j, p, T)\right)$ term yields

$$\mu_i(N_i, N_j, p, T) = \Delta G_{sim,i} - k_B T \ln\left\langle \frac{V(N_i, N_j, p, T)}{N_i \Lambda_i^3}\right\rangle$$
$$- k_B T \ln\left\langle \frac{V^*(N_i-1, N_j, p, T)}{V(N_i, N_j, p, T}\right\rangle. \quad (7.136)$$

The last term on the right side represents a correction to account for the change in volume due to the insertion of a particle of i. Although it could be calculated in free energy simulations, it is in general neglected, as the resulting correction to $\mu_i(N_i, N_j, p, T)$ is most often in the same magnitude of order than the uncertainties of the simulation results for the free energy differences $\Delta G_{sim,i}$. Therefore, this term is also omitted in the following discussions and derivations. The second term on the right side represent the chemical potential of the pure compound i in the ideal gas state at the same mean volume $V(N_i, N_j, p, T)$ as the $N_i + N_j$ system at full coupling. With this, the $\Delta G_{sim,i}$ corresponds to the residual chemical potential with regard to an ideal gas at the same volume. As before in Sect. 7.4.4, it is indicated by a subscript "V".

$$\mu_{i,V}^{res}(N_i, N_j, p, T) = \Delta G_{sim,i}. \quad (7.137)$$

The residual chemical potential $\mu_{i,p}^{res}$ related to the ideal gas of component i at the same temperature and pressure can be determined by (see Sect. 7.4.4)

$$\mu_{i,p}^{res}(N_i, N_j, p, T) = \mu_{i,V}^{res}(N_i, N_j, p, T) + \mu_{i,V}^{id} - \mu_{i,p}^{id}$$
$$= \Delta G_{sim,i} - k_B T \ln\left\langle \frac{V(N_i, N_j, p, T) p}{N_i k_B T}\right\rangle. \quad (7.138)$$

For a pure fluid, the chemical potential can accordingly be derived from

$$\mu_{0i}(N_{0i}, p, T) = \Delta G_{sim,0i} - k_B T \ln\left\langle \frac{V_{0i}(N_{0i}, p, T)}{N_{0i} \Lambda_i^3}\right\rangle, \quad (7.139)$$

wherein the $\Delta G_{sim,0i}$ again corresponds to the residual chemical potential $\mu_{0i,V}(N_{0i}, p, T)$ related to the chemical potential of the ideal gas at the same mean volume $V_{0i}(N_{0i}, p, T)$. The residual chemical potential $\mu^{res}_{0i,p}(N_{0i}, p, T)$ referring to the ideal gas at the same pressure is then given by

$$\mu^{res}_{0i,p}(N_{0i}, p, T) = \Delta G_{sim,0i} - k_B T \ln \left\langle \frac{V_{0i}(N_{0i}, p, T)p}{N_{0i} k_B T} \right\rangle. \qquad (7.140)$$

The index $0i$ in the expressions for the pure compound is now used to denote a property of the pure fluid i, whereas the simple index i refers to the properties of the component i in the mixture. The index $0i$ is introduced here to clearly distinguish between the property of the pure fluid and the component in the mixture, which is required for the discussion on fugacity and activity coefficients in the next sections.

7.4.7 Fugacity Coefficients

The concept of fugacity (coefficient) is used to describe the deviation of the chemical potential of a real component from its value in the ideal gas state at the same temperature. Based on the so called standard potential $\mu^{id}_{0i}(T, p^+)$ of the ideal gas at T and the standard pressure p^+, the chemical potential of a real pure fluid at T, p can be derived from

$$\mu_{0i}(T, p) = \mu^{id}_{0i}(T, p^+) + k_B T \ln \left(\frac{p}{p^+} \varphi_{0i}(p, T) \right), \qquad (7.141)$$

with $\varphi_{0i}(p, T)$ being the fugacity coefficient of the pure compound i at p, T. The product $p \cdot \varphi_{0i}(p, T) = f_{0i}$ yields the fugacity f_{0i}, which corresponds to a'corrected pressure' of the real gas. The term $k_B T \ln(p/p^+)$ describes the pressure difference of the ideal gas chemical potential between the system pressure p and the standard pressure p^+. Thus Eq. 7.141 can be rewritten as

$$\mu_{0i}(T, p) = \mu^{id}_{0i}(T, p) + k_B T \ln (\varphi_{0i}(p, T)), \qquad (7.142)$$

which defines the fugacity coefficient by the difference between the chemical potential of the real fluid and its ideal gas chemical potential at the same pressure and temperature. With this, it is directly related to the residual chemical potential $\mu^{res}_{0i,p}$.

When the chemical potential is computed by the Widom Method (see Sect. 7.4.4), the fugacity coefficient can be derived from the residual chemical potential $\mu^{res}_{0i,NpT}$, defined by Eq. 7.117

$$\begin{aligned} k_B T \ln (\varphi_{0i}(p, T)) &= \mu^{res}_{0i,NpT} \\ &= -k_B T \ln \left\langle \frac{p V_{0i}}{k_B T (N_{0i} + 1)} e^{-\Delta U_{conf,N_{0i}+1}/k_B T} \right\rangle_{NpT} \end{aligned} \qquad (7.143)$$

This then yields

$$\varphi_{0i}(p,T) = \frac{1}{\left\langle \frac{pV_{0i}}{k_BT(N_{0i}+1)} e^{-\Delta U_{conf,N_{0i}+1}/k_BT} \right\rangle_{NpT}}. \tag{7.144}$$

As the pressure and temperature are imposed constant values, they can be factored out of the ensemble averaging. This then results in the following expression for φ_{0i}, which was for instance used by Wierzchowski and Kofke in [92]

$$\varphi_{0i}(p,T) = \frac{1}{\frac{\beta p}{N_{0i}+1} \left\langle V_{0i} e^{-\Delta U_{conf,N_{0i}+1}/k_BT} \right\rangle_{NpT}}. \tag{7.145}$$

In order to deduce an expression for the fugacity coefficient when the chemical potential is determined by any free energy method (See Sect. 7.4.6), we only need to relate it to residual chemical potential $\mu^{res}_{0i,p}(N_{0i}, p, T)$ derived in Eq. 7.140

$$k_BT \ln\left(\varphi_{0i}(p,T)\right) = \mu^{res}_{0i,p}(N_{0i}, p, T) \tag{7.146}$$

$$= \Delta G_{sim,0i} - k_BT \ln \left\langle \frac{V_{0i}(N_{0i}, p, T)p}{N_{0i}k_BT} \right\rangle, \tag{7.147}$$

wherein $\Delta G_{sim,0i}$ represents the residual chemical potential $\mu^{res}_{0i,V}(N_{0i}, p, T)$ related to the ideal gas at the same volume, as discussed in Sect. 7.4.6. When Eq. 7.146 is solved for the fugacity coefficient of the pure compound, it yields

$$\varphi_{0i}(p,T) = \exp\left(\beta\mu^{res}_{0i,p}(N_{0i}, p, T)\right) \tag{7.148}$$

$$= \frac{\exp\left(\beta\mu^{res}_{0i,V}(N_{0i}, p, T)\right)}{\beta p} \left\langle \frac{N_{0i}}{V_{0i}(N_{0i}, p, T)} \right\rangle. \tag{7.149}$$

Based on the standard potential $\mu^{id}_{0i}(T, p^+)$, the chemical potential of a real compound in a mixture $\mu_i(T, p, x_i)$ can be derived by additive correction: the pressure term $k_BT\ln(p/p^+)$ to yield the system pressure, the mixing term $k_BT\ln(x_i)$, and finally the fugacity coefficient term to account for the deviation from an ideal gas

$$\mu_i(T, p, x_i) = \mu^{id}_{0i}(T, p^+) + k_BT \ln\left(\frac{p}{p^+}\right) + k_BT\ln(x_i) + k_BT\left(\varphi_i(p, T, x_i)\right),$$

$$= \mu^{id}_{0i}(T, p^+) + k_BT \ln\left(\frac{p}{p^+} x_i \varphi_i(p, T, x_i)\right) \tag{7.150}$$

$$= \mu^{id}_{0i}(T, p^+) + k_BT \ln\left(\frac{f_i}{p^+}\right). \tag{7.151}$$

The fugacity f_i of the component in the mixture is therefore defined as product $f_i = p \cdot x_i \cdot \varphi_i(p, T, x_i)$. The addition of the pressure term to the standard potential

$\mu_{0i}^{id}(T, p^+)$ gives the chemical potential in the ideal gas state at p. Adding the mixing term yields the chemical potential in the ideal gas mixture at T, p, x_i. Thus, the chemical potential of the real compound in the mixture can also be expressed by

$$\mu_i(T, p, x_i) = \mu_{0i}^{id}(T, p) + k_B T \ln\left(x_i \varphi_i(p, T, x_i)\right) \tag{7.152}$$
$$= \mu_i^{id}(T, p, x_i) + k_B T \ln\left(\varphi_i(p, T, x_i)\right). \tag{7.153}$$

Thus, the fugacity coefficient $\varphi_i(p, T, x_i)$ describes the deviation of the chemical potential of the real compound in a mixture from its value in an ideal gas mixture at the same values of temperature, pressure and composition. To derive the expression for the fugacity coefficient when the chemical potential is determined by free energy methods (Sect. 7.4.6), we only need to equate the second term on the right side of Eq. 7.152 with the expression for the residual chemical potential in the mixtures $\mu_{i,p}^{res}(N_i, N_j, p, T)$, we have derived in Eq. 7.138. This yields

$$k_B T \ln\left(x_i \varphi_i(p, T, x_i)\right) = \mu_{i,p}^{res}(N_i, N_j, p, T) \tag{7.154}$$
$$= \Delta G_{sim,i} - k_B T \ln\left\langle \frac{V(N_i, N_j, p, T) \cdot \beta p}{N_i} \right\rangle. \tag{7.155}$$

Again it should be noted that $\Delta G_{sim,i}$ corresponds to the residual chemical potential $\mu_{i,V}^{res}(N_i, N_j, p, T)$ related to an ideal gas at the same volume. Solved for the fugacity coefficient and inserting $x_i = N_i/(N_i + N_j)$, the fugacity coefficient is given by

$$\varphi_i(p, T, x_i) = \frac{\exp\left(\beta \mu_{i,V}^{res}(N_i, N_j, p, T)\right)}{\beta p} \left\langle \frac{N_i + N_j}{V(N_i, N_j, p, T)} \right\rangle, \tag{7.156}$$

wherein the expression in the ensemble average bracket represents the mean molar density ϱ_m of the mixture.

7.4.8 Activity Coefficients

In the concept of activity (coefficients), the reference system is not the ideal gas (as for φ_i) but the ideal mixture, whose components are real compounds. The chemical potential $\mu_i(T, p, x_i)$ of a real component in an ideal mixture can be derived from the chemical potential of the pure real compound $\mu_{0i}(T, p)$ at the same pressure and temperature by adding the mixing term $k_B T \ln(x_i)$. When the components do not form an ideal mixture, which is indicated by the occurrence of excess properties, $\mu_i(T, p, x_i)$ is corrected by the term involving the activity coefficient γ_i, according to

$$\mu_i(T, p, x_i) = \mu_{0i}(T, p) + k_B T \ln(x_i) + k_B T \ln\left(\gamma_i(p, T, x_i)\right). \tag{7.157}$$

The product of mol fraction and activity coefficient defines the so called activity $a_i = x_i \cdot \gamma_i$, which corresponds to a 'corrected mol fraction' of the component in the real mixture. In contrast to the fugacity, the activity coefficient is a corrective term that only occurs in the mixture, i.e. $\gamma_i = 1$ for $x_i = 1$.

In the previous Sect. 7.4.7, the relations were introduced to express both, the $\mu_i(T, p, x_i)$ and $\mu_{0i}(T, p)$ in terms of the fugacity coefficients $\varphi_i(p, T, x_i)$ and $\varphi_{0i}(p, T)$, respectively. When introducing Eqs. 7.152 and 7.142 into Eq. 7.157, we can therefore deduce the relation between the activity coefficients and the fugacity coefficients that is given by

$$ln\gamma_i(T, p) = \ln(\varphi_i(p, T, x_i)) - \ln(\varphi_{0i}(p, T)). \tag{7.158}$$

$$\gamma_i(T, p) = \frac{\varphi_i(p, T, x_i)}{\varphi_{0i}(p, T)} \tag{7.159}$$

When we use any energy method to determine the chemical potential or the fugacity coefficients, the corresponding expression for the activity coefficient can be derived by introducing the Eqs. 7.154 and 7.146 into Eq. 7.158

$$\begin{aligned} k_B T ln\gamma_i(T, p) &= k_B T \ln(\varphi_i(p, T, x_i)) - k_B T \ln(\varphi_{0i}(p, T)) \\ &= \mu_{i,p}^{res}(N_i, N_j, p, T) - k_B T \ln x_i - \mu_{0i,p}^{res}(N_{0i}, p, T). \end{aligned} \tag{7.160}$$

When expressing the residual chemical potential referring to an ideal gas at the same pressure in terms of the $\Delta G_{sim,i}$, which corresponds to the residual chemical potential at the same volume (see Sect. 7.4.6), it yields

$$\begin{aligned} ln\gamma_i(T, p) &= \beta\left(\mu_{i,V}^{res}(N_i, N_j, p, T) - \mu_{0i,V}^{res}(N_{0i}, p, T)\right) \\ &+ \ln\left\langle\frac{N_i + N_j}{V(N_i, N_j, p, T)}\right\rangle - \ln\left\langle\frac{N_{0i}}{V_{0i}(N_{0i}, p, T)}\right\rangle. \end{aligned} \tag{7.161}$$

The terms in the ensemble average brackets on the right side represent the mean values for the molar density of the mixture ϱ_m and the molar density $\varrho_{m,0i}$ of the pure compound i at p, T. With this, Eq. 7.161 can be rewritten to derive the activity coefficient from

$$\gamma_i(T, p) = \exp\left[\beta\left(\mu_{i,V}^{res}(N_i, N_j, p, T) - \mu_{0i,V}^{res}(N_{0i}, p, T)\right)\right] \cdot \frac{\varrho_m}{\varrho_{m,0i}}. \tag{7.162}$$

Palusch and Maginn [56] also derived a relation for the activity coefficient in terms of the residual chemical potential at the same volume as the ideal gas. Though they expressed the contribution from the pure fluid in terms of the fugacity $f_{0i} = p \cdot \varphi_{0i}$. Their expression for the activity coefficient can be deduced by using Eq. 7.156 to express the fugacity coefficient in the mixture $\ln(\varphi_i(p, T, x_i))$ and extracting the pressure

$$
\begin{aligned}
ln\gamma_i(T,p) &= \beta\mu_{i,V}^{res}(N_i,N_j,p,T) + \ln\left\langle\frac{k_B T(N_i+N_j)}{V(N_i,N_j,p,T)}\right\rangle \\
&\quad - \ln p - \ln\left(\varphi_{0i}(p,T)\right) \\
&= \beta\mu_{i,V}^{res}(N_i,N_j,p,T) + \ln\left\langle\frac{k_B T(N_i+N_j)}{V(N_i,N_j,p,T)}\right\rangle \\
&\quad - \ln\left(f_{0i}(p,T)\right). \qquad (7.163)
\end{aligned}
$$

The activity coefficient at infinite dilution γ_i^∞ provides valuable information for mixtures for which only limited data are available, and it is an important property to derive binary parameters for mixtures that are only partially miscible [73]. From molecular simulations, γ_i^∞ can be computed when $\Delta G_{sim,i} = \mu_{i,V}^{res}(N_i, N_j, p, T)$ is determined for the case that only one single molecule of component i ($N_i = 1$) is introduced in pure fluid j with $N_j \gg N_i$. Modifying Eq. 7.163 accordingly then yields [56]

$$
ln\gamma_i^\infty(T,p) = \beta\Delta G_{sim}^\infty(N_i=1,N_j) + \ln\left\langle\frac{k_B T(1+N_j)}{V(N_j,p,T)}\right\rangle - \ln f_{0i}. \qquad (7.164)
$$

When approximating $(1 + N_j) \approx N_j$, the term $N_j/V(N_j, p, T)$ gives the molar density of the pure compound j, $\varrho_{m,0j}(T, p)$. With this, the relation between the activity at infinite dilution and the simulated free energy difference ΔG_{sim}^∞ can be further combined to

$$
\Delta G_{sim}^\infty(N_i=1,N_j) = \ln\left(\frac{\gamma_i^\infty f_{0i}}{k_B T\,\varrho_{m,0j}}\right) \quad \text{at} \quad p, T. \qquad (7.165)
$$

This corresponds to the expression employed by Chang in [13].

7.4.8.1 Example: Activity Coefficients of R-32 in the R-32 + R-1234*yf* Mixture

In order to determine activity coefficients in the mixture R-32 + R-1234*yf*, we have performed free energy simulations in the *NpT* ensemble, using the molecular dynamics package GROMACS 5.0.4. [2], and employing our all-atoms force fields for R-32 [64] and R-1234*yf* [69]. The simulations on the free energy change for the coupling of an R-32 molecule in R-32 + R-1234*yf* mixtures were performed for different compositions ($x_{R-32} = 0.2, 0.4, 0.5, 0.6, 0.8$) and for the pure compound R-32, all at $T = 273$ K and $p = 1$ MPa. The systems consisted of $N = 400$ molecules for studies on the mixture, and $N = 600$ molecules for the pure compound R-32. 21 intermediate states were used to determined the free energy change for the transition from full coupled ($\lambda = 1$) to decoupled ($\lambda = 0$). For each composition studied, we first performed a short trial simulations (0.5 ns) with a equidistant λ spacing. Based on the analysis of these trial simulations, we optimized the λ spacings for each composition by imposing equal standard deviations of the partial free energies $\Delta G_{\lambda_i \rightarrow \lambda_{i+1}}$ as optimization criterion [47]. The systems with the optimized λ distributions were then equilibrated for 0.5 ns, followed by a production run of 2.5 ns. We then employed the pymbar package [77] to derive the change in the Gibbs free energies due to the insertion of an R-32 molecule by the MBAR method (see Sect. 3.5.3). From these free energies changes, we determined the activity coefficients of R-32 from Eq. 7.160.

Figure 7.6 shows the simulation results for the activity coefficients of R-32 in comparison with calculations by REFPROP. The REFPROP calculations are based on the equation of state (EOS) by Akasaka [3] for this mixture, and the activity coefficients were obtained from calculated fugacity coefficients using Eq. 7.158. For medium and higher R-32 concentrations, the predicted activity coefficients derived from the free energy simulations are in good agreement with the REFPROP calculations. Though at $x_{R-32} = 0.2$, the activity coefficient from molecular simulation considerably overestimates the EOS result. The same observations, i.e. increasing deviations in the predicted activity coefficients at low concentrations, were also made by Hempel et al. [24], for instance in water + methanol mixtures. They attributed this to inaccuracies in the force field models, and to the simple mixing rule used to determine unlike Lennard-Jones interactions. In their studies on activity coefficients in dilute aqueous solutions, Lazaridis and Paulaitis [36] stated that the prediction of γ_i^∞ by free energy simulations is sensitive to small variations in the partial charges. The large impact of the partial charges, i.e. the charge computing methods, on free energies of solvation was, for instance, also discussed by Jämbeck et al. [26] or Oostenbrink et al. [55]. Oostenbrink et al. observed that it is impossible to reproduce experimental bulk liquid properties and free energies of hydration in cyclohexane systems with the same set of partial charges. This is comprehensible as the fixed partial charges are usually determined to reflect the averaged polarization of the molecules in the bulk phase of the pure compound (see Sect. 6.1.1.2). The polarization of a solute molecule in a low concentrated solution, i.e. when it is mainly surrounded by solvent molecules, will in most cases be quite different due to the

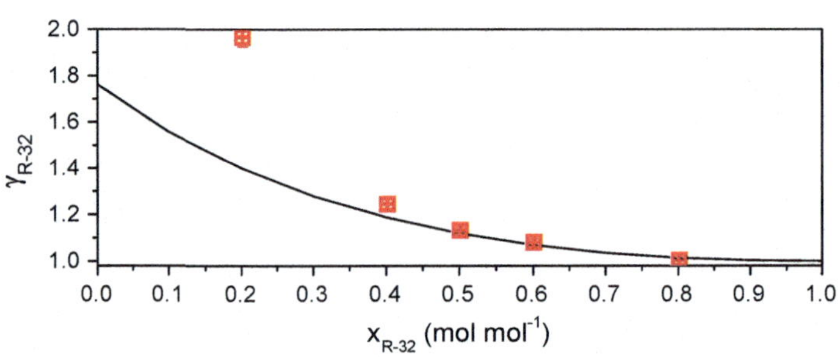

Fig. 7.6 Simulation results for the activity coefficient of R-32 in mixtures with R-1234*yf* at 273 K and 1 MPa (*red squares*) in comparison with calculations by REFPROP (*black line*) based on the EOS by Akasaka [3]

polarization effect of the solvent. Therefore, accurate predictions of free energies of solvation, i.e. activity coefficients at low concentrations, require different approaches to derive partial charges than simulation studies in bulk phases. The simultaneous reproduction of bulk properties and free energies at low concentrations might demand polarizable force fields (see Sect. 6.2). The simple approach to incorporate geometric polarization effects by using fully flexible molecular models (see Sect. 6.4.2) seems to be insufficient for the high sensitive free energy simulations at low concentrations. This will be subject of our future work.

7.4.9 Gibbs Duhem Equation

A phase at internal equilibrium consisting of N_c components is characterized by the N_c+2 intensive variables, i.e. the μ_i, p and T. The Gibbs Duhem equation represents an important relation in Thermodynamics as it states that the intensive variables are related. In its molar form (index m), the Gibbs Duhem equation is given by

$$S_m dT - V_m dp + \sum_{i=1}^{N_c} x_i d\mu_i = 0. \tag{7.166}$$

This means, that from the N_c+2 intensive variables, only N_c+1 are independent, and the last variable is fixed to obey the Gibbs Duhem relation. With this, the number of degrees of freedom of the system is reduced by one.

In the case of constant pressure ($dp = 0$) and constant temperature ($dT = 0$), the Gibbs Duhem equation reduces to a relation between the chemical potentials of the components in the mixture

$$\sum_{i=1}^{N_c} x_i d\mu_i = 0 \qquad p, T = const. \tag{7.167}$$

As the chemical potential can be expressed in terms of activity coefficients (see Sect. 7.4.8) or fugacity coefficients (see Sect. 7.4.7), the Gibbs Duhem equation also provides relations for these properties given by

$$\begin{aligned} \sum_{i=1}^{N_c} x_i d\ln\varphi_i &= 0 \qquad p, T = const. \\ \sum_{i=1}^{N_c} x_i d\ln\gamma_i &= 0 \qquad p, T = const. \end{aligned} \tag{7.168}$$

When simulation results for the chemical potentials, fugacity or activity coefficients are derived for all components of the mixture, these relations can be employed to check the thermodynamic consistency of the simulated properties.

In binary mixtures though, we can take advantage of the Gibbs Duhem relation to derive the chemical potential, activity or fugacity coefficient of one component from the known course of the values of the counter-compound. For a binary mixture with $x_2 = 1 - x_1$, the Gibbs Duhem equation for the activity coefficient becomes

$$x_1 d \ln \gamma_1 + (1 - x_1) d \ln \gamma_2 = 0. \tag{7.169}$$

The activity coefficients in binary mixtures are in general functions of the temperature, the pressure and the composition given by x_1. Thus, for $T, p = const.$, their total differential is given by

$$d \ln \gamma_1 = \frac{\partial \ln \gamma_1}{\partial x_1} dx_1, \quad d \ln \gamma_2 = \frac{\partial \ln \gamma_2}{\partial x_1} dx_1. \tag{7.170}$$

These expressions are introduced into Eq. 7.169, and the the resulting equation is integrated from $x_1 = 0$ (i.e. $x_2 = 1$ and $\ln \gamma_2 = 0$) up to an arbitrary composition x_1^*. This yields

$$\ln \gamma_2(x_1^*) = - \int\limits_{x_1=0}^{x_1^*} \frac{x_1}{1 - x_1} \left(\frac{\partial \ln \gamma_1}{\partial x_1} \right) dx_1. \tag{7.171}$$

Hempel et al. [24] for instance derived activity coefficients in aqueous solutions of alcohols and polymers by only determining the water activity coefficient directly from molecular simulations, whereas the counter-component activity coefficients were calculated by Gibbs Duhem integration.

7.5 Excess and Partial Molar Properties

Excess properties A^E are defined as difference of a property A in a real mixture and the property of an ideal mixture at the same pressure and temperature. For properties that are not related to the entropy, the ideal mixture property is derived as sum of the properties of the pure compounds

$$A^E = A - \sum_i^{N_c} A_{0i}. \tag{7.172}$$

For the entropy and the Gibbs and Helmholtz free energy, an additional mixing term has to be considered to account for the irreversibility of the mixing process. In the molar form, Eq. 7.172 yields

$$A_m^E = A_m - \sum_i^{N_c} x_i A_{m,0i} \quad \text{for} \quad A = U, H, V, C_p, C_v \text{ etc.} \tag{7.173}$$

Thus, excess properties can be determined by molecular simulation studies on both the mixture and the pure compounds, and evaluating Eq. 7.173. Though, as stated in Sect. 7.4.8, excess properties are related to the activity coefficient. This is due to the relation of the activity coefficient and the excess molar Gibbs free energy given by

$$G_m^E(T, p, x_i) = RT \sum_i^{N_c} x_i \ln \gamma_i(T, p, x_i). \tag{7.174}$$

As $G^E(T, p, x_i)$ represents a thermodynamic potential (see Sect. 2.2), all excess properties can be derived from $G_m^E(T, p, x_i)$ through algebraic transformations and derivations. Based on Eq. 7.174, all excess properties can be deduced from a known gradients of the activity coefficients, for instance

$$V_m^E(T, p, x_i) = RT \sum_i^{N_c} x_i \left(\frac{\partial \ln \gamma_i}{\partial p} \right)_{x,T} \tag{7.175}$$

$$H_m^E(T, p, x_i) = -RT^2 \sum_i^{N_c} x_i \left(\frac{\partial \ln \gamma_i}{\partial T} \right)_{x,p} . \tag{7.176}$$

Wierzchowski and Kofke [92] for instance made use of the relation between the activity coefficients and excess enthalpies to assess the thermodynamic consistency of their simulation results. An alternative approach to describe properties of real mixtures is through the concept of partial molar properties (index pm) that reflect the molar qualities of the components in a real mixture. Instead of accounting for real mixing effects by adding an excess property to the mixture property A_m, the correction is allocated to the single components within the mixture to account for the fact that their properties within a real mixture deviate from its value as pure compound. A partial molar property of a component in a mixture is defined by

$$A_{pm,i} = \left(\frac{\partial A_m}{\partial x_i} \right)_{T,p,x_j} , \tag{7.177}$$

whereby the molar property of the mixture A_m can be determined from

$$A_m = \sum_i^{N_c} x_i A_{pm,i} . \tag{7.178}$$

Partial molar properties can not be determined directly by experiments, as they only represent a quality within a mixture. They can only be derived by investigating the change of the mixture property A with composition at $p, T = const.$ For a binary mixture, the partial molar properties of the components "1" and "2" are given by

$$A_{pm,2} = A_m - x_1 \left(\frac{\partial A_m}{\partial x_1} \right)_{T,p} \tag{7.179}$$

$$A_{pm,1} = A_m + (1 - x_1) \left(\frac{\partial A_m}{\partial x_1} \right)_{T,p} . \tag{7.180}$$

Thus, partial molar properties can also be derived from molecular simulations by performing studies on the mixture property A_m at varying compositions to enable the evaluation of Eq. 7.179.

7.6 Fluid Phase Equilibria Properties

This section covers the determination of properties in fluid phase equilibria. This comprise mainly the analysis of Gibbs Ensemble, Gibbs Duhem or Histogram reweighting simulation studies for the vapor-liquid equilibria (VLE) properties of pure compounds and mixtures—including the estimation of critical properties of pure compounds. The Sect. 7.6.3 then addresses the description of gas solubilities in liquid by Henry's Law. Phase equilibria involving solid phases though are outside the scope of this section.

7.6.1 VLE of Pure Components

Fluid phase equilibria properties are usually determined by Monte Carlo simulations, for instance by Gibbs Ensemble (GEMC) simulations, the Gibbs-Duhem-integration or the Histogram Reweighting method as described in Sect. 3.6.

GEMC simulations of vapor-liquid phase equilibria (VLE) of pure components (See Sect. 3.6.1.1) are conducted in the *NVT*-ensemble, in which the global volume of both phases (subsystems), the total number of particles and the temperature is imposed. The volumes of the simulation boxes representing both phases, V' and V'', and the number of particles in the boxes, N' and N'' however will fluctuate. Thus, the molar saturated densities of the coexisting phases can be evaluated from the ensemble average of the instantaneous molar densities, sampled at regular intervals

$$\varrho'_m = \left\langle \frac{N'}{V'} \right\rangle, \quad \varrho''_m = \left\langle \frac{N''}{V''} \right\rangle . \tag{7.181}$$

The Gibbs Duhem-integration (GDI, see Sect. 3.6.2) does not require particle transfer, so the number of particles in both boxes are constant, whereas their volumes fluctuate. In the Histogram Reweighting techniques (Sect. 3.6.3) though, the volume is imposed and the number of particles will fluctuate. Thus, the ensemble averages for the saturated densities for these two simulation techniques are derived from similar expressions.

Although the Gibbs Ensemble method does not require the calculation of the pressures in both phases to evaluate the criteria of phase equilibria, comparing the results for the vapor pressure from both phases can prove if the equilibrium condition is fulfilled. The vapor pressure can be calculated from the virial expression Eq. 7.9 as described in Sect. 7.1, for both phases independently, though the pressure in the liquid phase is generally subject to large fluctuations. In the Gibbs-Duhem integration, the vapor pressure is computed as results of the numerical integration of the Clausius-Clapeyron differential equation. In the Histogram Reweighting technique, the vapor pressure is calculated using its relation to the grand canonical partition function

$$\beta pV = \ln Q_{\mu VT}(\mu, \beta) + C. \tag{7.182}$$

The additive constant C is derived from simulations in the ideal gas state i.e. at low densities, where $\beta pV = N$. Thus, the constant C is determined as intercept of the plot of $\ln Q_{\mu VT}$ versus N.

Finally, the thermodynamic criteria for phase coexistence demand for the equality of the chemical potential in both phases. As described in Sect. 7.4.4, the residual chemical potentials can be derived by inserting a test particle and determining the energy that is experienced by this particle due to its interactions with the molecules of the systems according to Eq. 7.107. Thus, in Gibbs Ensemble simulations, the chemical potentials are in general calculated within the particle transfer step. This allows for an independent verification that equilibrium has been attained. In the Gibbs-Duhem method, the thermodynamic integration along the saturation line ensures that the pressure changes with temperature in such a way that the conditions for phase equilibrium is satisfied. Thus, there is no particle exchange between the two phases so that the chemical potentials are usually not determined. In the Histogram Reweighting technique, the equality of the chemical potential in both phases is fulfilled by definition, as μ is an imposed variable in the simulations in the grand canonical ensemble.

Another property of interest in simulation studies on the VLE of pure compounds is the heat of vaporization that is the difference between the enthalpies of the vapor and the liquid phase at a given temperature

$$\Delta H_{m,vap} = H''_m - H'_m = U''_m - U'_m + p_s(V''_m - V'_m). \tag{7.183}$$

The internal energy U comprise both, the potential and the kinetic energy. However, MC simulations only provide the residual, configurational part of energies, whereas the kinetic, ideal gas energy can not be computed. Due to the thermal equilibrium between the phase, it can be assumed that the kinetic contribution is the same in both phases. Thus, the difference between the internal energies of both phases $\Delta U_{m,vap} = U''_m - U'_m$ can be calculated from the ensemble averages of their configuration energies, and the heat of vaporization can be derived from

$$\langle \Delta H_{m,vap} \rangle = \langle U''_{conf} \rangle - \langle U'_{conf} \rangle + \langle p_s \rangle (\langle V''_m \rangle - \langle V'_m \rangle) \tag{7.184}$$

$$= \langle U''_{conf} \rangle - \langle U'_{conf} \rangle + \langle p_s \rangle \left(\frac{1}{\langle \varrho''_m \rangle} - \frac{1}{\langle \varrho'_m \rangle} \right). \tag{7.185}$$

Assuming that the molar volume of the vapor phase is much larger than the volume of the liquid phase, and that the vapor phase obeys the ideal gas law, Eq. 7.184 becomes

$$\langle \Delta H_{m,vap} \rangle = \langle U''_{conf} \rangle - \langle U'_{conf} \rangle + RT. \tag{7.186}$$

This expression can also be used in MD studies to estimate the heat of vaporization. There, the ensemble average of the configurational energy $\langle U'_{conf} \rangle$ is determined during the course of a liquid phase simulation. The ensemble average of the configurational energy in the gas phase is simulated in the *NVT* ensemble in terms of a single molecule at the same temperature in a large enough simulation box.

7.6.1.1 Determination of the Critical Point

As discussed in Sect. 7.6.1, approaching the critical point by GEMC or GDI simulations is in general not possible due to the decreasing free energy barrier between the two phases, With this, it becomes difficult to maintain the identity of the simulation boxes, which shall each represent one of the two equilibrium phases. The critical temperature is therefore usually estimated by fitting the simulation results at sub-critical conditions to the scaling law

$$\varrho' - \varrho'' = A\,(T - T_c)^{\beta}\,. \tag{7.187}$$

The critical density is then derived by a fit to the law of rectilinear diameters

$$\frac{\varrho' + \varrho''}{2} = \varrho_c + B\,(T - T_c)\,. \tag{7.188}$$

The parameters A and B determined by the fit are dependent of the compound, and β is the critical exponent, for which a value of $\beta = 0.325$ (Ising exponent) is often used. The critical pressure can be estimated by extrapolating the vapor pressure curve using the Clausius-Clapeyron equation.

The Histogram Reweighting technique though is suitable for simulations in the near critical region, but here finite-size effects become relevant. At the critical point, the correlation length that measures the extent of density fluctuations diverges, and exceeds the size of the simulation box L. With this, the finite-size of the simulation systems suppresses the long-range fluctuation, and the critical properties derived from simulation $T_c(L)$, $\varrho_c(L)$ will deviate from the properties obtained for a system of infinite size $T_c(\infty)$, $\varrho_c(\infty)$. Finite-size scaling analysis can then be applied to correct for the size dependency of the thermodynamic properties. The analysis of collected histograms to obtain estimates for the critical parameters is in general based on the mixed-field size-scaling method by Wilding and Bruce [93]. Therein the ordering parameter M of the scaling field is a linear combination of number of particles N in the system and its configurational energy U_{conf} [58]

$$M \propto N - s_1 U_{conf} \tag{7.189}$$

with s_1 being the field mixing parameter. At the critical point, the normalized probability distribution of M has a universal form $\widetilde{p}_m^*(x)$. The principle of critical point universality states that the critical behavior of fluids is in the same universality class as the Ising model. Therefore, the critical point can be found by adjusting μ and s_1 such that the probability $P(N, U_{conf}, T, \mu)$ (see Sect. 3.6.3) matches the universal fixed-point ordering operator distribution $\widetilde{p}_m^*(x)$ of the 3D-Ising model. The deviation between the simulated critical properties $T_c(L)$, $\varrho_c(L)$ are expected to deviate from the values $T_c(\infty)$, $\varrho_c(\infty)$ of the infinite-system by

$$T_c(L) - T_c(\infty) \propto L^{-(\theta+1)/\nu} \tag{7.190}$$

$$\varrho_c(L) - \varrho_c(\infty) \propto L^{-(1-\alpha)/\nu}. \tag{7.191}$$

Therein, θ, α and ν are the exponents of the correction-to-scaling, heat capacity divergence, and correlation length, respectively. Approximate values of theses exponents for the 3D Ising universality class are $\theta = 0.54$, $\alpha = 0.11$ and $\nu = 0.629$ [58].

7.6.1.2 Example: VLE Properties of HFO-1216

The hexafluoropropene HFO-1216, also known as HFP, is an important intermediate in the fluorochemical industry. It is for instance oxidated to yield 2,2,3-trifluoro-3-(trifluoromethyl)oxirane, HFPO, an important perfluorinated epoxide [50]. We have employed our force field for hydrofluoroolefines (see Sect. 6.1.1.5) to study the VLE properties of HFO-1216 via GEMC simulations using the simulation code Towhee [46]. The system consisted of 256 molecules, and for each temperature it was equilibrated for 100.000–200.0000 cycles, before we performed production runs of 200.000–400.000 cycles. More simulation details are given in [63]. Figure 7.7 shows the GEMC simulation results for the vapor pressure and saturated densities in comparison with experimental data from Coquelet et al. [15] and calculations based on REFPROP [38]. It should be noted that no LJ parameters of our force field for hydrofluoroolefines have been adjusted to optimize agreement with experimental data

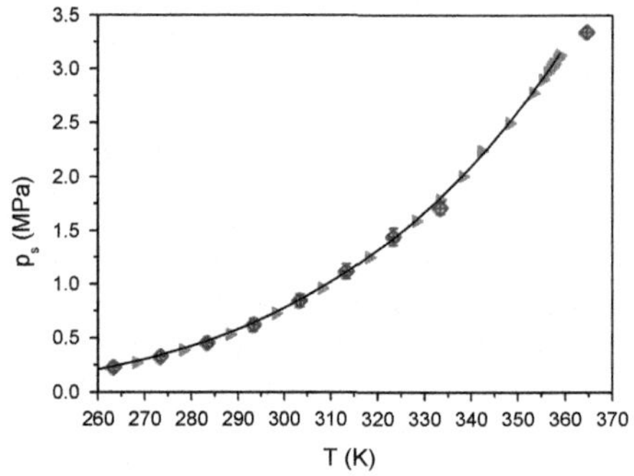

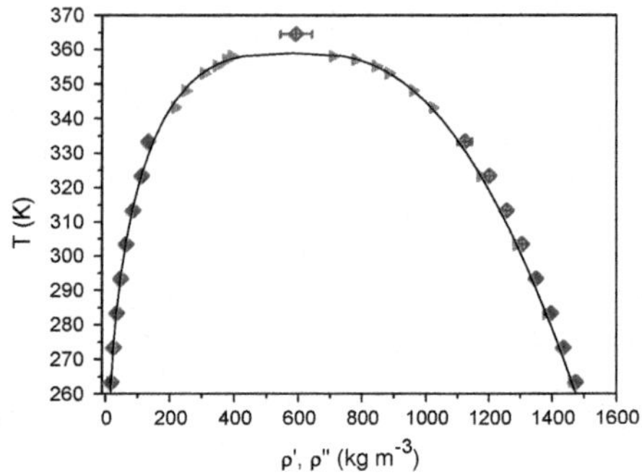

Fig. 7.7 GEMC simulation results (*blue rhombi*) for the vapor pressure p_s and the saturated densities ϱ', ϱ'' of HFO-1216 in comparison with experimental data (*grey triangles*, [15] and calculations by REFPROP (*line*)

Table 7.3 Estimated critical properties of HFO-1216 in comparison with experimental values [15]

	Experiment	Simulation
T_c (K)	358.93	364.3 ± 13.2
ϱ_c (kg m^{-3})	579.03	590 ± 49
p_c(MPa)	3.136	3.24 ± 1.13

for HFO-1216. Thus, our simulation studies for this compound are purely predictive. In spite of this, the agreement between simulation and experiment is quite satisfying.

By using the Towhee routine fitcoex [46], our simulation results for the saturation densities in the temperature range $T = (263 - 333)$ K were than fitted to the scaling law Eq. 7.187 and law of rectilinear diameters Eq. 7.188 to yield estimates for the critical temperature T_c and density ϱ_c of HFO-1216. The extrapolation of the fitted vapor pressure curve up to T_c gives the critical pressure p_c. In Table 7.3 we provide a comparison of the estimated critical properties of HFO-1216 with experimental data [15].

Due to the limited number of data points used in the fitting procedure, the estimated critical properties though are subject to large uncertainties. This is specially true for the critical pressure. The stated errors account for both uncertainties of the simulation results as input in the fitting procedure, and uncertainties of the fitting itself. According to the Eq. 7.184 we have also determined the heats of vaporization of HFO-1216, for which no experimental information exist. As illustrated by Fig. 7.8, our simulated $\Delta H_{m,vap}(T)$ tend to overestimate the predictions by REFPROP. Though the simulation results deviate by less than 1 kJ mol^{-1} from the calculations, which is still a good agreement for this property.

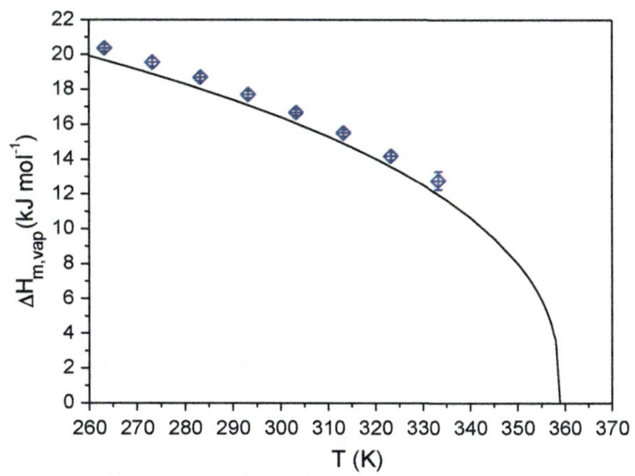

Fig. 7.8 GEMC simulation results (*blue rhombi*) for the molar heats of vaporization $\Delta H_{m,vap}$ of HFO-1216 in comparison with calculations from REFPROP (*line*)

7.6.2 *Phase Equilibria of Mixtures*

As described in Sects. 3.6.1.2 and 3.6.1.3, the Gibbs Ensemble simulation technique can be employed for simulation studies on multiphase equilibria of mixtures, whereby it is most commonly used for simulations on vapor-liquid equilibria, but it can also be applied for studies on liquid-liquid equilibria or VLLE. GEMC simulations on phase equilibria in mixtures are performed in the *NpT*-Ensemble in which the total number of particles, the equilibrium pressure and temperature are imposed. The total system volume now can fluctuate, and this also means that the volumes of the subsystems are allowed to change independently. The molar saturation densities of the coexisting phases can again be derived from the ensemble average of the instantaneous densities, calculated from the number of particles in the box divided by the volume as given by Eq. 7.181. Additionally, the partial molar density of a component i can be determined from its number of particles N_i in the box, divided by the volume of the phase. In case of studies on the VLE of mixtures, the partial molar densities of the different components i in the liquid and vapor phase are therefore given by

$$\varrho'_{m,i} = \left\langle \frac{N'_i}{V'} \right\rangle, \qquad \varrho''_{m,i} = \left\langle \frac{N''_i}{V''} \right\rangle, \tag{7.192}$$

The molar compositions of the phases are derived from

$$x'_i = \left\langle \frac{N'_i}{N'} \right\rangle, \quad x''_i = \left\langle \frac{N''_i}{N''} \right\rangle, \quad \text{with } N' = \sum_i N'_i, \; N'' = \sum_i N''_i. \tag{7.193}$$

The (partial) molar densities and compositions of the equilibrium phases of an LLE or VLLE are derived accordingly.

The residual chemical potentials of different components can be evaluated during the course of the simulation from the experienced interaction energy of an inserted particle of this compound. Though now, the fluctuation of the volume has to be taken into account (see Sect. 7.4.4, Eq. 7.114)

$$\mu_{i,NpT}^{res} = -k_B T \ln\left(\frac{\langle V exp\left(-\beta \Delta U_{conf,N_i+1}\right)\rangle}{\langle V\rangle}\right). \qquad (7.194)$$

7.6.2.1 Example: VLE in the Mixture CO_2 + R-32

In [64] we have proposed a new fully flexible all atoms force field model for R-32 difluoromethane to allow for studies on its mixtures with 2,3,3,3-tetrafluoropropenes R-1234yf and trans-1,3,3,3-tetrafluoropropenes R-1234$ze(E)$ (see Sect. 8.4.2). Here we present GEMC simulation results for the mixture of R-32 with carbon dioxide (CO_2, R-744), which represents a subsystem of some refrigerant blends that are currently proposed (see Sect. 8.5). For studies on this mixture, we have employed the TraPPE model by Potoff and Siepmann [62] for CO_2. No LJ parameters for the unlike interactions were adjusted, so the molecular simulation studies are purely predictive. The system consisted in the sum of $N = 400$ molecules, but the number of molecules of each component varied with the state point studied. For each state point, defined by the given temperature and pressure, the systems were equilibrated for 200.000 cycles, and production runs covered 300.000–400.000 cycles. The predictions from GEMC simulations for the compositions of the vapor and liquid phase and for the saturated densities of the 273 K and 293 K isotherms are presented in Fig. 7.9. Also shown are calculations using REFPROP and experimental data by Rivollet et al. [74].

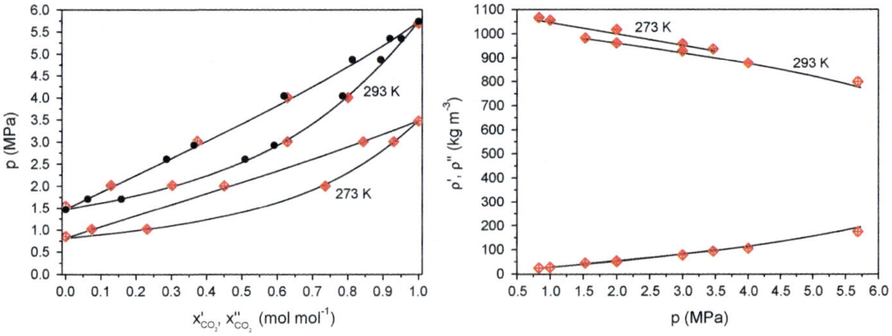

Fig. 7.9 GEMC simulation results (*red rhomb*) for the compositions and densities of the liquid and vapor phase of the VLE in the mixture CO_2 + R-32 at 273 and 293 K in comparison with REFPROP calculations (*line*). Also shown as *black dots* are experimental VLE data by Rivollet et al. [74]

The good agreement of our simulation results with both, REFPROP calculations and experimental data well illustrates the predictive capability of the molecular simulation studies.

7.6.3 Gas Solubilities

Gases in general only exhibit low solubility in liquids. In the limit of infinite solution, the Henry's law states that the molar fraction of the gas i in the liquid phase (of component j) is proportional to its partial pressure p_i in the gas phase

$$x_i' \cdot H_{i(j)} = p_i. \tag{7.195}$$

The proportionality factor is the Henry coefficient with the index marking the Henry coefficient of component i solved in component j. Following Henry's law, the gas solubility increases with its increasing partial pressure in the gas phase and decreasing Henry coefficient. The temperature dependence of the gas solubility is solely covered by the temperature dependence of the Henry coefficient, which is strong and nonlinear [21]. The Henry coefficient is defined by

$$H_{i(j)} = \lim_{x_i' \to 0} \frac{f_i}{x_i'}. \tag{7.196}$$

With the fugacity being defined by $f_i = p \cdot x_i \cdot \varphi_i$ (see Sect. 7.4.7), the Henry coefficient is determined by the limit of the fugacity coefficient at infinite solubility, denoted by φ_i^∞. As described for the activity at infinite dilution in Sect. 7.4.8, the limit of infinite solution can be represented in molecular simulations by only introducing one single molecule of component i in the pure fluid j with $N_j \gg N_i$. When the Eq. 7.154 is modified accordingly by setting $N_i = 1$, an expression for φ_i^∞ can be deduced

$$\ln\left(\varphi_i^\infty(p, T, x_i' \to 0)\right) = \beta \Delta G_{sim}^\infty(N_i = 1, N_j) - \ln\left\langle \frac{V(N_j, p, T) \cdot \beta p}{1 + N_j} \right\rangle. \tag{7.197}$$

Again, the term $(1 + N_j)/V(N_j, p, T)$ can be approximated by the molar density of the liquid j, $\varrho_{m,j}(p, T)$. This yields

$$\varphi_i^\infty(p, T, x_i' \to 0) = \exp\left(\beta \Delta G_{sim}^\infty\right) \varrho_{m,j} k_B T / p. \tag{7.198}$$

With this, the Henry coefficient can be derived from

$$H_{i(j)} = p \cdot \varphi_i^\infty(p, T, x_i' \to 0) = \exp\left(\beta \Delta G_{sim}^\infty\right) \varrho_{m,j} k_B T. \tag{7.199}$$

This illustrates that the Henry coefficient is directly related to the free energy of solvation.

7.7 Dielectric Properties

Dielectric properties are important in studies on polar or ionic compounds and allow conclusions regarding their solvent behavior, i.e. solubility for solvents and solvation dynamics. Dielectric properties of interest are for instance the static dielectric constant of a compound, its dipole moment, and the Debye relaxation for dipole fluctuations. The dipole moment of a molecule is derived from the positions and partial charges of its N_a atoms relative to the center of mass (COM) of the molecule by

$$\boldsymbol{\mu} = \sum_{i=1}^{N_a} q_i(\mathbf{r}_i - \mathbf{r}_{COM}), \qquad \mu = |\boldsymbol{\mu}| \tag{7.200}$$

In a liquid sample as collective of N molecular dipoles, a system's total dipole moment M can be computed from

$$\mathbf{M} = \sum_{j=1}^{N} \boldsymbol{\mu}_j. \tag{7.201}$$

An important quantity to characterize polar liquids is its dielectric or Debye relaxation for fluctuations of the dipoles. A normalized autocorrelation function can be formulated to describe the relaxation of the system's total dipole moment [95]

$$\Phi_{MM}(t) = \frac{\langle M(0)M(t)\rangle - \langle M\rangle^2}{\langle M^2(0)\rangle - \langle M\rangle^2}. \tag{7.202}$$

Assuming that $\Phi_{MM}(t)$ decays exponentially (Debye dielectric), a relaxation time τ_D can be determined from the time integral of the autocorrelation function

$$\tau_D = \int_0^\infty \Phi_{MM}(t)dt, \qquad \Phi_{MM}(t) = A\exp(-t/\tau_D). \tag{7.203}$$

This dielectric relaxation time τ_D might differ from the relaxation time τ_s of a single particle

$$\tau_s = \int_0^\infty \Phi_{\mu\mu}(t)dt, \qquad \text{with} \qquad \Phi_{\mu\mu}(t) = \frac{\langle \mu(0)\mu(t)\rangle - \langle \mu\rangle^2}{\langle \mu^2(0)\rangle - \langle \mu\rangle^2}. \tag{7.204}$$

This deviation is due to collective effects, i.e. the correlation of neighboring dipoles [85] in a liquid sample. Kirkwood [29] introduced a correction factor to account for the orientational correlation between neighboring dipolar molecules. For a finite system of N molecules, the Kirkwood factor G_K is given by [95]

$$G_K = \frac{\langle M^2 \rangle - \langle M \rangle^2}{N \langle \mu^2 \rangle}. \tag{7.205}$$

It is related to the infinite system Kirkwood factor g_K by [95]

$$g_K = \frac{2\varepsilon_r + 1}{3\varepsilon_r} G_K. \tag{7.206}$$

with the static dielectric constant ε_r. The Kirkwood factor relates the dielectric relaxation time of a liquid sample to that of a single molecule [51, 85]

$$g_K = \frac{\tau_D}{\tau_s}. \tag{7.207}$$

It is close to unity for weakly polar liquids, but deviates significantly from unity for hydrogen-bonding compounds such as water. Therefore, $g_K > 1$ indicates a strong orientational correlation between the molecular dipoles in the system. The static dielectric constant ε_r in Eq. 7.206 can be derived from the fluctuation of the systems's total dipole system by

$$\varepsilon_r = 1 + \frac{4\pi}{3Vk_BT}\left(\langle M^2 \rangle - \langle M \rangle^2\right) = 1 + \frac{4\pi\varrho\langle\mu^2\rangle}{3k_BT} G_K. \tag{7.208}$$

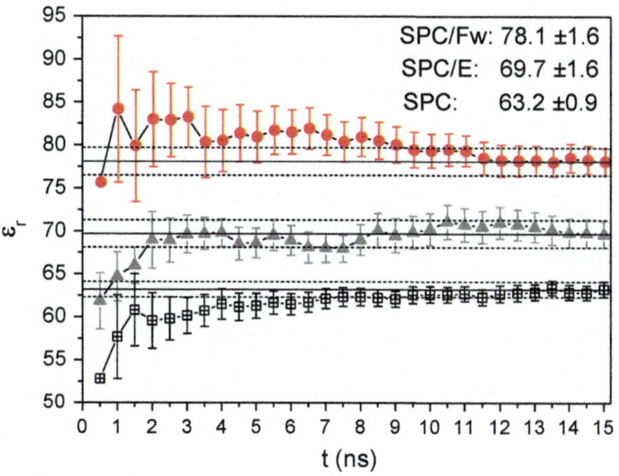

Fig. 7.10 Ensemble average of the dielectric constant of water at 298.15 K and 0.1 MPa, determined every 0.5 ns in a MD simulation run of 15 ns. Results for the SPC/Fw, SPC/E and the SPC model (see Sect. 6.4.2)

However, this quantity converges slowly, and simulation runs of several nanoseconds might be necessary until it reaches a stable value. This is exemplarily shown by the depiction of the running ensemble averages of the dielectric constant of water from different models in Fig. 7.10.

7.8 Structural Analysis

Apart from the ability to predict thermophysical properties, molecular simulation studies also provide insights into the behavior of the compounds on the molecular level, and allow for the direct analysis of the relationship between the microscopic structure and macroscopic properties. This section provides an introduction to analysis techniques for microstructural properties.

7.8.1 Distribution Functions

The structure of a fluid can basically be characterized by a set of radial pair distribution functions $g_{ij}(r)$ for two particles i and j in the fluid. The radial distribution function (RDF)

$$g_{ij}(r) = \frac{V}{N^2} \sum_i \sum_{j \neq i} \delta(\mathbf{r} - \mathbf{r}_{ij}) \tag{7.209}$$

yields the probability of finding the site j at distance r apart from the site i, relative to the expected probability for a uniform distribution at the same density [4]. In molecular simulation studies, $g_{ij}(r)$ is determined by using a histogram of discretized pair separation, in which the number of particles $N(b)$ within an interval $b = [r, r + \delta r]$ is related to the number of particles for a uniform distribution $N^*(b)$

$$g_{ij}(r + \frac{\delta r}{2}) = \left\langle \frac{N(b)}{N^*(b)} \right\rangle. \tag{7.210}$$

The volume of the corresponding spherical shell of thickness δr at a distance r is given by

$$V(b) = \frac{4}{3}\pi \left[(r + \delta r)^3 - r^3 \right] \approx 4\pi r^2 \delta r. \tag{7.211}$$

Thus, the average number of particles $N^*(b)$ in this spherical shell is

$$N^*(b) = \varrho V(b), \tag{7.212}$$

where ϱ is the number density in the bulk phase.

In a system with polyatomic molecules, the RDFs may either be determined for the center-of-masses of the molecules, or for characteristic sites. In this case, information on the orientational structure and intermolecular bonding with respect to specific domains of the molecules can be inferred from the combination of different RDFs.

In an ideal gas, the distribution of particles is random and $g_{ij}(r) = 1$. For real fluids $g_{ij}(r)$ tends towards 1 for larger distances, but at shorter distances, $g_{ij}(r)$ will differ from unity due to attractive ($g_{ij}(r) > 1$) or repulsive ($g_{ij}(r) < 1$) interactions. Thus, the analysis of RDFs might evidence preferred interactions between particles or sites, such as hydrogen bonding interactions (see Sect. 7.8.3)

The first peak of the RDF represents the first coordination shell and gives the most likely distance of the two sites i and j. Hence, the organization of the liquid can also be characterized by determining the coordination number for this first solvation shell. The coordination number N_{ij} of a site j around the site i is calculated by integrating $g_{ij}(r)$ from zero to the location of the first minimum

$$N_{ij}(r_{min}) = 4\pi\varrho_j \int_0^{r_{min}} g_{ij}(r)r^2 dr. \tag{7.213}$$

N_{ij} gives the average number of sites j within the sphere of radius r_{min} around a central site i. It can be used for instance to determine the ratio of solvent molecules around a solute molecule, or around one of its characteristic sites. However, it should be noted that the coordination number is sensitive to the choice of the integration limit r_{min}.

In fluids showing structural organization, another property may be used to quantitatively assess the characteristic length scale of this organization. This is the static structure factor $S(k)$ that is derived from the corresponding radial distribution function $g(r)$ by Fourier transformation [72]

$$S(k) = 1 + 4\pi\varrho \int_0^{\infty} \frac{\sin(kr)}{kr} g(r)r^2 dr. \tag{7.214}$$

The static structure factor may also be determined experimentally by neutron or X-ray scattering, allowing for a direct comparison of simulation results for the fluid microstructure with experiment.

Radial distribution functions in general only provide time averaged information on the local organization around a given site. For time dependent RDFs, an additional time dimension can be introduced to yield van Hove correlation functions (see Sect. 7.2.2.2).

In order to obtain an immediate image of three-dimensional density distributions, spatial distribution functions (SDF) are increasingly used in molecular simulation studies on local structures. These SDFs avoid the RDF's spherically averaging, and span both the radial and the angular coordinate. Spatial distribution functions are therefore able to provide details of anisotropies in spatial structures and insight into packing patterns. More details on the computation of SDFs can be found for instance in the contribution 'Spacial Structure in Molecular Liquids' by Kusalik et al. in [33]. SDFs are four-dimensional functions (probability, x, y, z) [9], consequently they are in general visualized by iso-surfaces (see Sect. 7.8.3 for examples).

7.8.2 Internal Structure

When investigating the structural features of systems comprising complex molecules such as polymers, not only the relative distribution of different molecules is of interest, but also the internal structure of the macromolecules. The configuration of a linear chain of N monomers can be measured by the internal mean square end-to-end distance $R(N)$

$$\langle R^2(N)\rangle = \langle(\mathbf{r}_1 - \mathbf{r}_N)^2\rangle. \tag{7.215}$$

For a branched molecule, an end-to-end distance between characteristic sites may be assessed. The size of the molecules consisting of N atoms or monomers can be characterized by the radius of gyration R_G

$$\langle R_G^2\rangle = \left\langle \frac{\sum_{i=1}^{N} m_i(\mathbf{r}_i - \mathbf{r}_{COM})^2}{m_{total}} \right\rangle, \qquad m_{total} = \sum_{i=1}^{N} m_i \tag{7.216}$$

where $\mathbf{r}_{COM}$ is the position of the center-of-mass of the molecules

$$\mathbf{r}_{COM} = \frac{\sum_{i=1}^{N} m_i \mathbf{r}_i}{m_{total}}. \tag{7.217}$$

7.8.3 Example: Structural Properties of Ionic Liquids

Is this section we exemplify the ability of molecular simulations to provide insight into structural features by our MD studies on the 1-alkyl-3-methyl-imidazolium-based ionic liquids (IL) $[amim][Cl]$ and $[amim][BF_4]$. In [67] we have analyzed the structural properties of the different ILs by SDFs and RDFs between specific sites [67] to investigate how they are influenced by the anion type, the cation size and temperature. Therefore we have employed an united atoms model from Liu et al. [39] for the 1-ethyl-3-methyl- $[emim^+]$ imidazolium cations, which we have extended for simulations in 1-hexyl-3-methyl-imidazolium $[hmim]$-ILs and combined with parameters of Canongia Lopes et al. [12] for the $[Cl^-]$ anion.

Figure 7.11 shows the spatial distribution of the $[Cl^-]$ and the $[BF_4^-]$ anion around the $[emim^+]$ and $[hmim^+]$ cation at a temperature of 363.15 K. The iso-surfaces of

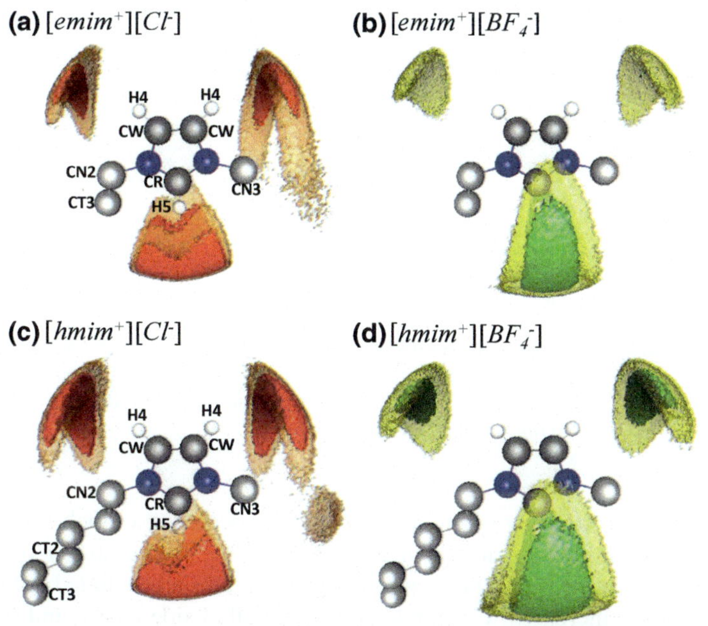

Fig. 7.11 Spatial distribution functions (SDFs) of the anion around the cation at 363.15 K as a function of the anion type and the size of the cation from molecular simulation [67]. The iso-surfaces are drawn at 14 (*green*, *red*) and 8 (*yellow*, *orange*) times of the average density, respectively. Also given is the notation for the atoms and sites for the united atoms model of Liu et al. $CN2$, $CN3$, $CT2$ and $CT3$ represent different CH_2 and CH_3 united atoms

the SDFs are visualized by the software package gOpenMol [34]. The figure also gives the notation for the different atoms and sites according to Liu et al.

For all four ILs three regions with a high probability of finding the anions exist: around the $CR - H5$ group, and between the $CW - H4$ group and the alkyl side chains, though the preferred distribution depends on both the anion type and cation size. The comparison of the SDFs for both $[emim^+]$-ILs at this temperatures shows that the very high density area (14 times the average) exists in all three regions for $[Cl^-]$ anion, but only in the region around CR for $[BF_4^-]$ anion. This can be ascribed to stronger coulombic interactions in the $[amim][Cl]$ ILs, whereas the charge is more dispersed in the $[amim][BF_4]$ ILs that weakens the electrostatic forces. The more dispersed charge and the special hindrances for the larger $[BF_4^-]$ anion also explains the difference in the shape of the high density regions around the $CR - H5$ group between both anions.

Increasing the alkyl side chain from $[emim^+]$ to $[hmim^+]$ has two different effects on the distribution of the anions. At first it can be observed that the high density region around the $CR - H5$ group is more extended towards the methyl side. Second, the highest density areas (14 times the average) now exist in all three regions. This indicates an enriched concentration of the anions in the vicinity of the imidazolium

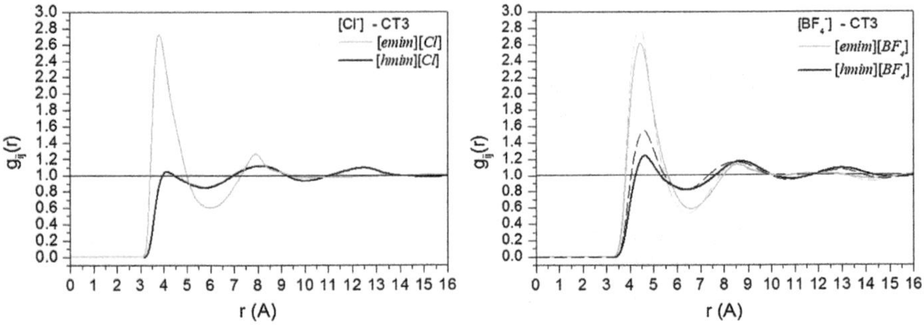

Fig. 7.12 Radial distribution functions (RDFs) for the anion around the end group $CT3$ of the alkyl side chain of the cation for the $[amim][Cl]$ and $[amim][BF_4]$ ILs (amim = emim and hmin) at 363.15 K. For the $[amim][BF_4]$ ILs, the corresponding RDFs were also determined at 298.15 K, depicted by *dash lines*, respectively

ring, whereas the extremity of the alkyl side chain becomes a much less favorable location for the anions the longer it gets. This is due to the ability of the alkyl side chain to freely rotate in space, which then generates an excluded volume. This excluded volume increases with the length of the alkyl side chain, and pushes the anions to distribute around the imidazolium ring.

Figure 7.12 gives the RDFs of the anion of both the $[amim][CL]$ and the $[amim][BF_4]$ IL around the $CT3$ end group of the alkyl chain at 363.15 K. The pronounced reduction of the first peak also well illustrates the decreasing concentration of the anions at the extremity of the alkyl side chain with their increasing lengths. For $[amim][BF_4]$, we additionally provide the corresponding RDFs at 298.15 K (dash lines). They show that the distribution of anions at the extremity of the alkyl side chain also decreases with increasing temperature, which can again be attributed to an increasing excluded volume.

We have further studied the structures of the $[hmim][CL]$ and $[hmim][BF_4]$ ILs in the vicinity of the alkyl side chain by determining the RDFs between its end groups $CT3$ and characteristic sites of the cation, i.e. the $H4$ and $H5$ hydrogens of the imidazolium ring, and the $CN3$ methyl side chain.

The depiction of the RDFs in Fig. 7.13 shows that only the $CT3 - CT3$ RDFs of both ionic liquids exhibit pronounced first maxima. This indicates a tendency of longer alkyl side chains to accumulate in the vicinity of the alkyl side chains of other cations. This phenomenon of alkyl side chains (tails) forming spatially heterogeneous domains is known as tail aggregation, and was first shown in molecular dynamics studies on $[amim][NO_3]$ ILs by Wang and Voth [88, 89], whereas the studies by Triolo [82] provided first experimental evidence for $[amin]$ based ionic liquids. A Special Issue of the Journal of Molecular Liquids summarizes studies on structural organizations in ionic liquids in the recent years [75].

In [66] we have studied mixtures of 1-alkyl-3-methylimidazolium chlorides $[amim][Cl]$ with ethanol and 1-propanol [66] to investigate how the structure of the ionic liquids changes upon mixing. In these simulation studies on the mixtures,

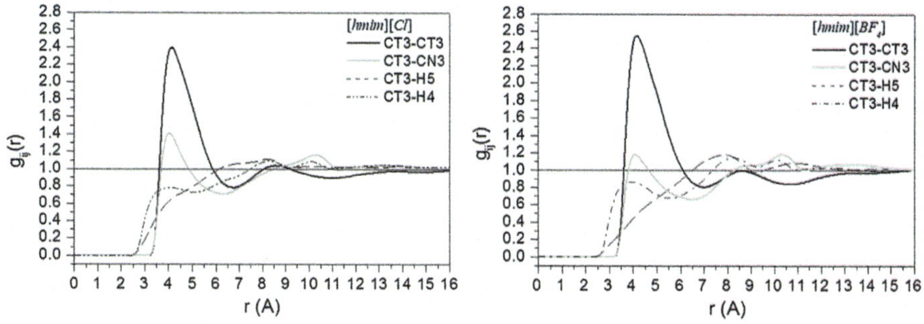

Fig. 7.13 Radial distribution functions (RDFs) for characteristic sites and atoms of the $[hmim^+]$ cation around the end group $CT3$ of the alkyl side chain of the cation for $[hmim][Cl]$ and $[hmim][BF_4]$ at 363.15 K. For $[hmim][BF_4]$, the corresponding RDFs were also determined at 298.15 K, depicted by *grey lines*, respectively

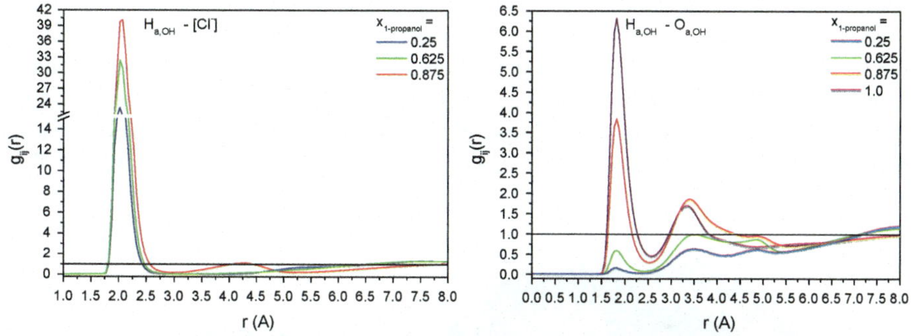

Fig. 7.14 RDFs of the H_{aOH} hydrogen (hydroxyl) of 1-propanol around the $[Cl^-]$ anion of the IL and around the O_{aOH} oxygen (hydroxyl) of other 1-propanol molecules in the mixture 1-propanol + $[hmim][Cl]$ at different mixture compositions. The $H_{a,OH} - [Cl^-]$ RDFs are determined at 298.15 K and the $H_{a,OH} - O_{a,OH}$ RDFs at 363.15 K

we have employed the NERD force field [27] for the alcohols, in which the CH_2 and CH_3 groups are represented as united atoms analogously to the molecular model for the ILs.

Figure 7.14 shows the RDFs of the hydrogen of the alcohol hydroxyl group $H_{a,OH}$ around the chloride anion $[Cl^-]$ of the IL (left), and around the oxygen atom of alcohol (right) in the mixture 1-propanol + $[hmim][Cl]$ at various compositions. The very high and sharp first peaks in the $H_{a,OH} - [Cl^-]$ RDF at about 2 Å clearly evidence the occurrence of hydrogen bonding interactions between the propanol and the anion. Increasing mole fractions of 1-propanol results in increasing peak heights and the establishment of a second solvation shell of alcohol molecules around the chloride anion, shown by the appearance of a second peak at about 4.2 Å for $x_{1-propanol} = 0.875$. The peak heights of the $H_{a,OH} - O_{a,OH}$ RDFs though clearly decrease with increasing mole fraction of the IL. This demonstrates that the propanol molecules prefer to hydrogen-bond to the chloride anion rather than to other alcohol molecules.

As discussed before for the pure ILs, the increasing length of the alkyl side chain causes the anion to increasingly concentrate around the imidazolium ring. This is also observed in their mixtures with alcohols. Due to the strong hydrogen bonding of the alcohol's hydroxyl and the anion, the OH_a group of the alcohol now accumulates in the same regions as the anions. This results in an enhanced concentration of polar groups around the polar domain of the cation. As also the alkyl chains (tails) of the cations and the alcohols tend to aggregate, an enhanced microphase separation into polar and nonpolar domains can be observed in mixtures of alcohol + $[amim][Cl]$-ILs with longer alkyl chains of the cation. This is visualized in Fig. 7.15 by the SDFs of the different sites around the $[hmim^+]$ cation in the mixture 1-propanol + $[hmim][Cl]$.

Increasing temperature even forward this separation, as the enlarged excluded volume at higher temperatures due to the intensified rotation of the alkyl side chain pushes the anions once more towards the imidazolium ring. Thus, the properties of alcohol + [amim][Cl] mixtures are determined by a significant local ordering due to strong coulombic interactions between both the hydroxyls of the alcohols and the chloride, and the polar domain of cation and the anion, and an aggregation of the non-polar groups with increasing cation size on the other side. The high solubility of the

Fig. 7.15 SDFs of the $[Cl^-]$ anion (*orange*), the 1-propanol hydroxyl group (*red*), 1-propanol $CH3a$ group (*black*) and of the cation $CT3$ group (*grey*) around the $[hmim^+]$ cation in the mixture 1-propanol + $[hmim][Cl]$, $x_{1-propanol} = 0.625$ at 298.15 K. *Orange and red surfaces* (polar) represent 10 times the average density, *black and grey ones* (nonpolar) 3 times the average density

alcohol can be attributed to its strong hydrogen bonding with the $[Cl^-]$ anion, but also to the interactions between its alkyl group and the nonpolar side chains of the ionic liquid's cation. These findings illustrate how structural analysis of the system can provide a molecular interpretation and a better understanding of the solvation process.

References

1. Aberg KM, Lyubartsev AP, Jacobsson SP, Laaksonen A (2004) Determination of solvation free energies by adaptive expanded ensemble molecular dynamics. J Chem Phys 120:1770–3776
2. Abraham MJ, Murtola T, Schulz R, Pall S, Smith JC, Hess B, Lindahl E (2015) GROMACS. High performance molecular simulation through multi-level parallelism from laptops to supercomputers. SoftwareX. 1–2:19–25
3. Akasaka R (2013) Thermodynamic property models for the difluoromethane (R-32) + trans-1,3,3,3-tetrafluoropropene (R-1234ze(E)) and difluoromethane + 2,3,3,3-tetrafluoropropene (R-1234yf) mixtures. Fluid Phase Equilib 358:98–104
4. Allen M, Tildesley DJ (1987) Computer simulation of liquids. Oxford Science Publication
5. Beahr HD, Kabelac S (2012) Thermodynamik. Springer 15. Auflage
6. Beutler TC, Mark AE, vanSchaik RC, Gerber PR, vanGunsteren WF (1994) Avoiding singularities and numerical instabilities in free energy calculations based on molecular simulations. Chem Phys Lett 222:529–539
7. Bhargava BL, Balasubramanian S (2005) Dynamics in a room-temperature ionic liquid: a computer simulation study of 1,3-dimethylimidazolium chloride. J Chem Phys 123:144505
8. Bhargava BL, Balasubramanian S, Klein ML (2008) Modelling room temperature ionic liquids. Chem Commun 3339–3351
9. Brehm M, Kirchner B (2011) TRAVIS—a free analyser and visualizer for Monte Carlo and molecular dynamics trajectories. J Chem Inf Model 51:2007–2023
10. Cadena C, Zhao Q, Snurr RQ, Maginn EJ, Theodorou DN (2006) Molecular modeling and experimental studies of the thermodynamic and transport properties of pyridinium-based ionic liquids. J Phys Chem B 110:2821–2832
11. Canongia Lopes JN, Costa Gomes MF, Pádua A (2006) Nonpolar, polar, and associating solute in ionic liquids. Chem Lett D 110:16816–16818
12. Canongia Lopes JN, Deschamps J, Pádua AAH (2004) Modeling ionic liquids using a systematic all-atom force field. J Phys Chem B 108:2038–2047
13. Chang J (2009) The calculation of the chemical potential of organic solutes in dense liquid phases by using expanded ensemble Monte Carlo simulations. J Chem Phys 132:074103
14. Chipot C, Pohorille A (2007) Free energy calculation, theory and application in chemistry and biology. Springer Series in Chemical Physics, vol 86
15. Coquelet C, Ramjugernath D, Madani H, Valtz A, Naidoo P, Meniai AH (2010) Experimental measurement of vapor pressures and densities of pure hexafluoropropylene. J Chem Eng Data 55:2093–2099
16. Daivis PJ, Evans DJ (1994) Comparison of constant-pressure and constant-volume nonequilibrium simulations of sheared model decane. J Chem Phys 100:541–547
17. Del Popolo MG, Voth GA (2004) On the structure and dynamics of ionic liquids. J Phys Chem B 108:1744–1752
18. Deschamps J, Padua AAH (2005) Interactions of gases with ionic liquids: molecular simulation. ACS Symp Series 901:150–158
19. Evans DJ, Morris GP (2007) Statistical mechanics of nonequilibrium liquids. Australian National University (ANU E) Press, 2 edn
20. Frenkel D, Smit B (1996) Understanding molecular simulation. Academic Press
21. Gmehling J, Kolbe B, Kleiber M, Rarey J (2012) Chemical thermodynamics for process simulation. Wiley-VCH, Weinheim

22. Greene G, Cho Y, Hartnett JP, Bor-Cohen A (2006) Advances in heat transfer. Academic Press by Elsevier
23. Hansen JP, McDonalds IR (1975) Statistical mechanics of dense ionized matter IV, Density and charge fluctuation in a simple molten salt. Phys Rev A 11
24. Hempel S, Fischer J, Paschek D, Sadowski G (2012) Activity coefficients of complex molecules by molecular simulations and Gibbs-Duhem-integration. Soft Mater 10:26–41
25. Hoover WG, Evans DJ, Hickman RB, Ladd AJC, Ashurst WT, Moran B (1980) Lennard-Jones triple-point bulk and shear viscosities. Green-Kubo theory, Hamiltonian mechanics, and nonequilibrium molecular dynamics. Phys Rev A 22(4):1690–1697
26. Jämbeck JPM, Mocci F, Lyubartsev AP, Laaksonen A (2013) Partial atomic charges and their impact on the free energy of solvation. J Comput Chem 34:187–197
27. Khare R, Sum AK, Nath SK, dePablo JJ (2004) Simulation of vapor-liquid phase equilibria of primary alcohols and alcohol-alkane mixtures. J Phys Chem B 108:10071–10076
28. Kirkwood JG (1935) Statistical mechanics of fluid mixtures. J Chem Phys 3:300–313
29. Kirkwood JG (1939) The dielectric polarization of polar liquids. J Chem Phys 7:911
30. Klimovich PV, Shirts MR, Mobley DL (2015) Guidelines for the analysis of free energy calculations. J Comput Aided Mol Des
31. Klomfar J, Souckova M, Patek J (2009) Buoyancy density measurements for 1-alkyl-3-methylimidazolium based ionic liquids with tetrafluoroborate anion. Fluid Phase Equilib 282:31–37
32. Kowsari MH, Saman A, Ashrafizaadeh M, Najafi B (2009) Molecular dynamics simulation of imidazolium-based ionic liquids II Transport coefficients. J Chem Phys 130:014703
33. Kusalik PG, Laaksonen A, Svishchev IM (1999) Spacial structure in molecular liquids. In: Balbuena PB, Seminario JM (eds) Molecular dynamics. Elsevier
34. Laaksonen L (1992) A graphics program for the analysis and display of molecular dynamics trajectories. J Mol Graph 10:33
35. Lagache M, Ungerer P, Boutin A, Fuchs AH (2001) Prediction of thermodynamic derivative properties of fluids by Monte Carlo simulation. Phys Chem Chem Phys 3:4333–4339
36. Lazaridis T, Paulaitis ME (1993) Activity coefficients in dilute aqueous solutions from free energy simulations. AIChE J 39:1051–1060
37. Leach A (1996) Molecular modelling: principles and applications, 1st edn. Prentice Hall
38. Lemmon EW, Huber ML, McLinden MO (2013) NIST Standard reference database 23: reference fluid thermodynamic and transport properties—REFPROP, Version 9.1, Standard Reference Data Program. National Institute of Standards and Technology: Gaithersburg, MD
39. Liu Z, Wu X, Wang W (2006) Modeling ionic liquids using a systematic all-atoms force field. Phys Chem Chem Phys 8:1096–1104
40. Lopes JNC, Deschamps J, Padua AAH (2005) Modeling ionic liquids of the 1-alkyl-3-methylimidazolium family using an all-atom force field. ACS Symp Series 901:134–149
41. Lustig R (1994) Statistical thermodynamics in the classical molecular dynamics ensemble. I. Fundamentals. J Chem Phys 100:3048–3059
42. Lyubartsev AP, Laaksonen A (1998) Parallel molecular dynamics simulations of biomolecular systems. Lect Notes Comput Sci 1541:296–303
43. Lyubartsev AP, Laaksonen A, Vorontsov-Volyaminov PN (1996) Determination of the free energy from chemical potential: application of the expanded ensemble method. Mol Simul 18:43–58
44. Lyubartsev AP, Martsinovskii AA, Shekunpv SV, Vorontsov-Volyaminov PN (1992) New approach to Monte Carlo calculation of the free energy; method of expanded ensemble. J Chem Phys 96:1176
45. Maginn EJ (2007) Atomistic simulation of the thermodynamic and transport properties of ionic liquids. Acc Chem Res 40:1200–1207
46. Martin MG (2013) MCCCS Towhee: a tool for Monte Carlo molecular simulation. Mol Simulat 39:1212–1222
47. Mecklenfeld A, Raabe G (2015) Predicting the solubility of drug candidates. In: Symposium on pharmaceutical engineering research. SphERe, Braunschweig, Germany

48. Meier K, Kabelac S (2006) Pressure derivatives in the classical molecular-dynamics ensemble. J Chem Phys 124:064104
49. Naden LN, Pham TT, Shirts MR (2014) Linear basis function approach to efficient alchemical free energy calculations: 1. Removal of uncharged atomic sites. J Chem Theo Comput 10:1128–1149
50. Nelson WM, Subramoney SC, Valtz A, Richon D, Naidoo P, Ramjugernath D (2011) Vapor-liquid equilibrium data for binary systems consisting of either hexafluoropropene (HFP) or 2,3,3-trifluoro-3-(trifluoromethyl)oxirane (HFPO) with carbon dioxide (R-744) or 2,2-dichloro-1,1,1-trifluoroethane (R-123). J Chem Eng Data 56:74–78
51. Neumann M (1986) Dielectric relaxation in water. Computer simulations with TIP4P potential. J Chem Phys 85(3):1567–1580
52. Ochterski JW (2000) Thermochemistry in Gaussian
53. Onsager L (1931) Reciprocal relations in irreversible processes I. Phys Rev 37:405
54. Onsager L (1931) Reciprocal relations in irreversible processes II. Phys Rev 38:2265
55. Oostenbrink C, Villa A, Mark AE, Gunsteren W (2004) A biomolecular force field based on free enthalpy of hydration and solvation. The GROMOS force-field parameter sets 53a5 and 53a6. J Comput Chem 25:1656–1676
56. Paluch AS, Maginn EJ (2013) Predicting the solubility of solid phenanthrene: a combined molecular simulation and group contribution approach. AIChE J 59:2647–2661
57. Paluch AS, Shah JK, Maginn EJ (2011) Efficient solvation free energy calculation of amino acid analogs by expanded ensemble. Mol Simul 7:1394–1403
58. Panagiotopoulos AZ (2000) Monte Carlo methods for phase equilibria of fluids. J Phys: Condens Matter 12(3):R25
59. Petravic J (2005) Thermal conductivity of ethanol. J Chem Phys 123:174503
60. Pham TT, Shirts MR (2011) Identifying low variance pathways for the free energy calculations of molecular transformations in solution phase. J Chem Phys 135:034114
61. Pham TT, Shirts MR (2012) Optimal pairwise and non-pairwise alchemical pathways for free energy calculations of molecular transformation in solution phase. J Chem Phys 136:124120
62. Potoff JJ, Siepmann JI (2001) Vapor-liquid equilibria of mixtures containing alkanes, carbon dioxide, and nitrogen. AIChE J 47:1676–1682
63. Raabe G (2012) Molecular modeling of fluoropropene refrigerants. J Phys Chem B 116:5744–5751
64. Raabe G (2013) Molecular simulation studies on the vapor-liquid-equilibria of binary mixtures of R-1234yf and R-1234ze(E) with R-32 and CO_2. J Chem Eng Data 58:1867–1873
65. Raabe G (2015) Molecular simulation studies on the vapor-liquid-equilibria of the cis- and trans-HCFO-1233zd and the cis- and trans-HFO-1336mzz. J Chem Eng Data 60:2412–2419
66. Raabe G, Köhler J (2008) Thermodynamical and structural properties of binary mixtures of imidazolium chloride ionic liquids and alcohols from molecular simulation. J Chem Phys 129:144503
67. Raabe G, Köhler J (2008) Thermodynamical and structural properties of imidazolium based ionic liquids from molecular simulations. J Chem Phys 128:154509
68. Raabe G, Maginn EJ (2010) A force field for 3,3,3-Fluoro-1-propenes, including HFO-1234yf. J Phys Chem B 114:10133–10142
69. Raabe G, Maginn EJ (2010) Molecular modeling of the vapor-liquid equilibrium properties of the alternative refrigerant 2,3,3,3-tetrafluoro-1-propene (HFO-1234yf). J Phys Chem Lett 1:93–96
70. Raabe G, Sadus RJ (2012) Molecular dynamics simulation of the effect of bond flexibility on the transport properties of water. J Chem Phys 127:104701
71. Rahman A (1964) Correlations in the motion of atoms in liquid argon. Phys Rev 136(2A):405–411
72. Rapaport DC (2004) The art of molecular dynamics simulation. Cambridge University Press
73. Reid RC, Prausnitz JM, Poling BE (1987) The properties of gases and liquids, 4th edn. McGraw-Hill Book Company

74. Rivollet F, Chapoy A, Coquelet C, Richon D (2004) Vapor-liquid equilibrium data for the carbon dioxide (CO_2) + difluoromethane (R32) system at temperatures from 283.12 to 343.25 K and pressures up to 7.46 MPa. Fluid Phase Equilib 218:95–101
75. Russina O, Schroer W, Triolo A (2015) Mesoscopic structural and dynamic organization in ionic liquids. J Mol Liq 210:161–163
76. Shing KS, Chung ST (1987) Computer simulation methods for the calculation of solubility in supercritical extraction systems. J Phys Chem 91:1674–1681
77. Shirts MR, Chodera JD A phyton implementation of the multistate Bennett Acceptance Ratio (MBAR)
78. Shirts MR, Mobley DL, Chodera JD (2007) Alchemical free energy calculations: ready for prime time? In: Annual reports in computational chemistry, vol 3, pp 41–59. Elsevier
79. Shirts MR, Pitera JW, Swope WC, Pande VS (2003) Extremely precise free energy calculations of amino acid side chain analogs: comparison and common molecular mechanics force fields for proteins. J Chem Phys 119:5740–5760
80. Smith TR, Forester TR, Todorov IT DL_POLY Classic User Manual, http://www.ccp5.ac.uk./DL_POLY_CLASSIC. Daresbury Laboratory, UK
81. Steinbrecher T, Joung I, Case DA (2011) Soft-core potential in thermodynamic integration comparing one- and two-step transformations. J Comput Chem 32:3253–3263
82. Triolo A, Russina O, Bleif HJ, Di Cola E (2007) Nanoscale segregation in room temperature ionic liquids. J Phys Chem B 111:4641–4644
83. Ungerer P, Nieto-Draghi C, Rousseau B, Ahunbay G, Lachet V (2007) Molecular simulation of the thermophysical properties of fluids: from understanding toward quantitative predictions. J Mol Liq 134:71–89
84. Van Hove L (1954) Time-dependent correlations between spin and neutron scattering in ferromagnetic crystals. Phys Rev 95(6):1374–1384
85. Volmari A, Weingärtner H (2002) Cross terms and Kirkwood factors in dielectric relaxation of pure liquids. J Mol Liq 98–99:293–301
86. Wang F, Landau DP (2001) Efficient, multiple-range random walk algorithm to calculate the density of states. Phys Rev Lett 86(10):2050–2053
87. Wang Y, Izvekov S, Yan T, Voth GA (2006) Multiscale coarse-graining of ionic liquids. J Phys Chem B 110:3564–3575
88. Wang Y, Voth GA (2005) Unique spatial heterogeneity in ionic liquids. J Am Chem Soc 127:12192–12193
89. Wang Y, Voth GA (2006) Tail aggregation and domain diffusion in ionic liquids. J Phys Chem B 110:18601–18608
90. Weigand B, Köhler J, vonWolfersdorf J (2008) Thermodynamik kompakt. Springer
91. Widom B (1963) Some topics in the theory of liquids. J Chem Phys 39:2308
92. Wierzchowski SJ, Kofke DA (2004) Liquid phase activity coefficients for saturated HF/H_2O mixtures with vapor-phase non-idealisties described by molecular simulations. Ind Eng Chem Res 43:218–227
93. Wilding NB, Bruce AD (1992) Density fluctuations and field mixing in the critical fluid. J Phys Condens Matter 4:3087–3108
94. Williams G (1964) Time-correlation functions and molecular motion. J Chem Phys 40(9):2527–2533
95. Wu Y, Tepper HL, Voth GA (2006) Flexible simple point-charge water model with improved liquid-state properties. J Chem Phys 124:024503
96. Yang L, Ahmed A, Sandler SI (2013) Comparison of two simulation methods to compute solvation free energies and partition coefficients. J Comput Chem 34:284–293
97. Zwanzig RW (1954) High-temperature equation of state by a perturbation method. I Nonpolar gases. J Chem Phys 22:1420–1426
98. Zwanzig RW (1964) Elementary derivation of time-correlation formulas for transport coefficients. J Chem Phys 40(9):2527–2533

Chapter 8
Applications of Molecular Simulations to Studies on Working Fluids

In the Great Soviet Encyclopedia [59] a working fluid is defined as "a gaseous or liquid substance by means of which an energy conversion is accomplished to obtain mechanical work (engine), cold (in refrigerating machines) or heat (heat pump)". Thus, this chapter covers molecular simulation studies on fluids that operate thermodynamic cycles to convert heat into work (Organic Rankine cycle), or serve as heat carrier in heating, ventilation, air-conditioning and refrigeration (HVAC&R) applications, whereby working fluids of HVAC&R cycles are generally referred to as refrigerants. The introduction of this chapter first provides a short overview of the historical development as well as of current regulations and trends that affect the progression of working fluids today. Air-conditioning and refrigeration have become essential in our society and economy, not least due to the need to refrigerate food or pharmaceuticals. Most thermodynamic cycles in HVAC&R applications are based on a vapor compression cycle that was introduced in mid 19th century. Since then there have been changing selection criteria for suitable refrigerants as working fluids in these cycles, so that the progression of refrigerants has now arrived in the 4th generation. The first generation of refrigerants in the 'whatever worked' period [10] comprised air, water, ammonia, carbon and sulphur dioxide, hydrocarbons (HC) and ethers. The increasing commercialization of refrigeration then entailed the demand for safer, more durable and efficient working fluids, resulting in the introduction of synthetic chlorofluorocarbons (CFC) and hydrochlorofluorocarbons (HCFC) as second generation of working fluids. From the 1930s onwards, vapor-compression systems using this second generation of refrigerants became the dominant refrigeration technology. The most widely used refrigerants then were dichlorodifluoromethane R-12 with its main application in automotive air-conditioning systems, and chlorodifluoromethane R-22 for residential air-conditioning and industrial refrigeration and process chillers. When it was found out that the chlorine of CFC and HCFC refrigerants reacts with the atmospheric ozone, these fluids got under phase-out schedules, following the Montreal Protocol from 1987 [82].

G. Raabe, *Molecular Simulation Studies on Thermophysical Properties*,
Molecular Modeling and Simulation, DOI 10.1007/978-981-10-3545-6_8

Due to their zero ozone depletion potential (ODP), hydrofluorocarbons (HFC) were then introduced as alternative refrigerants, representing the third generation of working fluids. First and foremost, tetrafluoroethane R-134a as replacement for R-12 in mobile air-conditioning systems (MAC) and also for medium to high-temperature refrigeration became the dominant refrigerant. With its lower chlorine content resulting in a smaller ODP value, R-22 has as longer phase-out period than R-12, and was therefore also regarded as interim alternative for it. However, countries of the European Union accelerated the phase out of R-22 with early regulations [5]. Thus, HFC refrigerant blends such as R-410A and R-407C are today used as replacement for R-22; R-410A is a binary, near azeotropic mixture of difluoromethane R-32 and pentafluoroethane R-125, while R-407C is a ternary zeotropic mixture of R-32, R-125 and R-134a. More details on the phase out-regulations and alternative refrigerants are, for instance, provided in the Bitzer Refrigerant Report [5], in the review by Calm [10], or in [80]. The hydrofluorocarbon refrigerants where later recognized as potent green house gases (GHG), i.e., they exhibit a significant global warming potential (GWP) due to their long atmospheric lifetimes. The atmospheric lifetime of the widely used R-134a, for instance, is 14 years. Its GWP(100-yr) value is 1300, which means that integrated over the specific time interval of 100 years, the radiative forcing from the release of 1 kg R-134a is by a factor of 1300 higher than that of 1 kg CO_2 as reference gas ($GWP_{CO_2} = 1$). The GWP(100-yr) value is the most commonly used metric for the climate impact of green house gas emission, and its definition is attributed to the Intergovernmental Panel on Climate Change (IPCC) that also provides accepted values of GWPs for different compounds [30]. Due to climate change concerns, the Kyoto Protocol sets binding targets for developed countries to limit their emission of greenhouse gas (GHG), which are based on these GWP values. Thus, phase-out regulations for high-GWP working fluids became effective, such as the F-gas regulation (No 517/2014, repealing No 842/2006) of the European Union that aims to reduce the EU's emission of fluorinated gases (HFCs, perfluorocarbons (PFCs) and sulphur hexafluoride (SF6)) by 2030 to 2/3 compared with levels from 2014. This results, for example, in a phase-out of HFC refrigerants with GWP > 150 in domestic refrigerators and freezers by 2015, or in commercial refrigeration by 2022. The EU has recently submitted an amendment proposal for the global phase-down of HFCs under the Montreal Protocol [81]. The phase-out of HFC refrigerants necessitates the adoption of alternative refrigerants with low GWP values as the forth generation of working fluids. This has also redirected attention to natural refrigerants of the first generation, such as ammonia (R-717), carbon dioxide (R-744) or hydrocarbons like propane (R-290). Ammonia, for example, is a common refrigerant in industrial refrigerant applications, whereas carbon dioxide is used in commercial refrigeration (supermarket) and water heating heat pumps, and is being tested in transport refrigeration. Moreover, carbon dioxide is proposed as alternative refrigerant in MAC systems. An overview of technical options for the use of these natural refrigerants is given in the UNEP '2014 Report of the Refrigeration, air-conditioning and Heat Pumps Technical Options Committee' [80].

Refrigerant manufacturers have responded to the projected ban of HFC compounds with the introduction of unsaturated, fluorinated compounds as alternative refrigerants, such as 2,3,3,3-tetrafluoropropene R-1234yf or trans-1,3,3, 3-tetrafluoropropene R-1234$ze(E)$. In the literature,these compounds are mostly named hydrofluoroolefines (HFO) to distinguish them from the saturated hydrofluorocarbons (HFC). This terminology is also used here. However, in the UNEP report [80], for instance, the prefix 'unsaturated HFC' is used for these compounds, as they are still composed of hydrogen, fluorine and carbons atoms as the saturated molecules. HFO compounds are chemicals that have at least one double bond between two carbon atoms. Due to the high reactivity of the carbon double bond, these compounds exhibit a short atmospheric lifetime resulting in a low GWP value. However, this in turn comes along with flammability, as the quick breakup in atmosphere also means that the substance breaks down just as easy when exposed to an ignition source [80]. Thus, blends of HFOs and non-flammable HFC compounds are proposed to yield refrigerants with reduced flammability compared to the pure HFOs, and lower GWPs than the pure HFCs.

Concerns on climate change and raising energy prices have also resulted in a growing interest in Organic Rankine Cycle (ORC) systems, which therefore represent another application for working fluids apart from the 'common' use in the vapor-compression cycles of HVAC&R systems. ORC systems are installed to generate electricity from low-temperature heat sources such as geothermic or solar energy and industrial waste heat. ORC power systems require working fluids with higher critical temperatures (up to 500 K) than those used in refrigeration applications (T_c = (300–400) K [46]). The thermophysical properties of working fluids required to match low-grade ORC applications are thereby similar to those for high-temperature heat pumps, which have also experienced growing interest for industrial waste heat recovery. Depending on the specific application, the selection of a suitable refrigerant or working fluid has to meet manifold criteria that also often represent tradeoffs and need to be balanced out [80]. Factors that have to be considered are:

- GWP and ODP values complying with legal requirements;
- safety (flammability, toxicity, operating pressure)
- performance (energy efficiency and capacity);
- costs (working fluid, equipment production and maintenance);
- stability (chemical degradation);
- compatibility (materials, lubricants);
- easy handling in service;
- recyclability, end-of-life issues.

Understanding, optimizing and controlling the performance of new working fluids in different applications requires a detailed knowledge of their thermodynamic and transport properties. Thereby molecular simulations are more and more applied as complement to experiment. A number of simulation studies and force fields are available in the literature for HFC refrigerants, for either fluoromethanes [19, 22, 58, 74] or fluoroethanes [19, 34, 43, 53, 74]. Stoll et al. [74], for instance, published force fields to study the vapor-liquid equilibria of low molecular weight hydrofluorocarbons that comprise several fluorinated derivatives of methane, ethane, but also ethene [11]. Kelkar and Maginn [34] have developed a transferable force field for

C1 and saturated C2 hydrofluorocarbons to quantitatively reproduce their vapor-liquid coexistence curves, vapor pressures and enthalpies of vaporization. Recently, Yang et al. [86] introduced a molecular model for 1,1-difluoroethane R-152a, which represents a low GWP, albeit flammable HFC refrigerant. In Sect. 6.1.1.5, we presented a transferable force field for HFO and HCFO compounds that enables reliable predictions for the thermophysical properties of this new class of working fluids.

In this chapter, we will present our molecular simulation results for different HFO and HCFO compounds, and also for refrigerant blends in the context of their potential technical applications, such as mobile air-conditioning, refrigeration, ORC and heat pump applications. This exemplifies the value of molecular simulations to provide predictions for poorly known compounds to allow for their evaluation in technical applications. When possible, the simulation results are compared with calculation or predictions from REFPROP, the REFerence fluid PROPerties program, developed by the National Institute of Standards and Technology (NIST) [41]. REFPROP is widely used in the refrigeration industry as it enables the calculation of thermophysical and transport properties for a large number of working fluids and mixtures [41]. In some cases, our molecular simulation results have already found their way into system simulation studies on the thermodynamic cycles, which are then mentioned in the respective text passage. However, a detailed description of the system simulations using the Modelica library TIL [69] is out of the scope of this book. More information on this subject is provided in the cited references.

8.1 Alternative Refrigerants for Mobile Air-Conditioning (MAC) Systems

Figure 8.1 shows a basic refrigerant cycle of an air-conditioning system based on the principals of a vapor-compression cycle. The refrigerant is evaporated on a low temperature and pressure level T_0, $p_s(T_0)$ and thereby absorbs heat from the vehicle interior. The (saturated) vapor is then compressed, and with the increase in pressure to the so called high-pressure, its temperature also increases to T. The hot vapor then flows through the condenser where it is condensed by rejecting heat to the ambient. Afterwards, the refrigerant passes the expansion device (valve) where its pressure is reduced to $p_s(T_0)$. It then flows to the evaporator to complete the cycle.

According to studies of the World Resources Institute, the transport section is responsible for about 22 % of the world wide (energy related) GHG emission—with increasing rate [40]. Again 10 % of this emission can be attributed to the use of mobile air-conditioning (MAC). The indirect GHG emission of MAC systems results from the additional fuel consumption to operate the system. The direct GHG emission is due to refrigerant leakage.

For the mitigation of GHG emissions, the European Union (EU) adopted in 2006 the Directive 2006/40/EC [52] that bans the use of refrigerants with GWP values higher than 150 from MAC systems. Thus, the phase-out of the presently used

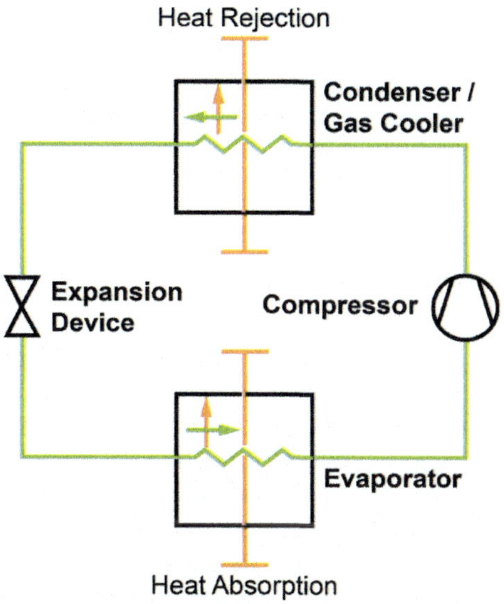

Fig. 8.1 Basic vapor-compression cycle of an air-conditioning system

tetrafluoroethane R-134*a* (GWP = 1430 [30]) has necessitated the introduction of alternative refrigerants that comply with the legal requirements of the EU and regulations placed by other countries. Possible candidates as MAC refrigerants, and the current status in the development of their systems are for, instance, summarized in the UNEP report [80].

German car manufacturers have decided to develop alternative MAC systems using CO_2 as refrigerant (R-744). It was shown that providing an appropriate system design and control, R-744 gives comparable cooling performance to R-134*a* systems [80]. Additionally, the process can be reversed so that the R-744 MAC system has good heat pump capabilities. Furthermore, the thermophysical properties of CO_2 are well-known [71], although they are quite different from those of R-134*a*. The low critical temperature of CO_2 ($T_c = 304.2\,\mathrm{K}$) means that at most conditions, the heat transfer to the ambient has to pass the supercritical region, yielding a so-called transcritical cycle. As the heat rejection occurs above the critical pressure, the high-pressure is no longer defined by the saturation pressure $p_s(T)$ but by the CO_2 charge [45]. In addition, the pressure level of the CO_2-transcritical cycle is significantly higher than that of the R-134*a* cycle. These special features of the CO_2 cycle entail the need for an appropriate redesign of the system with optimized heat exchangers, compressors with compatible lubricants, sealing, high-pressure controlling etc. Therefore, refrigerant suppliers have proposed alternative refrigerants that allow a near drop-in replacement of R-134*a*, i.e., without the need to change major components of the MAC system. Currently, 2,3,3,3-tetrafluoropropene R-1234*yf* [72] seems to be the "front runner of the race" [84] of potential MAC refrigerant candidates, and some car manufacturers have started to launch new car models with R-1234*yf* MAC systems. However, also blends are discussed as alternative

MAC refrigerants, and one blend that has attracted some attention is R-445A, which is a ternary zeotropic blend of CO_2, R-134a and trans-1,3,3,3-tetrafluoropropene (R-1234$ze(E)$) [54]. In the following subsections we will present and discuss our molecular simulation results for the thermophysical properties of both R-1234yf and R-445A.

8.1.1 R-1234yf

The HFO-1234yf (R-1234yf) was first introduced as component in an azeotropic blend with iodotrifluoromethane—referred to as "Fluid H" [73] that was proposed by the refrigerant suppliers Honeywell and DuPont in a joint collaboration as alternative low GWP MAC refrigerant. Later, R-1234yf alone was considered as suitable low GWP refrigerant (GWP = 4) for air-conditioning systems [72]. At this time, only limited information was available on its thermophysical properties. This originally motivated the development of the transferable force field for fluorinated propenes ([66, 67], see Sect. 6.1.1.5), in order to allow for predictions of these properties from molecular simulations to complement experimental data. As described in Sect. 6.1.1.5, the parameters for the intramolecular terms and the partial charges of the molecular model were derived from ab initio simulations, whereas the LJ parameters were adjusted to fine-tune agreement with the then available experimental data for the phase equilibria properties of different fluorinated compounds (see Appendix A).

Over the years, more experimental studies on the thermophysical properties of R-1234yf were published, and in 2011, Richter et al. presented an equation of state for this compound, which is now available in REFPROP [68]. Figure 8.2 shows results

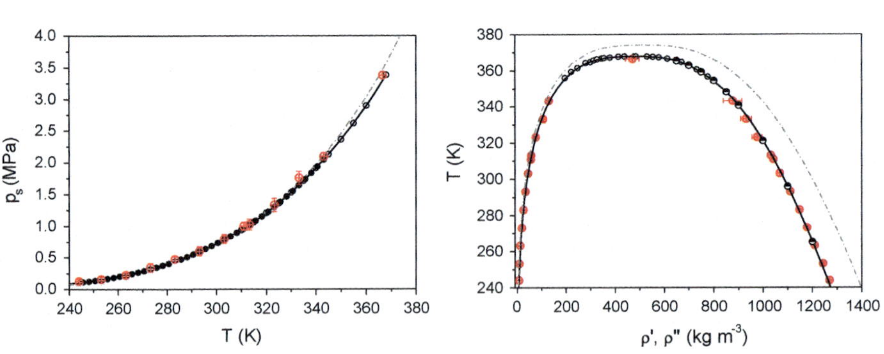

Fig. 8.2 GEMC simulation results (*red crossed circles*) for the vapor pressure and saturated densities of R-1234yf in comparison with experimental data by Tanaka et al. (*black circle* [75]), Fedele et al. (*black dot* [16]), Hulse et al. (*half filled black circle* [27]) and calculations by the REFPROP equation of state [41, 68] (*black line*). Also shown as *grey chain lines* are the vapor pressure curve and vapor-liquid coexistence curve of R-134a, calculated by its reference equation [79] in REFPROP

Table 8.1 Predicted critical properties and normal boiling point for HFO-1234yf in comparison with experimental data

	sim	exp	ref
$T_c\,(K)$	366.4 ± 9.5	367.85 ± 0.01	[75]
$p_c\,(MPa)$	3.376 ± 0.73	3.382 ± 0.003	[75]
$\varrho_c\,(kg\,m^{-3})$	469.8 ± 26	478 ± 3	[75]
$T_b\,(K)$	243.3 ± 3.5	243.7	[41, 68]

from our GEMC simulation studies on the vapor pressure and saturated densities of R-1234yf in comparison with experimental data [16, 27, 75] and calculations based on this EOS. Also shown as dot-dash line are REFPROP calculations for the vapor pressure curve and vapor-liquid coexistence curve (VLCC) of R-134a for comparison. Figure 8.2 illustrates that over the entire coexistence range, our GEMC simulation results for the vapor pressures and saturated densities of R-1234yf agree with the experimental data and EOS calculations within the error bars of the simulations. The good reproduction of both the vapor pressure curve and the VLCC also results in an excellent estimate of the critical point and the normal boiling point of R-1234yf as shown in Table 8.1.

We also performed MD simulation studies to derive liquid phase properties of R-1234yf in the temperature range from (263–313) K and at pressure up to 3.5 MPa [64, 66]. In Sect. 7.3.6 we already presented MD results for thermal expansivity α_P of R-1234yf. Now we will discuss its liquid densities and viscosities. The densities were determined from MD simulations in the *NpT*-ensemble, and we then performed additional simulations in the *NVT*-ensemble at these averaged densities to derive the viscosities from the Green-Kubo formula as described in Sect. 7.2.2.1 and in [64, 66]. Figure 8.3 shows exemplarily our MD simulation results for the liquid density of R-1234yf at different temperatures and $p = 2$ MPa, and for liquid viscosities at $p = 1$ MPa in comparison with experimental data [17, 35, 48] and calculations [41] for both R-1234yf and also R-134a.

Again, the simulated liquid densities are in very good agreement with both experiment and calculations by the REFPROP EOS for R-1234yf. Our simulation results using the Green Kubo formula for the viscosities of R-1234yf are predictive as no data for transport properties have been used to optimize the parameters of the force field. However, the simulated viscosities are also well confirmed by the experimental data by Meng et al. [48]. In REFPROP, the viscosities of R-1234yf are presently only described by an 'Extended Corresponding State (ECS)' approach with limited accuracy. Thus, the equation is increasingly overestimating the liquid viscosities of R-1234yf with decreasing temperatures. This is consistent with the findings by Cousins and Laesecke [13] who state that the ECS model describes a systematically different temperature dependence. The authors presented very accurate viscosity measurements for both the R-1234yf and R-1234$ze(E)$, but these data are

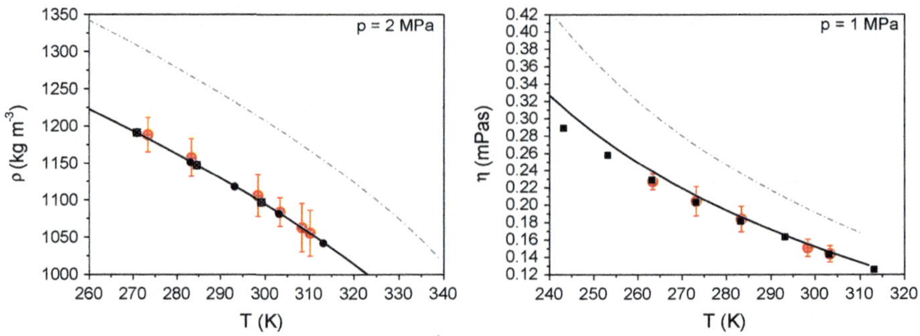

Fig. 8.3 MD simulation results (*red crossed circles*) for the liquid densities of R-1234*yf* at 2 MPa in comparison with experimental data by Komflar et al. (*black crossed circle* [35]) and Fedele et al. (*black dot* [17]) (*left*), and for the liquid viscosities at 1 MPa in comparison with experimental data by Meng et al. (*black squares* [48]) (*right*). Also shown are the calculated courses by REFPROP [41] for R-1234*yf* (*black line*) in comparison with the corresponding curves of R-134*a* (*grey chain line*)

not included in Fig. 8.3, as they were derived at saturation, i.e. at different pressures as shown in this figure.

The comparison of the vapor pressure curves of R-1234*yf* and R-134*a* in Fig. 8.2 demonstrates that both compounds have very similar vapor pressures. This suggests that R-1234*yf* can be used as drop-in replacement for R-134*a* as the cycles can operate at similar evaporating and condensing pressures. However, the critical temperature of R-1234yf is about 6.4 K below that of R-134*a* ($T_{c,R-134a} = 374.21$ K) and its critical pressure 0.68 MPa below the value for R-134*a* ($p_{c,R-134a} = 4.0593$ MPa [41]). Beyond that, there are also other key parameters that need to be considered to evaluate the performance of R-1234*yf* in MAC cycles [80, 84]. The Figs. 8.2 and 8.3 illustrate well that at the same temperature, the liquid density of R-1234*yf* is remarkably lower than for R-134*a*. When R-1234*yf* is used as drop-in replacement in R-134*a* cycles, i.e., the inner volume of the cycles is kept constant, this results in a different optimum refrigerant charge amount. It is thereby expected that the charge amount for R-1234*yf* is smaller than for R-134*a*. However, the refrigerant charge amount is not only determined by the density, but also by operation conditions, i.e., the criterion to hold an appropriate subcooling at the condenser output [84].

Figure 8.3 demonstrates that the liquid viscosities of R-1234*yf* are also smaller than those of R-134*a*. In general, this should positively affect the heat transfer during condensation. According to the Nusselt theory for the heat transfer of a (laminar) film condensation in a horizontal tube (diameter d), the heat transfer coefficient α is also determined by the density of the refrigerant, its heat of vaporization Δh_{vap}, and its thermal conductivity λ_T

$$\alpha = 0.728 \left[\frac{\varrho^L(\varrho^L - \varrho^V) g \Delta h_{vap} \lambda_T^L}{\eta^L (T_s - T_w)} \frac{1}{d} \right]^{1/4}. \tag{8.1}$$

The superscripts L and V identify the properties of the liquid and gaseous refrigerant, respectively, T_w is the temperature at the wall. As discussed before, the liquid densities of R-1234yf are significantly smaller than those of R-134a. Additionally, its heats of vaporization and thermal conductivities were also found to be lower [26, 75]. This results altogether in an inferior heat transfer performance in the condenser compared to R-134a [84]. In his recent review, Wang [84] provides an overview of studies addressing the evaluation of the system performance when R-134a is replaced by R-1234yf. He concludes that the capacities and coefficients of performance (COP) of current R-1234yf cycles are in general lower than those of R-134a systems. It should be noted that a lower COP, i.e. lower energy efficiency, in turn results in an increased indirect emission of green house gases due to higher energy consumption to operate the cycle.

8.1.2 R-1234$ze(E)$ and Blend R-445A

As another possible replacement for R-134a in MAC systems, a zeotropic refrigerant blend composed of 85 % R-1234$ze(E)$, 9 % R-134a, and 6 % CO_2 was proposed [54]. It was first named AC6 and is now designated R-445A by ASHRAE [4]. The GWP of R-445A is about 135, i.e. below the limit set by the EU regulation, and it is considered to be non-flammable under transport and handling conditions, non-toxic, affordable, and available [54, 55].

In [61] we have presented the extension of the force field model to the trans-tetrafluoro-1-propene HFO-1234$ze(E)$ (R-1234$ze(E)$). Therefore, we derived its partial charges from ab initio simulations as described in Sect. 6.1.1.5. All LJ-parameters used in the modeling of R-1234ze(E) are the same as for HFO-1234yf with the exception of the introduction of the AMBER $H1$ parameter [12], as the hydrogen in R-1234$ze(E)$ is bonded to the same CM atom as the electron withdrawing FCM fluorine (see Sect. 6.1.1.5). As no LJ parameters were adjusted to experimental data of R-1234$ze(E)$, the simulations for this compound are purely predictive.

Figure 8.4 shows our GEMC simulation results for the vapor pressure and vapor-liquid coexistence curve of R-1234$ze(E)$ in comparison with experimental data [25, 47, 77], and calculations by the REFPROP EOS [47]. The agreement of the simulations with the experimental data is quite satisfying, although the simulations tend to overestimate the vapor pressures for $T > 300$ K, and the saturated liquid densities for $T < 320$ K. Figure 8.4 also illustrates that at the same temperatures, the vapor pressures of R-1234$ze(E)$ are lower than those of R-134a and R-1234yf, whereas its critical temperature ($T_{c,R-1234ze(E)} = 382.51$ K [25]) is higher than those of both other refrigerants. The saturated liquid densities of R-1234$ze(E)$ are higher than those of R-1234yf and with this closer to the R-134a values. The higher saturated liquid densities of R-1234$ze(E)$ compared to R-1234yf might be surprising as both compounds are composed of the same atoms. However, the different positions of the fluorine atoms change the charge distribution within the molecules and result in

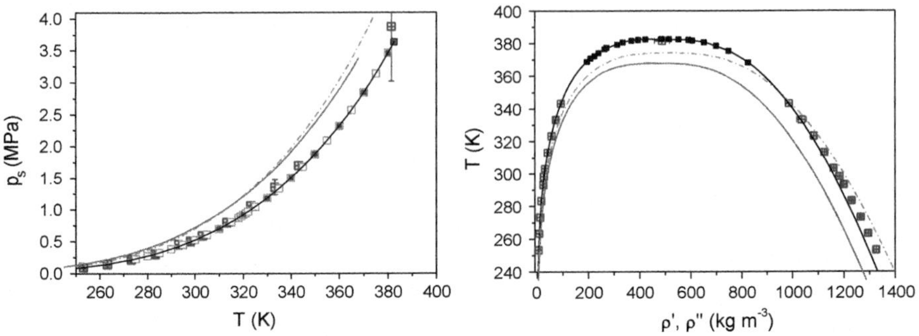

Fig. 8.4 GEMC simulation results (*blue crossed squares*) for the vapor pressure and saturated densities of R-1234$ze(E)$ in comparison with experimental data by McLinden et al. (*white squares* [47]) and Tanaka et al. (*black squares* [25, 77]), and calculations by the REFPROP equation of state [41, 47] (*black line*). Also shown as *grey chain lines* and *grey solid lines* are the vapor pressure curves and vapor-liquid coexistence curves of R-134a and R-1234yf, calculated by their reference equations in REFPROP

different dipole moments for both compounds. Thereby, R-1234yf exhibits a higher dipole moment ($\mu = 2.57\ D$ [64]) than R-1234$ze(E)$ ($\mu = 1.31\ D$ [64]), and it is expected that the liquid densities then scale with the dipole moments—but this is not the case here. In [64] we analyzed the contributions of LJ and electrostatic interactions to the interaction potential of different fluoropropenes, and found that the highest contribution from electrostatic interactions occurs in R-1234$ze(E)$. We then employed an empirical equation [15] for estimating effective molecular quadrupole moments of nonsymmetric molecules from the diagonal elements of their traceless quadrupole tensor obtained from ab initio simulations

$$Q_{eff}^2 = \frac{2}{3}\left(Q_{xx}^2 + Q_{yy}^2 + Q_{zz}^2\right). \tag{8.2}$$

Based on this equation, we derived estimates of the effective molecular quadrupole moments of different fluoropropenes, and found that R-1234$ze(E)$ with $Q_{eff} = 3.43 D\dot{A}$ exhibits a significantly higher quadrupole moment compared to the other compounds ($Q_{eff,R-1234yf} = 2.10 D\dot{A}$, for instance). From this, we concluded that the relatively high densities of the R-1234$ze(E)$ result from strong coulombic interactions that are due to its high quadrupole moment. For simulation studies on the ternary blend R-445A in [62], we employed our force field model for R-1234$ze(E)$ together with common models from the literature for CO_2 [57] and R-134a [53]. Prior to studies on the ternary blend, we performed GEMC simulations on the VLE of the binary subsystems CO_2 + R-134a, CO_2 + R-1234$ze(E)$ and R-134a + R-1234$ze(E)$. Thereby, all LJ parameters for interactions between unlike atoms were obtained from the Lorentz-Berthelot combining rule without adjustable interaction parameters. With this, simulation studies on all mixtures are purely predictive. Figure 8.5 shows our simulation result for the two binary subsystems with CO_2. For the system

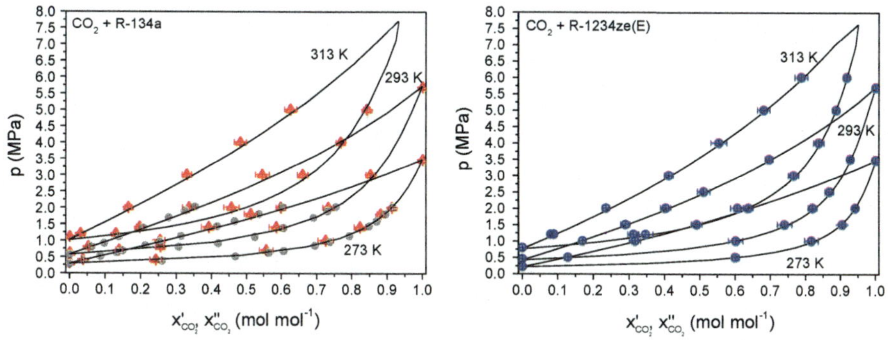

Fig. 8.5 GEMC simulation results for the VLE in the binary mixture CO_2 + R-134*a* (*red crossed triangle*) and CO_2 + R-1234*ze*(*E*) (*blue crossed circles*) at 273, 293 and 313 K, in comparison with REFPROP calculations. Also shown are experimental data for the CO_2 + R-134*a* mixture (*grey dots*, [14])

CO_2 + R-134*a*, there are some experimental VLE data available [14], which allow for a validation of our predictive simulation results. EOS calculations for mixtures in REFPROP on the other hand require the derivation of binary interaction parameters. For the subsystem CO_2 + R-134*a*, these REFPROP interaction parameters were adjusted to the experimental data. The good agreement between the simulation results and both the experimental data by Duran-Valencia et al. [14] and REFPROP calculations illustrates that the two force fields from literature for CO_2 and R-134*a* are well suited to describe this mixture.

The right graph in Fig. 8.5 shows our predictions for the VLE in the binary subsystem CO_2 + R-1234*ze*(*E*) [62, 63]. Due to the lack of experimental VLE data for this mixture, the binary interaction parameters for the REFPROP calculations were fitted to these molecular simulation results. Figure 8.5 also demonstrates that the $p - x$-diagrams of the VLE in the systems CO_2 + R-134*a* and CO_2 + R-1234*ze*(*E*) resemble each other, which could be expected. From this we conclude that our simulation studies also yield reliable predictions for the VLE in the mixture CO_2 + R-1234*ze*(*E*).

Figure 8.6 shows our predictive simulation results for the VLE in the mixture R-134*a* + R-1234*ze*(*E*) for which—to our best knowledge—no experimental data are available in the literature. For this mixture, the binary interaction parameters for the REFPROP calculations were derived from an estimation scheme [41]. The calculations by REFPROP (shown as lines) indicate that this mixture is azeotropic with narrow phase envelopes. The depiction of our GEMC simulation results in Fig. 8.6 shows good agreement with the predictions by REFPROP.

We also performed MD simulations at $T = (270\text{–}315)\,\text{K}$ and pressures up to 1.6 MPa to provide predictions for the densities and viscosities of the ternary blend R-445*A* in the liquid phase. Thereby, the viscosities were derived by the Green-Kubo formula (Eq. 7.27) from MD simulations in the *NVT*-ensemble at the densities $\varrho^L(p, T)$, which we determined in preceding simulations in the *NpT*-ensemble.

Fig. 8.6 GEMC simulation results for the VLE in the binary mixture R-134*a* + R-1234*ze*(*E*) (*violet crossed circles*) at 273, 293 and 313 K, in comparison with REFPROP predictions

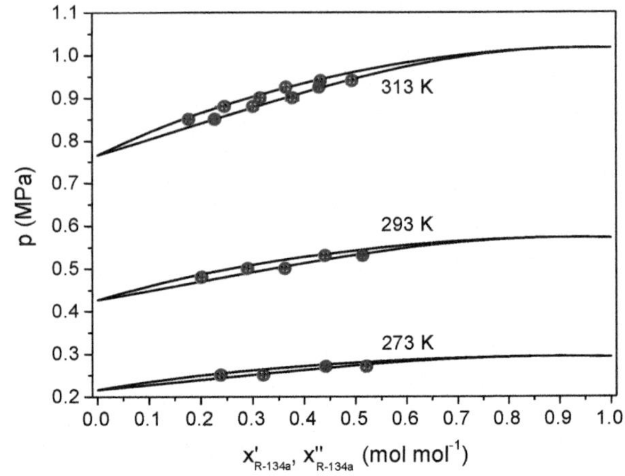

Figure 8.7 shows our MD simulation results for the liquid density and liquid viscosity of R-445*A* in comparison with REFPROP predictions for the mixtures, and also with calculated densities and viscosities of the pure compounds R-134*a* and R-1234*yf*.

The predictions for the liquid densities and viscosities of R-445*A* are generally in good agreement with the REFPROP calculations, and yield ϱ^L and η values for R-445*A* that are higher than the corresponding values of R-1234*yf*. With this, they are closer to the properties of R-134*a*, albeit still below them. The higher liquid densities and viscosities of R-445*A* compared to R-1234*yf* not only result from its R-134*a* content, but are mainly caused by the fact that the pure compound R-1234*ze*(*E*) offers already higher liquid densities and viscosities than R-1234*yf*, due to its high quadrupole moment as discussed before [64].

The liquid densities and viscosities of R-445*A* determine the optimum refrigerant charge amount in the cycle and the heat transfer condition in the condenser as discussed in Sect. 8.1.1. However, the feasibility of using R-445*A* as replacement for R-134*a* is significantly determined by its phase behavior. The depiction of the $p - x$-diagrams of the VLE in the binary subsystems in Fig. 8.5 illustrates that CO_2 has significantly different boiling points from R-134*a* and R-1234*ze*(*E*), i.e. at the same temperature, the vapor pressures of the pure compounds differ considerably. This means in turn that at evaporating and condensing pressure, the saturation temperatures differ accordingly. With this, the refrigerant blend exhibits a remarkable temperature glide in the two-phase region, which affects the temperature distribution in the heat exchangers. Due to this, the use of refrigerant blends with temperature glide generally requires the modification/optimization of the heat exchangers [80].

The use of R-445*A* in MAC systems additionally raises the question of the selective leakage of CO_2. For this reason, we also studied the CO_2 composition in the equilibrium phases during vaporization and condensation. Figure 8.8 shows GEMC simulation results for the CO_2 composition in the coexisting phases at isobaric conditions of vaporization ($p = 0.3$ MPa) and condensation ($p = 1.6$ MPa). The

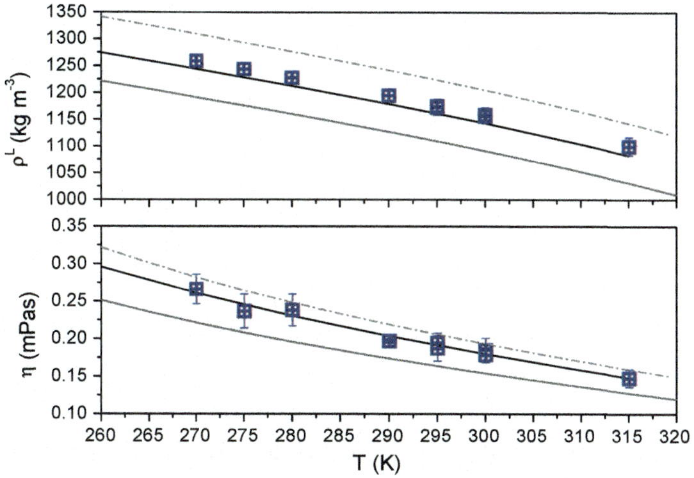

Fig. 8.7 MD simulation results for the liquid density and liquid viscosity of $R-445A$ (*blue crossed squares*) at $p=$ (1–1.6) MPa in comparison with REFPROP predictions at $p=1,6$ MPa (*black line*). Also shown are the calculated [41] densities and viscosities of R-134*a* (*grey chain line*) and R-1234*yf* (*grey solid line*) at $p=1,6$ MPa

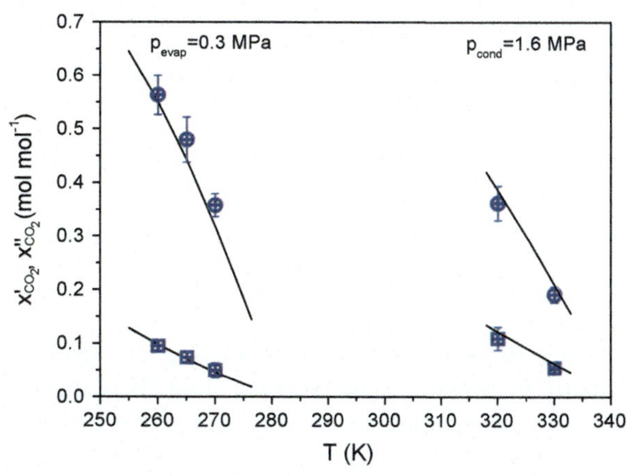

Fig. 8.8 GEMC simulation results for the mole fraction of CO_2 in the vapor phase (*blue crossed circle*) and the liquid phase (*blue crossed squares*) during evaporation ($p=0.3$ MPa) and condensation ($p=1.6$ MPa) of $R-445A$ in comparison with REFPROP calculation (*black line*)

simulation results agree well with the REFPROP calculations (black line) in describing a decreasing CO_2 composition in both the liquid and vapor phase for increasing temperatures. Figure 8.8 also illustrates that the highest CO_2 compositions in the vapor phase are observed at conditions of evaporation at low temperatures and pressures. Therefore, leakage of refrigerant vapor in the evaporator results in highest possible CO_2 loss in the system.

In [70], we performed system simulation studies to evaluate R-445A as drop-in replacement for R-134a in a MAC system of a city bus. Figure 8.9 shows the $p-h$-diagrams of both refrigerants derived from REFPROP calculations. For the

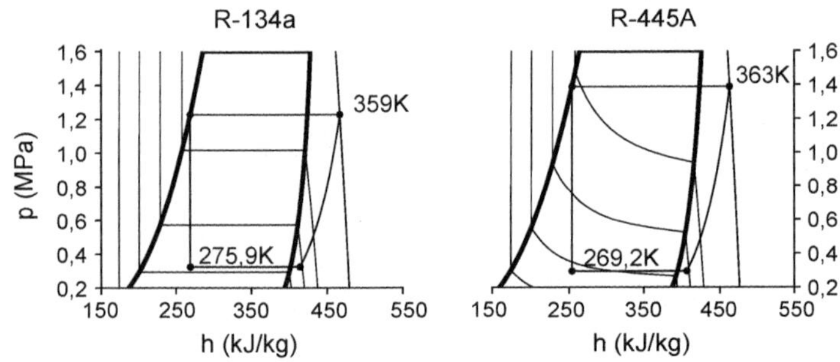

Fig. 8.9 $p - h$-diagrams with typical states of the R-134a and R-445A MAC systems at the condition of same cooling capacities

blend R-445A, the REFPROP calculations employ interaction parameters for the CO_2 + R-1234$ze(E)$ derived from our molecular simulation results. Also depicted in Fig. 8.9 are typical states of the R-134a and R-445A systems, assuming same cooling capacities of both cycles. The comparison of the two cycles well illustrates the effect of the drop-in of R-445A: it results in increased high pressures, and also in higher compressor outlet temperatures, which influences the compressor lifetime. Due to the temperature glide of R-445A during evaporation, the evaporator inlet temperature is below $0\,°\mathrm{C} = 273.15\,\mathrm{K}$, which may cause ice formation. For the conditions we studied in [70], we found reduced maximum cooling capacities and lower COPs for the R-445A system.

8.2 Working Fluids for Organic-Rankine Cycles and High Temperature Heat Pumps

Figure 8.10 shows a basic Organic-Rankine cycle in which a waste heat is used directly to evaporate the working fluid. It is then expanded—in most systems in turbines—to generate shaft-work. Often, the turbine is directly connected to a generator to yield electrical power. After expansion, the superheated working fluid is cooled and liquefied in the condenser. The saturated liquid working fluid is then pressurized in the pump and enters the evaporator, where it is again heated and evaporated.

Thus, ORC processes can be used to convert low-grade waste heat, or more generally heat from any low temperature heat sources—into work or electric power. Another option to make use of low-grade heat is the application of high temperature heat pumps (HTHP) that convert the low grade heat into higher grade heat by employing vapor-compression cycles corresponding to the system shown in Fig. 8.1. The working fluid of an ORC process or HTHP application needs to be selected to match the specific application, i.e. it has to allow for evaporation at heat source temperatures, and it has to be chemically stable in the temperature range covered

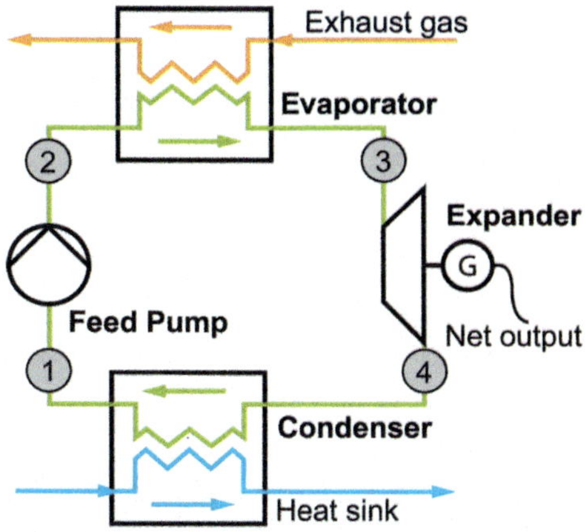

Fig. 8.10 Configuration of an Organic-Rankine cycle with direct evaporation of the working fluid by the waste heat

by the application. Therefore, both applications require working fluids with higher critical temperatures than those used in typical refrigeration applications. A number of possible working fluids have been proposed in the literature for low-grade ORC processes, and possible candidates have been reviewed, for example, by Chen et al. [11], Tchanche et al. [78] or Quoilin et al. [60]. Assessments of different working fluids for HTHP applications are amongst others provided by Pan et al. [51] or Kondou and Koyama [36]. This section though covers only HFO and HCFO compounds that were recently proposed as working fluids for low-grade ORC or HTHP applications: cis-1,3,3,3-trifluoropropene HFO-1234$ze(Z)$, cis-1,1,1,4,4,4-hexafluoro-2-butene, HFO-1336$mzz(Z)$ and trans-1-chloro-3,3,3-trifluoropropene HCFO-1233$zd(E)$. Although HCFO-1233zd(E) is chlorine containing, its ozone-depletion potential (ODP) is expected to be negligible due to its short atmospheric lifetime [29]. The performance of both HCFO-1233$zd(E)$ and HFO-1336$mzz(Z)$ as working fluids in low-grade ORC processes was studied by Molés et al. [49], however, the evaluation of HFO-1336mzz(Z) was based on data from DuPont, which are not available in the open literature. Studies by Fukuda et al. [21], Longo et al. [44] or Brown et al. [9] focused on the evaluation of R-1234ze(Z) as HTHP working fluid, whereas we have studied its performance in low-grade ORC processes [56].

In the following, we will discuss our molecular simulation results for the vapor-liquid phase equilibria properties of HFO-1234$ze(Z)$, HFO-1336$mzz(Z)$ and HCFO-1233$zd(E)$. We will thereby compare their thermophysical properties with those of 1,1,1,3,3-pentafluoropropane R-245fa, which is a widely used ORC-working fluid [56], but is also considered for HTHP applications [51]. However, in this evaluation, we will mainly focus on the ORC application.

When we performed simulation studies on the trans-isomere R-1234$ze(E)$ in [61], we also provided predictions for the cis-isomere HFO-1234$ze(Z)$. As described above in Sect. 8.1.2 for the trans-compound, we only derived its partial charges from ab initio simulations, whereas all LJ-parameters were transferred from the HFO-1234yf modeling with the additional introduction of the AMBER $H1$ parameter

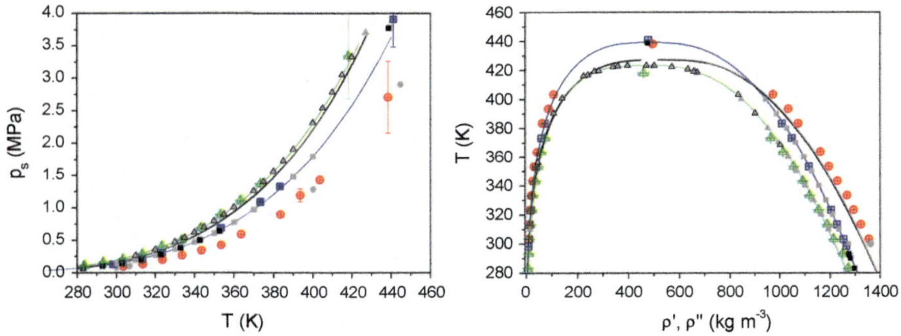

Fig. 8.11 GEMC simulation results for the vapor pressure and saturated densities of HFO-1234*ze*(*Z*) (*green crossed triangle*), HCFO-1233*zd*(*E*) (*blue crossed square*) and HFO-1336*mzz*(*Z*) (*red crossed circle*) in comparison with experimental data by Higashi et al. (*white triangle* [24]) and Tanaka et al. (*grey triangle* [76]) for 1234*ze*(*Z*), Tanaka et al. (*grey squares*) and Hulse et al. (*black squares*, [29]) for 1233*zd*(*E*), and Moles et al. (*grey dot*, [49]) for HFO-1336*mzz*(*Z*). Also shown are correspondingly *colored lines* representing calculations with REFPROP with the new EOS for 1234*ze*(*Z*) [2] and 1223*zd*(*E*) [50]. The *black line* shows the calculated vapor pressure curve and VLCC of R-245*fa* [42]

[12]. Only recently, experimental data for the cis-isomere became available [18, 24, 33, 76], which confirm our predictions from molecular simulations, as illustrated by the depiction of the vapor pressure curve and the VLCC of R-1234*ze*(*Z*) in Fig. 8.11. In 2014, Akasaka et al. [2] published a fundamental equation of states for R-1234*ze*(*Z*), which is now available in REFPROP.

To enable simulation studies on the chlorinated compound HCFO-1233*zd*(*E*), we extended the force field model for HFOs by deriving Lennard-Jones parameters for the Cl atom type and parameters for the stretching of the Cl-CM bond and bending of the Cl-CM-CM and Cl-CM-H1 angles as described in [65] and Sect. 6.1.1.5. Figure 8.11 also shows our GEMC simulation results in comparison with experimental data [29] and with calculations using the new EOS by Mondejar et al. [50].

For simulation studies on the butene compounds HFO-1336*mzz*(*Z*), we have transferred all force field parameters for LJ interactions and the intramolecular terms from our model for fluoropropenes. Thereby we also employed the slightly modified LJ parameters for the fluorine atoms in compounds with five or more fluorine atoms. This is described in more detail in [65], and in Sect. 6.1.1.5. As no LJ parameters were adjusted for HFO-1336*mzz*(*Z*), our simulation studies on its phase behavior are purely predictive. The simulation results for HFO-1336*mzz*(*Z*) are also shown in Fig. 8.11 in comparison with experimental data, which are still very limited.

From our molecular simulation results for the vapor pressure and VLCC of the different compounds we also derived estimates of their critical properties and normal boiling points, which are summarized and compared with experimental data in Table 8.2.

As illustrated in Fig. 8.11, the HFO-1234*ze*(*Z*) compound provides a vapor pressure curve that is most similar to that of R-245*fa*, which means that the ORC or

Table 8.2 Predicted critical properties and normal boiling point for HFO-1234$ze(Z)$, HCFO-1233$zd(E)$ and HFO-1336$mzz(Z)$ in comparison with experimental data

		HFO-1234$ze(Z)$	HCFO-1233$zd(E)$	HFO-1336$mzz(Z)$
$T_c\ (K)$	sim.	417.9 ± 9.2	440.8 ± 1.9	438.0 ± 6.7
	exp.	423.3 [24]	439.6 [50]	444.45 [37]
$p_c\ (MPa)$	sim	3.34 ± 0.66	3.90 ± 0.43	2.70 ± 0.56
	exp	3.533 [24]	3.624 [50]	2.9 [37]
$\varrho_c\ (kg\,m^{-3})$	sim	457 ± 23	475.6 ± 4.0	492.3 ± 20.0
	exp	470 ± 5 [24]	480.2 [50]	
$T_b\ (K)$	sim	283.6 ± 3.5	290.9 ± 3.5	305.5 ± 5.7
	exp	282.9 [2]	293.41 [50]	306.5 [37]

HTHP process can operate at similar evaporating and condensing pressures. Furthermore, the critical temperature HFO-1234$ze(Z)$ is closest to that of R-245fa with $T_{c,R-245fa} = 427.16$ K [42]. Both HCFO-1233$zd(E)$ and HFO-1336$mzz(Z)$ show lower vapor pressures than R-245fa. The predicted critical pressure of HFO-1336$mzz(Z)$ of (p_c = (2.7–2.9) MPa) is significantly lower than the corresponding value of R-245fa with $p_{c,R-245fa} = 3.651$ MPa [42]. The lower vapor pressures of both HCFO-1233$zd(E)$ and HFO-1336$mzz(Z)$ allow for higher operation temperatures, or result in lower pump power consumption when their ORC cycles operate at the same temperature as the R-245fa process [37, 49]. The saturated liquid densities of HFO-1336$mzz(Z)$ are very similar to that of R-245fa, whereas HFO-1234$ze(Z)$ and HCFO-1233$zd(E)$ exhibit lower densities.

Figure 8.12 shows molecular simulation results for the heats of vaporization of HFO-1234$ze(Z)$, HCFO-1233$zd(E)$ and HFO-1336$mzz(Z)$ in comparison with calculations by their EOS in REFPROP, when available, and the corresponding curve $\Delta H_{m,vap}(T)$ of R-245fa. To the best of our knowledge, at present there is no experimental information available for the heats of vaporization of the HFO and HCFO compounds studied here.

Although the molecular simulation results overestimate the heats of vaporization calculated by the REFPROP equations, they agree with the calculations in yielding values for the $\Delta H_{m,vap}$ of HCFO-1233$zd(E)$, which are closest to the values of R-245fa, whereas HFO-1234$ze(Z)$ exhibits the smallest heats of vaporization. Our molecular simulations predict heats of vaporization of cis-1336mzz(E) that are significantly higher than those of R-245fa, which should result in higher turbine output in ORC applications [11].

In [6, 7] Brown et al. proposed a methodology for evaluating 'not-so-well-described' working fluids for ORC processes, i.e. for fluids with limited experimental data available. This method is based on the knowledge of the working fluid's molar mass M, its critical properties T_c, p_c, the acentric factor ω and its molar isobaric ideal gas heat capacity $C^{id}_{p,m}(T)$. As characteristic value for $C^{id}_{p,m}$, they employed its value at the critical point, $C^{id}_{p,m,c}$, and presented a simple correlation

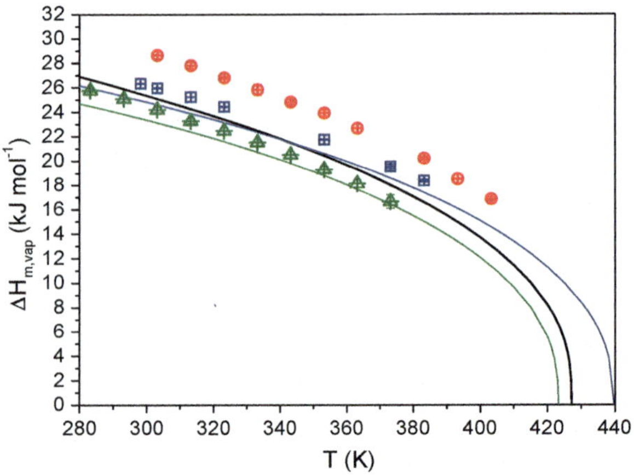

Fig. 8.12 GEMC simulation results for the heats of vaporization of HFO-1234$ze(Z)$ (*green crossed triangle*), HCFO-1233$zd(E)$ (*blue crossed square*) and HFO-1336$mzz(Z)$ (*red crossed circle*). Also shown are correspondingly colored lines representing calculations with REFPROP with the new EOS for 1234ze(Z) [2] and 1223$zd(E)$ [50], and as *black line* the calculated heats of vaporization for R-245fa [42]

Table 8.3 Comparison of M, ω and $C^{id}_{p,m,c}$ values of HFO-1234$ze(Z)$, HCFO-1233$zd(E)$, HFO-1336$mzz(Z)$ and R-245fa

	1234$ze(Z)$	1233$zd(E)$	1336$mzz(Z)$	245fa
$M\ (g\ mol^{-1})$	114.04	130.5	164.06	134.05
$\omega\ (-)$	0.3274	0.3035	0.4065	0.3776
$C^{id}_{p,m,c}$ (kJ kmol$^{-1}K^{-1}$)	119.88	127.92	176.42	142.85

$$\frac{\varrho_c}{mol\ dm^{-3}} = 224.77\left(\frac{C^{id}_{p,m,c}}{kJ\ kmol^{-1}K^{-1}}\right)^{-0.834} \tag{8.3}$$

to estimate $C^{id}_{p,m,c}$ from the critical density of the working fluid. Table 8.3 summarizes the values for M, ω and $C^{id}_{p,m,c}$ of the working fluids HFO-1234$ze(Z)$, HCFO-1233$zd(E)$, HFO-1336$mzz(Z)$ in comparison with the R-245fa values. For HFO-1336$mzz(Z)$ we derived an estimate for the acentric factor from our molecular simulation results for the vapor pressures, whereas $C^{id}_{p,m,c}$ was calculated using Eq. 8.3 from our estimate for ϱ_c. All other properties listed in Table 8.3 were derived from REFPROP.

The comparison of the molar masses and $C^{id}_{p,m,c}$ values in Table 8.3 illustrates well that an increasing molecular complexity, reflected by the molar masses, generally involves increasing $C^{id}_{p,m,c}$ values. This in turn affects the shape of the saturation dome on the T-S-diagram as illustrated in Fig. 8.13 that shows the REFPROP calculations for the saturation curves of HFO-1234$ze(Z)$, HCFO-1233$zd(E)$ and R-245fa. With

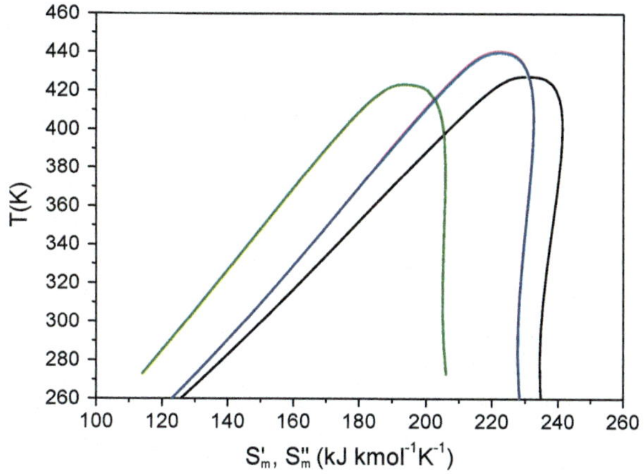

Fig. 8.13 Saturation dome on the *T*-*S*-diagram of HFO-1234*ze*(Z) (*green line*), HCFO-1233*zd*(*E*) (*blue line*) and R-245*fa* (*black line*) calculated by REFPROP

increasing molar masses and $C^{id}_{p,m,c}$ values, the slope of the dew line in the *T*-*S*-diagram is possessing a more and more positive slope (i.e. $dS_m/dT > 0$). HFO-1234*ze*(Z) represents an isentropic fluid with nearly zero slope of the dew line, whereas HCFO-1233*zd*(*E*) and R-245*fa* exhibit positive slopes, which characterize them as dry working fluids. Both, dry and isentropic fluids are considered suitable working fluids for ORC processes as they do not require superheating to prevent condensation in the turbine during the isentropic expansion. From the fluids compared here, HFO-1336*mzz*(Z) exhibits the highest molar mass and $C^{id}_{p,m,c}$ value, and with this, it can be expected that it also represents a dry working fluid. In [6] Brignoli and Brown have parametrically varied the critical properties, ω and $C^{id}_{p,m,c}$ values to investigate their influence on the ORC cycle efficiency and volumetric work output. Thereby they found that both parameters decrease with increasing $C^{id}_{p,m,c}$ values. With regard to this, the high $C^{id}_{p,m,c}$ value of HFO-1336*mzz*(Z) is less favorable.

The afore discussed properties only enable a first evaluation and screening of the performance potentials of the working fluids. A detailed evaluation of a working fluid for a specific ORC application though has to account for several operational conditions, such as heat source and heat sink temperature, chosen operating point and pinch point, heat transfer effects, pressure drops etc.

8.3 Alternative Refrigerants for Chillers

A chiller is a plant that absorbs heat from a fluid—most commonly via a vapor-compression cycle, as shown in Fig. 8.1. The chilled fluid (often water) is then distributed to heat exchangers, either to cool air in air-conditioning applications, or to control the cooling of machines or products in industrial processes. The chiller represents a so-called indirect system, as the cooling is not directly provided by the refrigerant in the chiller cycle, but by the chilled fluid (water). Chillers employ different compressor types in the vapor-compression cycle, depending on the required cooling capacity. The choice of compressor in turn limits the options for suitable refrigerants in the cycle. An overview of chiller types and typical refrigerants is, for instance, provided in the UNEP Report [80]. Air-conditioning in large commercial building complexes represents a typical application for chillers with large cooling capacities, and these chillers generally employ centrifugal compressors. In a centrifugal compressor, the kinetic energy of the refrigerant is increased through an impeller, and afterwards the refrigerant flow is slowed down in the diffuser section so that the kinetic energy is converted into pressure. Thus, the impeller and diffusor geometry need to be designed specifically for a chosen refrigerant. The HCFC refrigerant R-123, i.e. 2,2-dichloro-1,1,1-trifluoroethane, was found to be highly energy-efficient in these chiller applications, but it is destined for phase-out by 2020 under the Montreal Protocol. Trans-1-chloro-3,3,3-trifluoropropene HCFO-1233$zd(E)$ was identified as suitable alternative to R-123 as it provides high cooling capacities in chillers [28, 83]. As stated in Sect. 8.2, HCFO-1233zd(E) is also chlorine containing, but due to its short atmospheric lifetime of 26 days [29], its ozone-depletion potential (ODP) is negligible, hence it is not affected by the phase-out regulations for HCFC compounds. In [83] Horn et al. also examined mixtures of cis- and trans-1233zd for chiller application. Thus, this section deals with a comparison of the properties of cis- and trans-1233zd to exemplarily illustrate the effect of conformational isometry in HFO/HCFO compounds. The different positions of the chlorine atom in cis- and trans-1233zd result in quite different charge distributions in both compounds, and with this in different dipole moments. This is well illustrated by Fig. 8.14 that shows the charges on the atomic sites and the resulting dipole moments derived from ab initio simulations for isolated molecules of both compounds [65]. The cis-isomer of HCFO-1233zd exhibits a remarkably higher dipole than the trans-isomer, though the difference in the dipole moments $\Delta\mu_{iso,1233zd} = 1.91D$ is smaller than for the cis- and trans-HFO-1234ze ($\Delta\mu_{iso,1234ze} = 2.23D$), [66], where the chlorine atom is replaced by a fluorine atom. This is comprehensible in consideration of the higher electronegativity of the fluorine compared to the chlorine (see for instance CRC Handbook of Chemistry [85]).

In [65] we provided GEMC simulation results for the VLE properties of the cis-1233zd. These are based on the same LJ parameters as for the trans-isomer, but on individually derived partial charges according to Fig. 8.14. In the molecular simulation studies, we predicted a difference in the normal boiling point between cis- and trans-HCFO-1233zd of $\Delta T_{b,iso,1233zd} \approx 21$ K. This difference in the normal boiling points is again smaller as for the cis- and trans-HFO-1234ze ($\Delta T_{b,iso,1234ze} =$ 28 K), which can be attributed to the smaller difference in the dipole moments for the HCFO-1233zd compounds.

While an accurate equation of state for HCFO-1223$zd(E)$ [50] is available in REFPROP, no such model exists for the cis-isomer until now. To allow for calculations of vapor-liquid phase equilibria properties of both pure compounds, and also of their mixture, we employed the PC-SAFT equation of state (EOS) [23]. The PC-SAFT EOS uses a hard-sphere chain (hc) fluid as reference fluid, in which the chains are formed by equal spherical segments. The dispersive attractive interactions

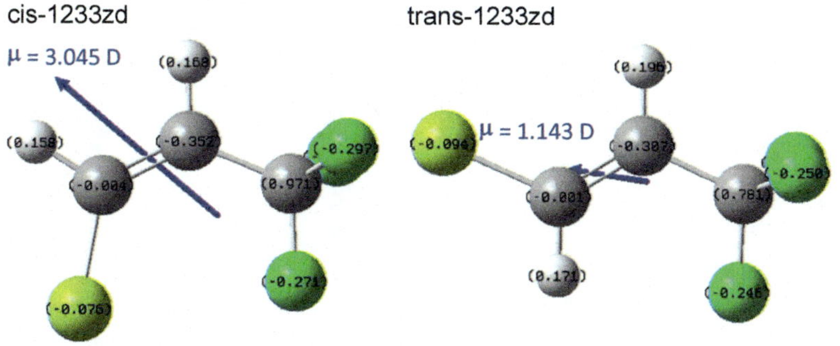

Fig. 8.14 Partial charges and resulting dipole moments of cis- and trans-1233*zd* from ab initio simulations by the ESP approach with the CHELPG fitting scheme at the HF/6-31G* level of theory, using the Gaussian 03 package

Table 8.4 PC-SAFT parameters for cis- and trans-1223*zd*

Component	$m(-)$	$\sigma^*(\AA)$	$\varepsilon^*/k_B(K)$
trans-HCFO-1233*zd*	3.1368	3.3909	202.51
cis-HCFO-1233*zd*	3.4853	3.2749	205.36

between molecules are then treated as a perturbation to the reference fluid, so that the residual free energy F^{res} is given by

$$\frac{F^{res}}{Nk_BT} = \frac{F^{hc}}{Nk_BT} + \frac{F^{pert.}}{Nk_BT}. \tag{8.4}$$

Each pure component is modeled by three parameters: the number of segments m, the segment diameter σ^* and the attraction parameter ε^*. Please note that the EOS parameters σ^* and ε^* are here marked with an asterisk to distinguish them from the LJ parameters used in the molecular simulations. The parameters m, σ^* and ε^* need to be adjusted for each compound. For the trans-isomer R-1233$zd(E)$, the PC-SAFT parameters were determined by fitting calculated vapor pressures and saturated densities to experimental data [29, 50]. For the cis-isomer, the parameters were fitted to the GEMC simulation results [65]. The PC-SAFT parameters for both compounds are listed in Table 8.4.

Figure 8.15 shows the calculated vapor-liquid coexistence curves and vapor pressure curves for cis- and trans-1233*zd* using the PC-SAFT EOS in comparison with the experimental data and GEMC simulation results.

Apart from the fact that the PC-SAFT equation overestimates the critical points, it yields a good reproduction of the VLCCs and vapor pressure curves of both compounds. Also shown as grey chain lines are REFPROP calculations for R-123 for

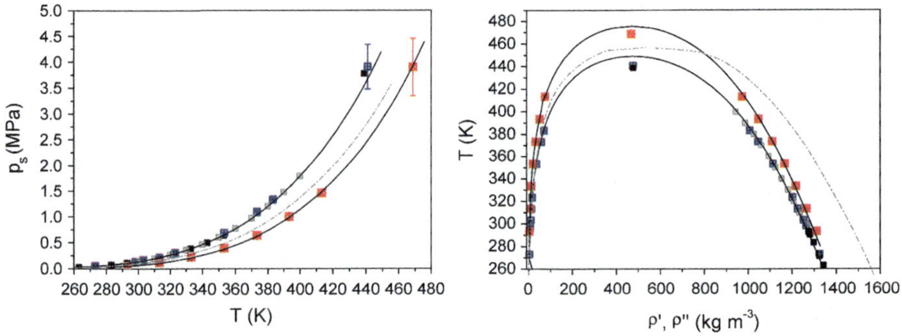

Fig. 8.15 Calculated vapor-liquid coexistence curves and vapor pressure curves for cis- and trans-1233*zd* using the PC-SAFT EOS (*black line*) together with GEMC simulation results for the cis- (*red squares*) and trans-isomer (*blue squares*), and experimental data for $R - 1233zd(E)$ (*grey squares*, [50]). Also shown as *grey chain line* are REFPROP calculations for R-123

the sake of comparison. This illustrates that the vapor pressure curve of R-123 lies between those of the cis- and trans-1233*zd*, which might give reason to use a mixture of the cis- and trans-isomer as R-123 replacement. The densities of R-123 are remarkably higher than those of both cis- and trans-1233*zd*. This is due to the higher molar mass of R-123 ($M_{R-123} = 152.93\,\mathrm{g\,mol^{-1}}$) compared to the 1233*zd* compounds ($M_{1233zd} = 130.5\,\mathrm{g\,mol^{-1}}$). Whereas the molar densities of R-123 and trans-1233*zd* are quite similar, the cis-isomer exhibits higher densities. This is exemplified by comparing the saturated liquid densities of the compounds at $T = 313\,\mathrm{K}$: the saturated liquid densities of R-123 and HCFO-1233*zd*(*E*) are $\varrho'_{m,R-123} = 9,32\,\mathrm{kmol\,m^{-3}}$ and $\varrho'_{m,R-1233zd(E)} = 9,39\,\mathrm{kmol\,m^{-3}}$, i.e. they differ by 0.75 %, whereas the molar density of the cis-isomer with $\varrho'_{m,1233zd(Z)} = 9,67\,\mathrm{kmol\,m^{-3}}$ is about 3.8 % higher than the R-123 value. The use of mixtures of cis- and trans-1233*zd* as refrigerant blend in chiller applications also requires knowledge of their phase behavior. Thus, we also performed GEMC simulation studies on mixtures of both compounds. Figure 8.16 shows our molecular simulation results together with calculations by the PC-SAFT EOS. For PC-SAFT-calculations on mixtures, the parameters m, σ^* and ε^* are calculated from

$$\varepsilon^*_{ij} = \sqrt{\varepsilon^*_{ii}\varepsilon^*_{jj}}(1 - k_{ij}), \quad \sigma^*_{ij} = \frac{\sigma^*_{ii} + \sigma^*_{jj}}{2}, \quad \overline{m} = \sum x_i m_i, \tag{8.5}$$

with using an adjustable interaction parameter k_{ij} for the attraction parameter ε^*_{ij}. For predictive calculations of the VLE in the mixtures of the cis- and trans-isomer, k_{ij} was set to 0. Figure 8.16 illustrates that both, the PC-SAFT calculations and the GEMC simulation results agree in predicting that the cis- and trans-1233*zd* form a zeotropic mixture. This observation that the mixtures of cis- and trans-1233*zd* fractionate zeotropically is thereby in accordance with the experimental results by Horn et al. [83].

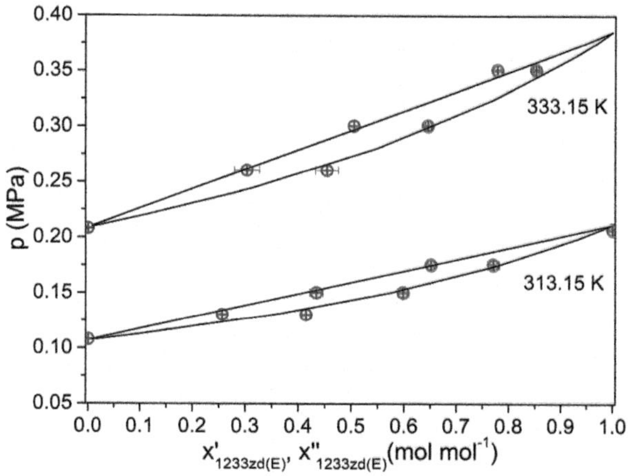

Fig. 8.16 GEMC simulation results for the VLE in the mixture trans-1233*zd* + cis-1233*zd* (*blue circles*) in comparison with PC-SAFT calculations for $k_{ij} = 0$ (*black line*)

8.4 Refrigerant Blends

Regulations to phase-out HFC refrigerants with high GWP values also initialized the development of alternative refrigerants for stationary refrigeration applications. Table 8.5 lists some possible candidates to replace R-410*A*, R-404*A*/ R-507*A*, R-22/R-407*C* and R-134*a* in commercial refrigeration, stationary air-conditioning systems, water heating heat pumps, supermarket refrigeration etc. In order to reduce the GWP, HFO compounds such as the R-1234yf or R-1234$ze(E)$ have been introduced as blend components. The HFOs though offer lower volumetric refrigerant capacities than the conventional refrigerants, so that HFC compounds are added to improve the cooling capacities [5]. Moreover, R-1234yf and R-1234$ze(E)$ are flammable (A2L safety class) so that HFC compounds with high fluoric content such as R-134*a* or the pentafluoroethane R-125 are often added to suppress the flammability. These compounds, however, in turn exhibit high GWP values (1300/2800, [30]), and this results in a trade-off between low GWP and low flammability when formulating the blend composition.

The evaluation of the performance of the proposed novel blends by detailed system simulations requires the calculation of their thermophysical and transport properties at the different state points of the refrigeration cycle. This information can be provided, for instance, by the REFPROP program. The modeling of thermophysical properties of the mixtures in REFPROP is thereby based on highly accurate EOS for the pure compounds, and the application of mixing rules to the Helmholtz free energy of the mixture. These mixing rules contain (binary) mixing parameters that have to be fitted to experimental data for the binary subsystems. Experimental data for mixtures of the conventional HFC compounds are generally available in literature, but the number of experimental studies on mixtures with HFO compounds is still limited. In this context, we can take advantage of the predictive capability of molecular simulations to determine relevant properties of refrigerant blends to

Table 8.5 Proposed blends of HFO/HFC compounds as alternatives for HFC refrigerants according to [5, 8]. The GWP values of the blends are also taken from [5, 8] unless otherwise noted

ASHRAE #	Trade name	Composed of	GWP
	Replacement for R-410*A* (GWP = 1725 [30])		
R-447*A*	Solstice L-41	R-32/125/1234$ze(E)$	582
	Opteon® XL41	R-32/1234yf	466
	HPR1D	R-32/1234$ze(E)/CO_2$	407
	L-41b	R-32/1234$ze(E)$	494
	ARM-70a	R-32/1234$yf/134a$	482
	Replacement for R-404*A*/R-507*A* (GWP = 3260/3300 [30])		
R-448*A*	Solstice N-40	R-32/125/1234yf/1234$ze(E)/134a$	1386
R-449*A*	Opteon® XP40	R-32/125/1234$yf/134a$	1397
R-452*A*	Opteon® XP44	R-32/125/1234yf	2140
	L-40	R-32/1234$ze(E)$/1234$yf/152a$	623
	ARM-31a	R-1234$yf/32/134a$	492
	Replacement for R-22/R-407*C* (GWP = 1500/1525 [30])		
R-444*B*	Solstice L-20	R-32/152a/1234$ze(E)$	294
	LTR4X	R-1234$ze(E)/32/125/134a$	1295
	Replacement for R-134*a* (GWP = 1300) [30]		
R-450*A*	Solstice N-13	R-1234$ze(E)/134a$	601
R-513*A*	Opteon® XP10	R-1234$yf/134a$	631
	AC5X	R-1234$ze(E)/134a/32$	622
	ARM-42a	R-1234$yf/152a/134a$	117

complement experimental data. Thus, the following sections cover simulation studies on binary mixtures containing R-1234yf or R-1234$ze(E)$.

It should be noted that we are not claiming that the list in Table 8.5 is complete. Actually, different blends that are composed of the same compounds, but at different compositions, often exist. Therefore, we only listed one example, because the information needed from the binary subsystems in order to model the blend is always the same.

8.4.1 Binary Mixtures of CO_2 and HFOs

The molecular simulation results for the VLE properties of the mixtures CO_2 + R-1234$ze(E)$ are shown and discussed in Sect. 8.1.2. Therefore, we here only present simulation results for the VLE of the binary mixture CO_2 + R-1234yf [63]. For simulation studies on this mixture we also employed a CO_2 molecular model from literature [57] as described in Sect. 8.1.2. Again, all LJ parameters for unlike interaction were derived from the common Lorenz-Berthelot combining rule (Eq. 6.2)

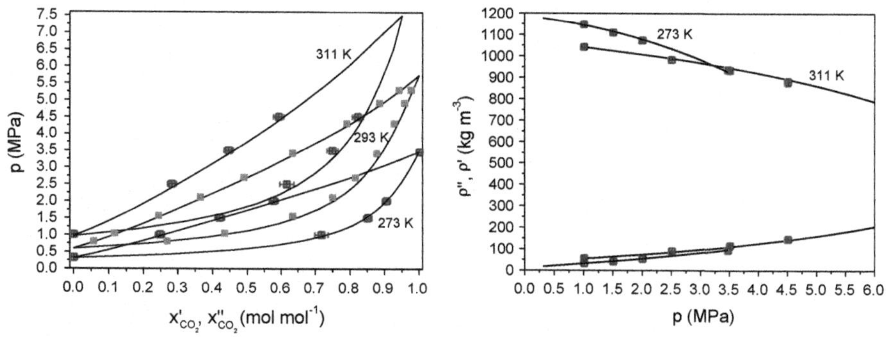

Fig. 8.17 GEMC simulation results for the VLE in the mixture CO_2 + R-1234*y* (*violet square*) at 273 and 311 K: compositions (*left*) and densities (*right*) of the liquid and vapor phase. Also shown are experimental data at 293 K [31] and REFPROP calculations (*solid line*) based on interaction parameters fitted to our simulation results

without any adjustable parameters. With this, the molecular simulation studies for this mixture are purely predictive. Figure 8.17 shows our predictions for the $p - x$-diagrams and saturation densities of the VLE in the binary system CO_2 + R-1234yf together with calculations by REFPROP. As discussed before, EOS calculations in REFPROP require the fitting of binary interaction parameters in the mixing rule. Due to the lack of experimental data back then, the binary interaction parameters for the REFPROP calculations were fitted to our molecular simulation results. In 2014, experimental pTx-data for the VLE of this binary mixture were published by Juntarachat et al. [31]. Figure 8.17 also shows experimental data for the VLE at 293 K in comparison with REFPROP calculations based on interaction parameters derived from our molecular simulation results. It is apparent that the saturated liquid line is well described, whereas the calculated saturated vapor line deviates from the experimental data at high CO_2 mole fractions. However, the same can be observed when comparing the REFPROP calculation for the saturated vapor line at 311 K and our molecular simulation results. The alikeness of the simulated and measured pTx-relations of the VLE illustrates that the molecular simulations provide reliable predictions for this mixture. At the same time, the GEMC simulation studies also yield information for the saturated densities of the system, for which no experimental data are available until now.

We did not limit our simulation studies to the VLE in the mixtures of CO_2 + R-1234yf and CO_2 + R-1234$ze(E)$, but we also performed MD simulations to provide predictions for their liquid densities and viscosities [64]. Thereby, the viscosities were derived from the Green-Kubo formula, as discussed before. Figure 8.18 shows exemplarily our simulation results for $\varrho^L(x_{CO_2})$ and $\eta(x_{CO_2})$ at 273.15 K and 3.5 MPa. The MD results for the liquid densities in the CO_2 + R-1234yf agree very well with the REFPROP calculations, whereas the simulations overestimate the densities in the CO_2 + R-1234$ze(E)$ mixture for high R-1234$ze(E)$ contents. This is due to the fact that the molecular modeling for the pure R-1234$ze(E)$ tends to yield

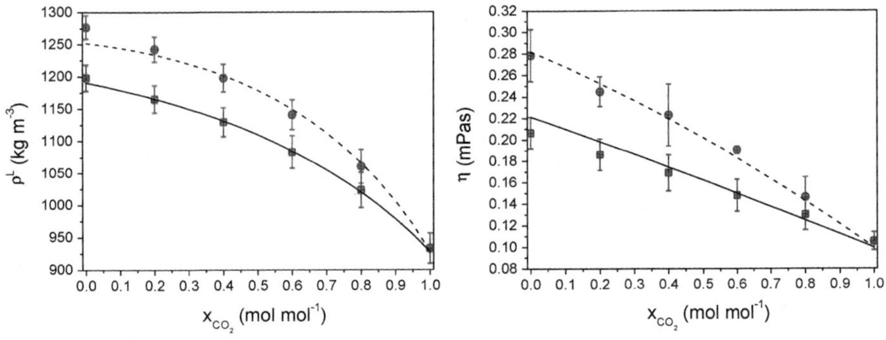

Fig. 8.18 MD simulation results for the liquid densities and viscosities in the mixture CO_2 + R-1234y (*violet square*) and CO_2 + R-1234$ze(E)$ (*blue circle*) at 273.15 K and 3.5 MPa. Also shown are REFPROP calculations (*solid and chain line*, respectively) based on interaction parameters fitted to our simulation results

too high liquid densities at lower temperatures as described in Sect. 8.1.2. Though for the liquid viscosities in the mixture CO_2 + R-1234$ze(E)$ both the MD simulations and REFPROP predictions are in good agreement. In REFPROP, the calculation of the viscosities of the pure HFO compounds is at the moment still limited to an Extended Corresponding State (ECS) approach. As discussed in Sect. 8.1.1, the ECS model for R-1234yf is overestimating its liquid viscosities at lower temperatures. This explains the increasing deviations between our MD results and the REFPROP calculations in the mixture CO_2 + R-1234yf with increasing R-1234yf content.

To gain insight into how the different locations and also numbers of fluorine atoms in the HFO compounds influence their interactions with the CO_2 molecule, we have also analyzed the local ordering in the mixtures [64]. We therefore performed MD simulations in which we determined the radial distribution functions of the center of mass (CoM) of CO_2 molecules around the CoM of different HFO molecules and also other CO_2 molecules in binary CO_2+ HFO mixtures with $x_{CO_2} = 0.4$ at $T = 273$ K and $p = 3.5$ MPa, as shown in Fig. 8.19. The highest first peak in the CO_2–HFO RDFs is found in the mixture of CO_2 with the hexafluoropropene R-1216, whereas the interaction of CO_2 with R-1234$ze(Z)$ seems to be less favorable. The $CO_2 - CO_2$ RDFs in mixtures with different fluoropropenes show very similar peak heights that are notably higher than in the pure CO_2 liquid, and also higher than the peaks in the CO_2–HFO RDFs. This is in agreement with studies by Zhang and Siepmann [87] on mixtures of CO_2 with n-alkanes and perfluoroalkanes. For these mixtures they also found the highest peak for the like $CO_2 - CO_2$ pairs, and concluded that the CO_2 molecules tend to form clusters. This also seems to apply for mixtures of CO_2 with fluoropropenes. The relatively small first peaks in all CO_2-HFO RDFs further indicate that there are no hydrogen-bonding interactions between the HFO hydrogens and oxygen atoms of CO_2.

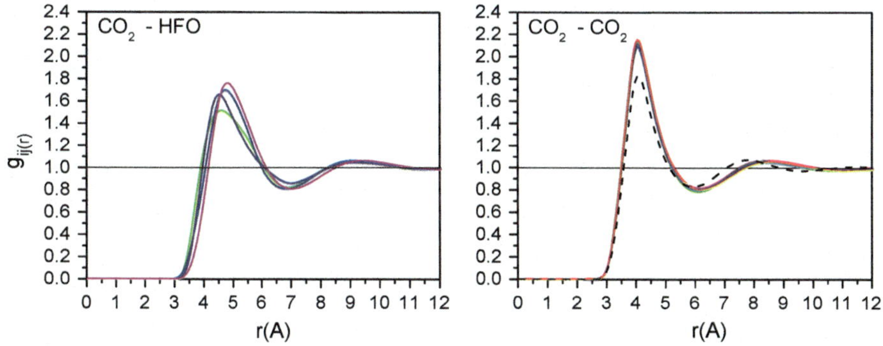

Fig. 8.19 CoM-RDF for CO_2 in mixtures with different HFO molecules at $x_{CO_2} = 0.4$, $T = 273$ K, $p = 3.5$ MPa: HFO-1234*ze*(*Z*) (*green line*), HFO-1234*ze*(*E*) (*blue line*), HFO-1234*yf* (*violet line*) and HFO-1216 (*red line*). *Left* CoM-RDF of CO_2 around the HFO molecules, *right* CoM-RDF of $CO_2 - CO_2$ pairs in mixtures with HFOs, and in the pure fluid (*black dashed line*)

8.4.2 Binary Mixtures of R-32 and HFOs

As illustrated by the compilation in Table 8.5, difluoromethane R-32 is a popular component in novel refrigerant blends, but it is also a component in the currently used blends R-407*A* and R-410*A*. Its popularity is based on its good volumetric cooling capacities [5] and its comparatively low GWP value of 650 [30]. However, R-32 exhibits low flammability and is classified as A2L, i.e. it is assigned the same safety class as R-1234*ze*(*E*) and R-1234*yf*. Despite their flammability, mixtures of R-32 with R-1234*ze*(*E*) or R-1234*yf* have been proposed as candidates to replace R-410*A* in domestic heat pump or air-conditioning systems [20, 39]. In order to allow for simulation studies on these mixtures, we have developed a new all-atoms force field for R-32, which is compatible with our HFO molecular model (see Sect. 6.1.1.5 or [63]). For simulation studies on the mixtures, all LJ parameters for interactions are obtained from the Lorentz-Berthelot combining rule (Eq. 6.2) without using any adjustable interaction parameters. Therefore, all our simulations present pure predictions. Figure 8.20 shows our GEMC simulation results for the VLE in the mixtures R-32 + R-1234*yf* and R-32 + R-1234*ze*(*E*) in comparison with experimental data [32, 38] and calculations by the EOS model for these mixtures by Akasaka [1].

The simulation results generally agree well with both the EOS calculations and the experimental data. This attests the predictive ability of the molecular models and simulation studies. The depiction of the experimental data [38] for R-32 + R-1234*ze*(*E*) reveals some scattering. Therefore, the molecular simulation results for this mixture were also incorporated in the fitting procedure of the EOS development by Akasaka. Furthermore, the GEMC simulations not only provide pTx-relations of the mixture VLE, but also data for the saturated densities that are shown in Fig. 8.21. Experimental information on saturated densities of these mixtures in literature are limited to data near the critical point for mixtures of R-1234*yf* + R-32 with fixed

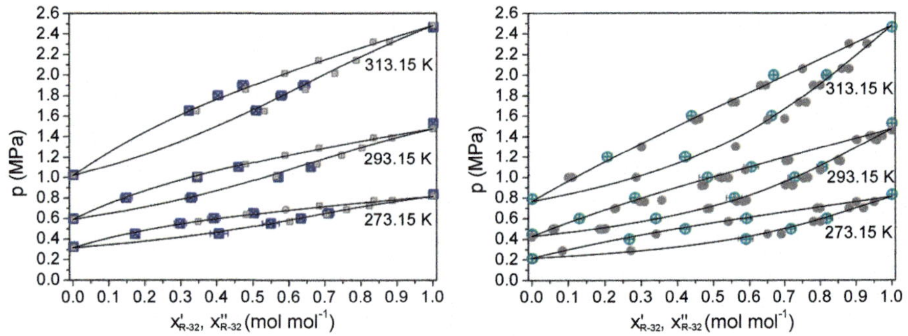

Fig. 8.20 GEMC simulation results for the VLE in the mixtures R-32 + R-1234*yf* (*left, blue squares*) and R-32 + R-1234*ze*(*E*) (*right, cyan circles*) in comparison with calculations based on the EOS model by Akasaka [1] and experimental data (*grey squares* Kamiaka et al. [32], *grey dots* Koyama et al. [38]

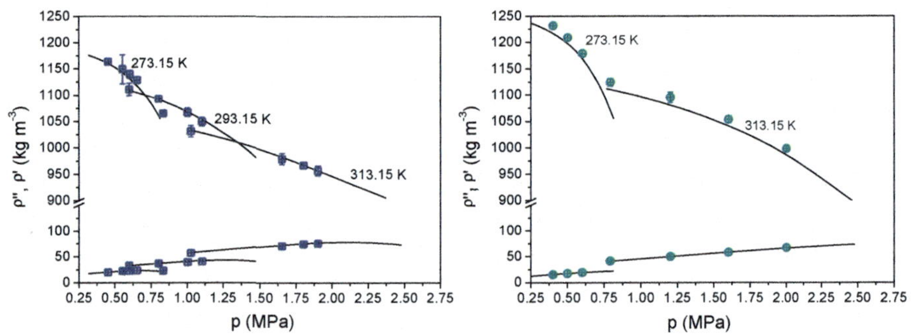

Fig. 8.21 GEMC simulation results for the saturated densities in the mixtures R-32 + R-1234*yf* (*left, blue squares*) and R-32 + R-1234*ze*(*E*) (*right, cyan circles*) in comparison with calculations based on the EOS model by Akasaka [1]. The 293 K data for the R-32 + R-1234*ze*(*E*) were omitted for the sake of clarity

compositions [3]. The predicted saturated densities for the R-32 + R-1234*yf* mixture are in excellent agreement with the EOS calculations although they were not used in the EOS fitting. The GEMC simulation results for the saturated liquid densities of the R-32 + R-1234*ze*(*E*) mixture, however, slightly deviate from the EOS calculations due to the fact that the molecular modeling for the pure R-1234*ze*(*E*) tends to overestimate its liquid densities. The deviation between simulation and EOS calculation at 313.15 K, 2 MPa for example is only 1.4 %, and this still represents a good agreement.

As described before in Sect. 8.4.1 for the CO_2 mixtures, we also performed MD simulations to analyze the local structure of different binary R-32 + HFO mixtures by radial distribution functions in order to gain insight into preferred interactions. The left depiction in Fig. 8.22 shows exemplarily the CoM-RDFs of the R-32 around different fluoropropenes with $x_{R-32} = 0.4$ at 273 K and $p = 3.5$ MPa. The

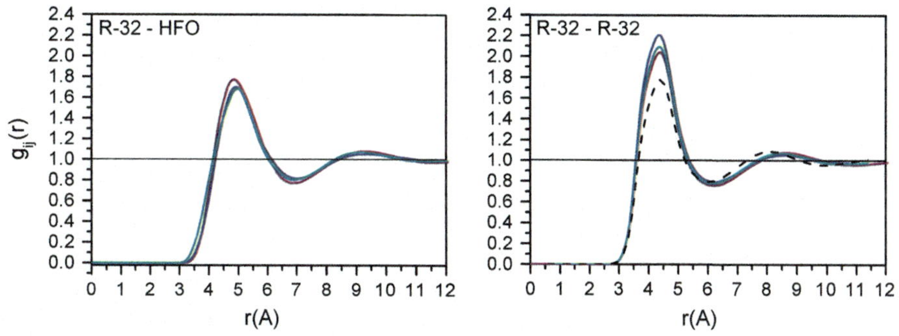

Fig. 8.22 CoM-RDF for R-32 in mixtures with different HFO molecules at x_{R-32}, $T = 273$ K, $p = 3.5$ MPa: HFO-1234$ze(Z)$ (*purple line*), HFO-1234$ze(E)$ (*dark cyan line*) and HFO-1234yf (*blue line*). *Left* CoM-RDF of R-32 around the HFO molecules, *right* CoM-RDF of R-32–R-32 pairs in mixtures with HFOs, and in the pure fluid (*black dashed line*)

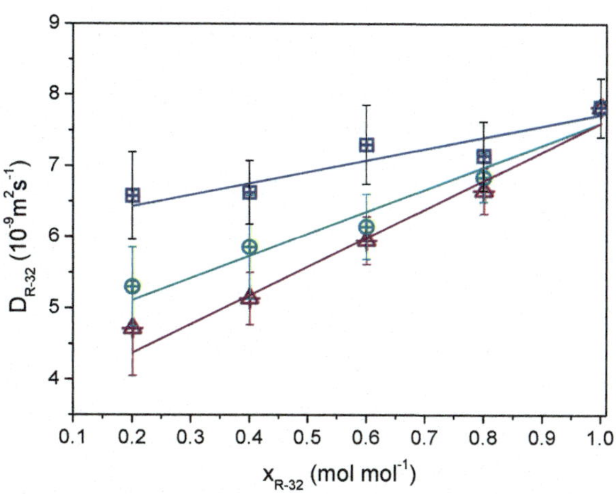

Fig. 8.23 MD results for the diffusion coefficient of R-32 in mixtures with different HFO molecules at $T = 273$ K, $p = 3.5$ MPa: HFO-1234$ze(Z)$ (*purple triangle*), HFO-1234$ze(E)$ (*dark cyan circles*) and HFO-1234yf (*blue squares*). Also shown are linear fits of the MD simulation results as correspondingly *colored lines*

right depiction provides the R-32–R-32 RDF in the same mixtures as well as in the pure compound. The first peaks in the R-32—HFO RDFs are neither high nor sharp, indicating that hydrogen bonding interactions do not play an important role in these mixtures. The R-32–R-32 RDFs in mixtures with different fluoropropenes are notably higher than in the pure R-32 liquid. This suggests that R-32 molecules prefer to form clusters in mixtures with HFOs, similarly to CO_2 in mixtures with HFOs as discussed before.

The highest peak in the R-32–HFO RDF is observed for the mixture with R-1234$ze(Z)$, and for this mixture also the clustering of R-32 is less pronounced. This hints at stronger interactions of R-32 with HFO-1234$ze(Z)$ compared to the other fluoropropenes. This is also supported by the fact that R-32 exhibits the lowest diffusivity in mixtures with this compound, which is well illustrated in Fig. 8.23

showing MD simulation results for the diffusivity of R-32 in mixtures with different HFO compounds at $T = 273$ K, $p = 3.5$ MPa and different compositions.

References

1. Akasaka R (2013) Thermodynamic property models for the difluoromethane (R-32) + trans-1,3,3,3-tetrafluoropropene (R-1234ze(E)) and difluoromethane + 2,3,3,3-tetrafluoropropene (R-1234yf) mixtures. Fluid Phase Equil 358:98–104
2. Akasaka R, Higashi Y, Miyara A, Koyama S (2014) A fundamental equation of state for cis-1,3,3,3-tetrafluoropropene (R-1234ze(Z)). Int J Refrig 44:168–176
3. Akasaka R, Tanaka K, Higashi Y (2013) Measurements of saturated densities and critical parameters for the binary mixture of 2,3,3,3-tetrafluoropropene (R-1234yf) + difluoromethane (R-32). Int J Refrig 36(4):1341–1346
4. Ashrae S (2013) Designation and safety classification of refrigerants. American Society of Heating, Refrigeration and Air-Conditioning Engineers Inc
5. Bitzer KG (2014) Bitzer refrigerant report 18. Bitzer Kühlmaschinenbau GmbH
6. Brignoli R, Brown JS (2015) Organic Rankine cycle model for well-described and not-so-well-described working fluids. Energy 86:93–104
7. Brown JS, Brignoli R, Quine T (2015) Parametric investigation of working fluids for organic Rankine cycle applications. Appl Therm Eng 90(5):64–74
8. Brown JS, Zilio C, Brignoli R, Cavallini A (2014) Thermophysical properties and heat transfer and pressure drop performance potentials of hydrofluoroolefines, hydrochlorofluoroolefines and their blends. HVAC&R Res 20:203–230
9. Brown JS, Zilio C, Cavallini A (2009) The fluorinated olefin R-1234ze(Z) as a high-temperature heat pumping refrigerant. Int J Refrig 32:1412–1422
10. Calm JM (2008) The next generation of refrigerants: historical review, considerations, and outlook. Int J Refrig 31:1123–1133
11. Chen H, Goswami Y, Stefanakow EK (2010) A review of thermodynamic cycles and working fluids for the conversation of low-garde heat. Renew Sustain Energy Rev 14:3059–3067
12. Cornell WD, Cieplak P, Bayly CI, Gould IR, Merz KM Jr, Ferguson DM, Spellmeyer DC, Fox T, Caldwell JW, Kollmann PA (1995) A second generation force field for the simulation of proteins, nucleic acids, and organic molecules. J Am Chem Soc 117:5179–5197
13. Cousins DS, Laesecke A (2012) Sealed gravitaional capillary viscometry of dimethyl ether and two next-generation alternative refrigerants. J Res Natl Inst Stand Technol 117
14. Duran-Valencia C, Pointurier G, Valtz A, Guilbot P, Richon D (2002) Vapor-liquid-equilibrium (VLE) data for the carbon dioxide (CO2) + 1,1,1,2-Tetrafluoroethane (R134a) system at temperatures from 252.95 K to 292.95 K and pressures up to 2 MPa. J Chem Eng Data 47:59–61
15. Eubank P (1972) Estimation of effective molecular quadrupole moments. AIChE J 18:454–456
16. Fedele L, Bobbo S, Groppo F, Brown JS, Zilio C (2011) Saturated pressure measurements of 2,3,3,3-tetrafluoroprop-1-ene (R1234yf) for reduced temperatures ranging from 0.67 to 0.93. J Chem Eng Data 56:2608–2612
17. Fedele L, Brown JS, Colla L, Ferron A, Bobbo S, Zilio C (2012) Compressed liquid density measurements for 2,3,3,3-tetrafluoroprop-1-ene (R1234yf). J Chem Eng Data 57:482–489
18. Fedele L, DiNicola G, Brown JS, Bobbo S, Zilio C (2014) Measurement and correlations of cis-1,3,3,3-tetrafluoroprop-1-ene (R1234ze(Z)) saturation pressure. Int J Thermophys 1:1–12
19. Fermeglia M, Ferrone M, Pricl S (2003) Development of an all-atoms force field from ab initio calculations for alternative refrigerants. Fluid Phase Equil 210:105–116
20. Fujitaka A, Tsutomu S, Shigehiro S, Kawabe Y (2010) Application of low global warming Potential refrigerants for room air conditioner. In: International symposium on next-generation air conditioning and refrigeration technology, Tokyo, Japan

21. Fukuda S, Kondou C, Takata N, Koyama S (2014) Low GWP refrigerants R1234ze(E) and R1234ze(Z) for high temperature heat pumps. Int J Refrig 40:161–173
22. Gough CA, Debolt SE, Kollmann P (1992) Derivation of fluorine and hydrogen atom parameters using liquid simulations. J Comp Chem 13:963–970
23. Gross J, Sadowski G (2001) Perturbed-chain-SAFT: an equation of state based on a prturbation theory for chain molecules. Ind Eng Chem Res 40:1244–1260
24. Higashi Y, Hayasaka S, Shirai C, Akasaka R (2015) Measurement of $p\varrho T$ properties, vapor pressures, saturated densities, and critical parameters R-1234ze(Z) and R-245fa. Int J Refrig 52:100–108
25. Higashi Y, Tanaka K, Ichikawa T (2010) Critical parameters and saturated densities in the critigal region for trans-1,3,3,3-tetrafluoropropene (HFO-1234ze(E)). J Chem End Data 55:1594–1597
26. Huber ML, Perkins RA (2011) Measurement and correlation of the thermal conductivity of 2,3,3,3-tetrafluoroprop-1-ene (R1234yf) and trans-1,3,3,3-tetrafluoropropene (R1234ze(E)). J Chem Eng Data 56:4868–4874
27. Hulse R, Singh RR, Pham H (2009) Physical properties of HFO-1234yf. In: 3rd IIR conference on thermophysical properties and transfer processes of refrigerants, Boulder, USA
28. Hulse R, Singh RR, Spatz MW (2013) Trans-chloro-3,3,3-trifluoropropene for use in chiller application. US Patent No US 8.574,451,B2
29. Hulse RJ, Rajat SB, Singh RR, Thomas RHP (2012) Physical properties of HCFO-1233zd(E). J Chem Eng Data 57:3581–3586
30. IPCC Working Group I TSB (2001) 3rd assessment report: climate change 2001 (TAR). IPCC
31. Juntarachat N, Valtz A, Coquelet C, Privat R, Jaubert JN (2014) Experimental measurements and correlation of vapor-liquid equilibrium and critical data for the CO_2 + R1234yf and CO_2 + R1234ze(E) binary mixtures. Int J Refrig 47:141–152
32. Kamiaka T, Dang C, Hihara E (2013) Vapor-liquid equilibria measurements for binary mixtures of R1234yf with R32, R125, and R134a. Int J Refrig 36:965–971
33. Kayukawa Y, Tanaka K, Kano Y, Fujita Y, Akasaka R, Higashi Y (2012) Experimental evaluation of the fundamental properties of low-GWP refrigerant R-1234ze(Z). In: International symposium on new refrigerants and environmental technology, Kobe, Japan
34. Kelkar MS, Shiflett MB, Yokozeki A, Maginn EJ (2008) Development of force fields for hydrofluorocarbons. AIChE Annu, Meeting, Philadelphia, USA
35. Klomfar J, Souckova M, Patek J (2012) Liquid-phase $p - \varrho - T$ data for 2,3,3,3-tetrafluoroprop-1-ene (R-1234yf) and 1,1,2,3,3,3-hexa-fluoroprop-1-ene (R-1216) at temperatures from (208 to 353) K under pressures up to 40 MPa. J Chem Eng Data 57:3283–3289
36. Kondou C, Koyama S (2015) Thermodynamic assessment of high-temperature heat pumps using Low-GWP HFO refrigerants for heat recovery. Int J Refrig 53:126–141
37. Kontomaris K (2014) HFO-1336mzz-Z: high temperature chemical stability and use as s working fluid in organic rankine cycles. In: 15th international refrigeration and air conditioning conference, Purdue, USA
38. Koyama S, Matsuo Y, Fukuda S, Akasaka R (2010) Measurement of vapor-liquid equilibrium of HFO-1234ze(E)/HFC-32. In: Proceedings of JSRAE annual conference, Kanazawa, Japan
39. Koyama S, Takada N, Matsuo Y, Yoshitake D, Fukuda S (2010) Possibility to introduce HFO-1234ze(E) and its mixtures with HFC-32 as low-GWP alternative for heat pump/refrigeration systems. In: International symposium on next-generation air conditioning and refrigeration technology, Tokyo, Japan
40. Lefevre B, Enrique A (2014) transport sector key to closing the world's emission gap. http://www.wri.org
41. Lemmon EW, Huber ML, McLinden MO (2013) NIST standard reference database 23: reference fluid thermodynamic and transport properties—REFPROP, Version 9.1, standard reference data program. National Institute of Standards and Technology, Gaithersburg, MD
42. Lemmon E, Span R (2006) Short fundamental equations of state for 20 industrial fluids. J Chem Eng Data 51:785–850

43. Lisal M, Vacek V (1996) Molecular dynamic simulations of fluorinated ethanes. Mol Phys 87:167–187
44. Longo GA, Zilio C, Righetti G, Brown JS (2014) Experimental assessment of the low GWP refrigerant HFO-1234ze(Z) for high temperature heat pumps. Exp Thermal Fluid Sci 57:293–300
45. Malvicino C, Seccardini R, Markowitz M, Schuermann K, Bergami A, Arnaud C, Haller R, Petitjean C, Strupp C, Lemke N, Clodic D, Post C, Hafner A (2009) B-COOL project—Ford ka and Fiat Panda R-744 MAC systems. SAE Technical Papers 2009-01-0967
46. McLinden MO, Kazakov AF, Brown JS, Domanski PA (2014) A thermodynamic analysis of refrigerants: possibility and tradeoffs for low-GWP refrigerants. Int J Refrig 38:80–92
47. McLinden MO, Thol M, Lemmon EW (2010) Thermodynamic properties of trans-1,3,3,3-tetrafluoropropene [R1234ze(E)]: measurements of density and vapor pressure and a comprehensive equation of state. In: International refrigeration and air conditioning conference, Purdue, USA
48. Meng X, Qju G, J W, Abdulagatov IM (63) Viscosity measurement for 2,3,3,3-tetrafluoroprop-1-ene (R1234yf) and trans-1,3,3,3-tetrafluoropropene (R1234ze(E). J Chem Thermodyn 2013:24–30
49. Molés F, Navarro-Esbri J, Peris B, Mota-Babiloni A, Baragán-Cervera A, Kontomaris K (2014) Low GWP alternatives to HFC-245fa in organic rankine cycles for low temperature heat recovery: HCFO-1233zd-E and HFO-1336mzz-Z. Appl Therm Eng 71:2014–2212
50. Mondejar M, McLinden M, Lemmon E (2015) Thermodynamic properties of trans-1-chloro-3,3,3-trifluoropropene (R1233zd(E)): vapor pressure, p-ϱ-T data, speed of sound measurements and equation of state. J Chem Eng Data 60(8):2477–2489
51. Pan L, Wang H, Chen Q, Chen C (2011) Theoretical annd experimental study on several refrigerants of moderately high temperature heat pump. Appl Therm Eng 31:1886–1893
52. Parliament E, Council (2006) Directive 2006/40/EC of the European parliament and of the council of 17 May 2006 relating to emissions from air-conditioning systems in motor vehicles and amending Council Directive 70/156/EEC. Official Journal of the European Union. L161/12
53. Peguin RPS, Kamath G, Potoff JJ, daRocha SRP (2008) All-atom force field for the prediction of vapor-liquid equilibria and interfacial properties of HFA 134a. J Phys Chem 113:178–187
54. Peral-Antunez E (2012) MAC refrigerant blend cooperative research program update. In: SAE thermal management systems symposium, Scottsdale, USA
55. Peral-Antunez E, L'Huillier J (2011) Renault experiences with zeotropic blend for MAC. In: SAE alternative refrigerant and system efficiency symposium, Scottsdale, USA
56. Petr P, Raabe G (2015) Evaluation of R-1234ze(Z) as drop-in replacement for R-245fa in organic rankine cycles—from thermophysical properties to cycle performance. Energy 93:266–274
57. Potoff JJ, Siepmann JI (2001) Vapor-liquid equilibria of mixtures containing alkanes, carbon dioxide, and nitrogen. AIChE J 47:1676–1682
58. Potter SC, Tildesley DJ, Burgess AN, Rogers SC (1997) A transferable potential model for the liquid-vapor equilibria of fluoromethanes. Mol Phys 92(5):825–833
59. Prokhorov A (1969–1978) Great soviet encyclopedia, 3rd edn. Soviet State, Moscow
60. Quoilin S, Van Den Broek M, Declaye S, Dewallef P, Lemort V (2013) Techno-economic survey of organic rankine cycle (ORC) systems. Renew Sustain Energy Rev 22:168–186
61. Raabe G (2012) Molecular modeling of fluoropropene refrigerants. J Phys Chem B 116:5744–5751
62. Raabe G (2013) Molecular simulation studies on the thermophysical properties of the refrigerant blend R-445A. J Chem End Data 58:3470–3476
63. Raabe G (2013) Molecular simulation studies on the vapor-liquid-equilibria of binary mixtures of R-1234yf and R-1234ze(E) with R-32 and CO_2. J Chem Eng Data 58:1867–1873
64. Raabe G (2014) Molecular dynamics studies on liquid-phase dynamics and structures of four different fluoropropenes and their binary mixtures with R-32 and CO_2. J Phys Chem B 118:240–254
65. Raabe G (2015) Molecular simulation studies on the vapor-liquid-equilibria of the cis- and trans-HCFO-1233zd and the cis- and trans-HFO-1336mzz. J Chem Eng Data 60:2412–2419

66. Raabe G, Maginn EJ (2010) A force field for 3,3,3-fluoro-1-propenes, including HFO-1234yf. J Phys Chem B 114:10133–10142
67. Raabe G, Maginn EJ (2010) Molecular modeling of the vapor-liquid equilibrium properties of the alternative refrigerant 2,3,3,3-tetrafluoro-1-propene (HFO-1234yf). J Phys Chem Lett 1:93–96
68. Richter M, McLinden MO, Lemmon EW (2011) Thermodynamic properties of 2,3,3-tetrafluoroprop-1-ene (R1234yf): p-ϱ-t-measurements and an equation of state. J Chem Eng Data 56:3254–3264
69. Schulze C, Gräber M, Huhn M, Grätz U (2011) Real-time simulation of vapour compression cycles. In: Proceeding of the 8th international modelica conference, Dresden, Germany
70. Schulze C, Raabe G, Tegethoff WJ, Köhler J (2015) Transient evaluation of a city bus air conditioning system with R-445A as drop in—from the molecules to the system. Int J Therm Sci 96:355–361
71. Span R, Wagner W (1996) A new equation of state for carbon dioxide covering the fluid region from the triple-point temperature to 1100 K at pressures up to 800 MPa. J Phys Chem Ref Data 25(6):1509–1596
72. Spatz M, Minor B (2008) HFO-1234yf. A low GWP refrigerant for MAC. Honeywell/DuPont Joint Collaboration. In: VDA alternative refrigerant winter meeting, Saalfelden, Austria
73. Spatz MW (2007) Update on a low GWP refrigerant: fluid H. In: SAE alternative refrigerant system symposium, Phoenix, USA
74. Stoll J, Vrabec J, Hasse H (2003) A set of molecular models for carbon monoxide and halogenated hydrocarbons. J Chem Phys 119:11396–11407
75. Tanaka K, Higashi Y (2010) Thermodynamic properties of HFO-1234yf (2,3,3,3-tetrafluoropropene). Int J Refrig 33:474–479
76. Tanaka K, Maruko K, Fujimoto Y, Tanaka M (2013) PVT properties of R1234ze(Z). In: 4th conference on thermophysical properties and transfer processes of refrigerants, paper TP-072, Delft, The Netherlands
77. Tanaka K, Takahashi G, Higashi Y (2010) Measurements of the vapor pressure and $p\varrho T$ properties for trans-1,3,3,3-tetrafluoropropene (HFO-1234ze(E)). J Chem End Data 55:2169–2172
78. Tchanche BF, Lambrinos G, Frangoudakis A, Papadakis G (2011) Low-garde heat conversation into power using organic rankine cycles—a review od various application. Renew Sustain Energy Rev 15:3963–3979
79. Tillner-Roth R, Baehr HD (1994) An international standard formulation of the thermodynamic properties of 1,1,1,2-tetrafluoroethane (HFC-134a) for temperatures from 170 to 455 K at pressures up to 70 MPa. J Phys Chem Ref Data 23:657–729
80. UNEP (2014) 2014 report of the refrigeration. Air Conditioning and Heat Pumps Technical Options Committee
81. Union E, itsMember States (2015) Proposed amendment to the Montreal Protocol, UNEP/OzL.Pro.WG.1/36/5. European Union and its Member States
82. United Nations U (1987) Montreal protocol on substances that deplete the ozone layer. United Nations (UN), New York, NY, USA
83. Van Horn BL, Bonnett P, Abbas L (2015) Use of R-1233 in liquid chillers. Patent WO 2015(130589):A1
84. Wang CC (2014) System performance of R-1234yf refrigerants in air-conditioning and heat pump systems—an overview of current status. Appl Therm Eng 73:1412–1420
85. West RC (2005) CRC handbook of chemistry. CRC, Boca Raton
86. Yang Z, Gong M, Dong X, Li X, Wu J (2015) Molecular modeling and simulation of vapor-liquid equilibrium of the refrigerant R152a and its mixture R152a + R32. Fluid Phase Equil 294(93–100)
87. Zhang L, Siepmann IJ (2005) Pressure dependence of the vapor-liquid-liquid phase behavior in ternary mixtures containing of n-alkanes, n-perfluoroalkanes, and carbon dioxide. J Phys Chem B 109:2911–3010

Chapter 9
Conclusion and Outlook

The intention of this thesis is to summarize the fundamentals of molecular simulation and the available know-how for its application to derive information on thermophysical properties. Simulation studies on hydrofluoroolefines (HFO) and hydrochlorofluoroolefins (HCFO) as new class of working fluids exemplify the value of molecular simulations to provide predictions for the thermodynamic and transport properties of poorly known compounds to complement experimental data. However, reliable predictions of thermophysical properties require an adequate intermolecular potential functions ("force fields"), i.e. analytical models that describe the conformational energy of the system as sum of contributions from intra- and intermolecular interactions.

Considerable progress has been made over the past years to derive highly accurate pair potential from ab initio simulations. However, these ab initio potentials represent pure two-body potentials, and thus they are not able to describe properties of dense bulk phases or phase equilibria, as these are determined by multi-body interactions. Therefore, additional multi-body potentials are required when ab initio two-body potentials shall be applied in simulations on bulk phase properties. First attempts are found in literature to develop ab initio three-body interactions, but their inclusion in molecular simulation studies is computationally very expensive. In our studies on mercury, we proposed an empirical effective term into the ab initio pair potential to account for non-additive contributions of multi-body effects in a computationally more feasible way. However, ab initio potentials are still limited to small and medium-sized atoms and molecules such as nitrogen, carbon dioxide, or mercury. With this, the determination of ab initio pair and multi-body potentials for more complex components such as the HFOs or HCFOs is still far from reality. Therefore, intermolecular van der Waals interactions between complex molecules are usually treated by additive interactions between their atomic sites, which are modeled by Lennard-Jones potentials. The Lennard-Jones parameters of the different atom (site) types are then adjusted to represent experimental data for bulk phases to include implicitly the effect of multi-body interactions.

G. Raabe, *Molecular Simulation Studies on Thermophysical Properties*,
Molecular Modeling and Simulation, DOI 10.1007/978-981-10-3545-6_9

In molecules that consists of atoms with unequal electronegativities, a significant contribution to the configurational energy arises from electrostatic interactions. The common approach to account for electrostatic interactions is the calculation of coulomb interactions between fixed partial charges on the atom sites. The partial charges are generally derived in such a way that they reproduce the averaged polarization of the molecules in bulk phases. Consequently, the assigned partial charges can not account accurately for changing polarizability effects in the condensed phase with varying state points. Polarizable force fields that explicitly account for electronic polarization by determining the redistribution of charges due to an electrostatic field are computationally highly expensive. Our studies on the simple SPC/Fw water model though illustrated that the introduction of bond flexibility allows the geometry—and with this the distribution of the fixed charges—to change in response to the thermodynamic state point. This can be regarded as 'geometric polarization' as it mimics the change of the molecular interactions in response to the local environment. It therefore represents a computationally less demanding way to incorporate polarizability effects. Based on these results, the molecular model for HFO and HCFO compounds was developed as fully flexible all-atoms force fields.

The commercialization of this new class of working fluid is in its early stage, and hence, experimental data for most compounds in literature are limited. Thus, we intended to provide a transferable force field for HFO/HCFO compounds, in which Lennard-Jones parameters are fine-tuned to reproduce experimental data for the well known components, in order to enable reliable predictions for components with limited experimental data available. The molecular model was developed in the functional form of classical force fields such as AMBER, OPLS etc. to allow for combination with models from the literature for simulation studies on mixtures. Currently, the force field covers

- 3,3,3-trifluoro-1-propene HFO-1243zf;
- 2,3,3,3-tetrafluoro-1-propenes HFO-1234yf, 1,3,3,3-tetrafluoro-1-propenes HFO-1234$ze(Z)$ and HFO-1234$ze(E)$;
- 1,2,3,3,3-pentaflouro-1-propene HFO-1225$ye(Z)$;
- hexafluoro-1-propene HFO-1216
- 1,1,1,4,4,4 hexafluoro-2-butenes HFO-1336$mzz(Z)$ and HFO-1336$mzz(E)$
- 1-chloro-3,3,3-trifluoro-1-propene HCFO-1233$zd(E)$ and HCFO-1233$zd(Z)$

The Lennard-Jones parameters were adjusted to experimental data for phase equilibria properties (saturated densities and vapor pressure) of the model compound C_2F_4, HFO-1243zf, HFO-1234yf, HFO-1225$ye(Z)$ and HCFO-1233$zd(E)$. The force field was validated by molecular simulations on the vapor-liquid phase equilibria of these compounds, and the simulation results were found to agree well with available experimental data and correlations. For HFO-1234yf for instance, the simulated saturated densities and vapor pressures agree with experimental data within the error bars of the simulations. Studies on liquid phase viscosities prove the predictive capability of the molecular model to yield reliable predictions for properties that were not included in the parametrization. The molecular modeling of the compounds HFO-1234$ze(Z)$ and HFO-1234$ze(E)$, HFO-1216, HFO-1336$mzz(Z)$,

HFO-1336$mzz(E)$ and HCFO-1233$zd(Z)$ is based on transferred intra- and intermolecular parameters, and individually derived partial charges. As no Lennard-Jones parameters were adjusted for these compounds, the simulation studies for them are purely predictive. The simulation results for the vapor pressure and vapor-liquid coexistence curve of these compounds were also generally in good agreement with experimental data (if available). For the compounds HFO-1234$ze(Z)$ and HFO-1336$mzz(Z)$, which have recently attracted attention as working fluids for ORC processes, the predictions from molecular simulations provided first information on their VLE properties, before experimental data became available. The reliable predictions for compounds that were not included in the parametrization attests the good transferability of the force field parameters. However, we found that the transferability of the Lennard-Jones parameters depends on the number of fluorine atoms in the molecule. The parameters used to model compounds with four fluorine atoms describe compounds with five and six fluorines as too high boiling compounds. For these compounds, slightly modified fluorine parameters needed to be established.

The detailed modeling of the HFO and HCFO compounds by an all-atoms force field allows for studies on the effect of conformational isometry, as illustrated exemplarily for the cis- and trans-isomers of HCFO-1233zd, but also for the cis- and trans-isomer of HFO-1234ze. The different positions of the chlorine, or respectively the fluorine atoms, in cis- and trans-conformation yield different charge distributions, and with this different dipole moments. The higher dipole moments of the cis-isomers result in higher normal boiling points compared to the trans-isomers. The difference in the dipole moment, and consequently the normal boiling points, is less pronounced for HCFO-1233zd than for HFO-1234ze, which can be attributed to the higher electronegativity of the fluorine compared to the chlorine.

Simulation studies on mixtures of HFO compounds with "conventional" refrigerants such as R-32, CO_2 or R-134a have proven the predictive capability of molecular simulations to determine relevant properties of refrigerant blends to complement experimental data. The prediction of properties of the refrigerant blend R-445A from molecular simulations has enabled the first independent system simulation study on the performance of the refrigerant in mobile air-conditioning applications [3]. The simulation results for the VLE properties of binary blends of HFO compounds with R-32 and CO_2 have also found application by other researchers: the binary interaction parameters in REFPROP [2] for the mixtures CO_2 + R-1234yf and CO_2 + R-1234$ze(E)$ were derived from the molecular simulation results, and the GEMC simulation results for R-32 + R-1234$ze(E)$ were incorporated in the fitting of an EOS for this mixture by Akasaka [1].

Molecular simulations do not only yield information on thermophysical properties, they also provide insight into the systems on a molecular level to allow for molecular interpretations of these properties. This is well illustrated by the comparative simulation studies on the tetrafluoro-1-propenes HFO-1234yf and HFO-1234$ze(E)$. Although HFO-1234yf has the higher dipole moments, HFO-1234$ze(E)$ exhibits higher densities and viscosities. An analysis of contributions from Lennard-Jones and electrostatic interactions to the configurational energy of different fluoropropenes revealed that the highest contribution from electrostatic interactions occurs in HFO-1234$ze(E)$. We found that this can be attributed to an effective quadrupole moment of HFO-1234$ze(E)$ that is significantly higher than that of the other HFO compounds studied. Simulation studies on the local ordering of ionic liquids, but also in mixtures of HFO compounds with R-32 and CO_2 have additionally illustrated the value of molecular simulations to gain insight into preferred interactions. For the R-32 + HFO systems, for instance, the stronger interactions of R-32 with HFO-1234$ze(Z)$ could be related to the comparatively low diffusivity of R-32 in this mixture.

Future molecular simulations on HFO/HCFO based working fluids will cover studies on further mixtures with "conventional" refrigerants, but also on mixtures of different HFO or HCFO compounds. Furthermore, the force field model will be extended to other compounds, for example fluorinated ethenes, that are currently considered as refrigerants. The motivation for this future work is well reflected in this citation from the UNEP report [5]: "The perfect refrigerant does not exist, and is unlikely to come into existence. Choices will therefore include existing very low GWP refrigerants (e.g. R-717, R-744 or HCs) and the newly applied or developed chemicals. Many new alternatives are proposed which creates a challenge of finding the right refrigerant for each application...". Therefore, our work is aimed at providing information on the thermophysical properties of new refrigerants and blends to allow for the evaluation of their performance in potential applications.

To make a statement on future research trends in the wide field of molecular simulations is difficult, and is surely affected by personal research interests. In recent years, a surge in simulation studies on free energy calculations could be observed, which can be ascribed to both methodological improvement in the simulation techniques and enhanced computational power [4]. It can be expected that this research area will further grow due to the importance of free energies to define equilibrium and stability in mixtures at conditions relevant for experiments and processes. However, studies in the literature, and also our simulation studies on the activity coefficient of R-32 in R-32 + R-1234yf mixtures, reveal a high sensitivity of the computed free energies on the partial charges, and on the method to derive them. Therefore, future work to improve the accuracy of free energy calculations will have to include studies on rigorous methods to assign fixed partial charges, or on the use of polarizable force fields. The need for alternative approaches to assign partial charges was also revealed in simulation studies on ionic liquids. Therefore, studies on different approaches to account for polarization effects in force field models might represent a general research trend in the molecular modeling.

Another research field that experiences a growing interest is the application of reactive force fields such as ReaxFF that allow for a bridging between the system size and time scale of ab initio and classical molecular simulations. The reaction system of carbothermal reduction of silica is one application for which the reactive simulations might yield insight the underlying reaction mechanisms. This will also be the subject of future work in our research group.

References

1. Akasaka R (2013) Thermodynamic property models for the difluoromethane (R-32) + trans-1,3,3,3-tetrafluoropropene (R-1234ze(E)) and difluoromethane + 2,3,3,3-tetrafluoropropene (R-1234yf) mixtures. Fluid Phase Equilib 358:98–104
2. Lemmon EW, Huber ML, McLinden MO (2013) NIST standard reference database 23: reference fluid thermodynamic and transport properties—REFPROP, Version 9.1, Standard Reference Data Program. National Institute of Standards and Technology: Gaithersburg, MD
3. Schulze C, Raabe G, Tegethoff WJ, Köhler J (2015) Transient evaluation of a city bus air conditioning system with R-445A as drop in—from the molecules to the system. Int J Therm Sci 96:355–361
4. Shirts MR, Mobley DL, Chodera JD (2007) Alchemical free energy calculations: ready for prime time? In: Annual reports in computational chemistry, vol 3. Elsevier, pp 41–59
5. UNEP (2014) 2014 Report of the refrigeration. Air conditioning and heat pumps technical options committee

Appendix A
Force Field Parameters

Parameters of the all-atoms force fields for hydrofluoroolefines (HFO) and hydrochlorofluoroolefines with the following functional form (Tables A.1, A.2)

$$U_{conf} = \sum_{bonds} k_r (r - r_0)^2 + \sum_{angles} k_\theta (\theta - \theta_0)^2 + \sum_{dihedrals} k_\phi \left[1 + cos(n\phi - \delta)\right]$$
$$+ \sum_i \sum_{j>i} \left[4\varepsilon_{ij} \left[\left(\frac{\sigma_{ij}}{r_{ij}}\right)^{12} - \left(\frac{\sigma_{ij}}{r_{ij}}\right)^{6} \right] + \frac{q_i q_j}{4\pi \varepsilon_0 r_{ij}} \right]$$

To distinguish between the conformational (cis- and trans-) isomers and to account for different equilibrium geometries in the HFOs/HCFOs, the phase angles δ and multiplicities n in the torsion terms in Table A.3 need to be assigned attentively as summarized in the following tables (Tables A.4, A.5). The Tables A.6, A.7 and A.8 give the partial charges of the different HFO/HCFO compounds. Table A.9 provides the force field parameters for R-32.

Table A.1 LJ Parameters for the force field for HFO and HCFO compounds

Atom type	ε ($\frac{kJ}{mol}$)	σ (Å)	Source, 'fitted' to exp. data from
CM	0.41000	3.40	C_2F_4
CT	0.31091	3.40	HFO-1243zf/1234yf
FCM	0.23617	2.90	C_2F_4
FCM^h	0.21784	2.92	HFO-1225$ye(Z)$
FCT	0.23617	2.94	HFO-1243zf/1234yf
FCT^h	0.21784	2.94	Transferred from ε_{FCM^h}
HC	0.06570	2.65	AMBER [1]
$HC1$	0.06570	1.85	HFO-1243zf
$H1$	0.06570	2.47	AMBER [1]
Cl	1.09752	3.55	HCFO-1233$zd(E)$

G. Raabe, *Molecular Simulation Studies on Thermophysical Properties*, Molecular Modeling and Simulation, DOI 10.1007/978-981-10-3545-6

Table A.2 Parameters for the bond stretching and angle bending of the force field for HFO and HCFO compounds, derived from ab initio simulations

Bond	k_r ($\frac{\text{kJ}}{\text{mol Å}^2}$)	r_0 (Å)	Derived from
$CM = CM$	2831.69	1.331	C_2F_4, HFO-1234yf
$CM - CT$	1328.84	1.511	HFO-1243$zf/1216$
$CT - FCT$	1554.61	1.353	CF_4
$CM - HC, CM - H1$	1627.07	1.086	C_2H_4
$CM - FCM$	1864.73	1.330	C_2F_4, HFO-1234$yf/1216$
$CM - Cl$	953.80	1.734	HCFO-1233zd
Angle	k_θ ($\frac{\text{kJ}}{\text{mol rad}^2}$)	θ_0 (°)	
$HC - CM = CM$,	152.09	120.6	C_2H_4
$H1 - CM = CM$			
$FCT - CT - FCT$	367.61	107.5	CF_4
$CM - CT - FCT$	313.17	111.3	HFO-1243$zf/1216$
$HC - CM - CT$	135.27	115.1	HFO-1243zf
$HC - CM - HC$	122.63	118.7	C_2H_4
$CM = CM - FCM$	211.38	122.6	C_2F_4
$FCM - CM - FCM$	357.23	112.6	C_2F_4
$FCM - CM - CT$	319.57	112.5	HFO-1234$yf/1216$
$CM = CM - CT$	209.70	124.1	HFO-1243$zf/1216$
$H1 - CM - FCM$	214.36	113.7	HFO-1234$ze(E)/1225ye(Z)$
$Cl - CM = CM$	280.80	122.9	HCFO-1233zd
$Cl - CM - H1$	152.60	113.8	HCFO-1233zd

Table A.3 Torsion parameters of the force field for HFO and HCFO compounds, derived from ab initio simulations

Dihedral	k_ϕ ($\frac{\text{kJ}}{\text{mol}}$)	n	δ (°)	Derived from
$X - CM = CM - X$	27.84	$1/2^a$	$0/180$	AMBER [1]
$HC - CM - CT - FCT$	0.745	3	$0/180$	Kelkar et al. [2]
$H1 - CM - CT - FCT$				
$FCM^{(h)} - CM - CT - FCT$	1.0445	3	$0/180$	HFO-1234yf
$CM = CM - CT - FCT$	0.5951	3	$0/180$	HFO-1243zf

$^a n = 1$ for cis- and trans- isomers, $n = 2$ for all molecules without configurational isomers (HFO-1234yf, HFO-1243zf, HFO-1216))

Table A.4 Phase angles δ of the dihedrals in different HFO and HCFO compounds

Dihedral	δ (°)	In
$HC - CM - CT - FCT$	0	HFO-1234$ze(E)$/1336$mzz(Z)$, HCFO-1233$zd(E)$/1233$zd(Z)$
	180	HFO-1243zf/1234$ze(Z)$/1336$mzz(E)$
$FCM^{(h)} - CM - CT - FCT$	0	HFO-1234yf/1225$ye(Z)$
	180	HFO-1216
$CM = CM - CT - FCT$	0	HFO-1243zf/1216/1234$ze(Z)$, HCFO-1233$zd(E)$/1233$zd(Z)$
	180	HFO-1234yf/1234$ze(Z)$/1225$ye(Z)$

Table A.5 Phase angles δ assignment to distinguish between cis- and trans-isomers of different HFO and HCFO compounds ($n = 1$)

Dihedral		δ (°)	In
$H1 - CM = CM - HC$	cis	180	HFO-1234$ze(Z)$, HCFO-1233$zd(Z)$
	trans	0	HFO-1234$ze(E)$, HCFO-1233$zd(E)$
$FCM^h - CM = CM - FCM^h$	cis	180	HFO-1225$ye(Z)$
$FCM^{(h)} - CM = CM - CT$	cis	180	HFO-1234(Z)
	trans	0	HFO-1234(E)/1225$ye(Z)$
$H1 - CM = CM - CT$	cis	180	HFO-1225$ye(Z)$/1234$ze(E)$, HCFO-1233$zd(E)$
	trans	0	HFO-1234ze, HCFO-1233$zd(Z)$
$FCM - CM = CT - HC$	cis	180	HFO-1234$ze(E)$
	trans	0	HFO-12$ze(Z)$
$H1 - CM = CM - FCM^{(h)}$	trans	0	HFO-1225$ye(Z)$
$Cl - CM = CM - HC$	cis	180	HCFO-1233$zd(E)$
	trans	0	HCFO-1233$zd(Z)$
$Cl - CM = CM - CT$	cis	180	HCFO-1233$zd(Z)$
	trans	0	HCFO-1233$zd(E)$
$HC - CM = CM - HC$	cis	180	HFO-1336$mzz(Z)$
	trans	0	HFO-1336$mzz(E)$
$CT - CM = CM - CT$	cis	180	HFO-1336$mzz(Z)$
	trans	0	HFO-1336$mzz(E)$
$CT - CM = CM - HC$	cis	180	HFO-1336$mzz(E)$
	trans	0	HFO-1336$mzz(Z)$

Table A.6 Partial charges q_i for fluoropropenes HFO-1234yf, HFO-1243zf and HFO-1216

	HFO-1234yf		HFO-1243zf		HFO-1216	
#	Atom	q_i (e)	Atom	q_i (e)	Atom	q_i (e)
1	CM	−0.41911	CM	−0.23610	CM	0.46789
2	CM	0.19743	CM	−0.30512	CM	−0.14276
3	CT	0.63064	CT	0.84964	CM	0.87049
4	HC	0.20473	HC	0.16423	FCM	−0.14577
5	HC	0.20473	HC	0.16423	FCM	−0.14577
6	FCM	−0.18254	HC	0.17204	FCM	−0.13272
7	FCT	−0.21196	FCT	−0.26964	FCT	−0.25712
8	FCT	−0.21196	FCT	−0.26964	FCT	−0.25712
9	FCT	−0.21196	FCT	−0.26964	FCT	−0.25712

Table A.7 Partial charges q_i for fluoropropenes HFO-1234$ze(E)$, HFO-1234$ze(Z)$ and HFO-1225$ye(Z)$

	HFO-1234$ze(E)$		HFO-1234$ze(Z)$		HFO-1225$ye(Z)$	
#	Atom	$q_i(e)$	Atom	q_i (e)	Atom	q_i (e)
1	CM	0.25325	CM	0.23034	CM	0.07582
2	CM	−0.48504	CM	−0.52066	CM	0.05606
3	CT	0.77614	CT	0.98095	CT	0.53681
4	FCM	−0.19161	$H1$	0.13221	FCM^h	−0.14148
5	$H1$	0.13249	FCM	−0.16993	$H1$	0.15570
6	HC	0.24464	HC	0.20221	FCM^h	−0.12905
7	FCT	−0.24329	FCT	−0.28504	FCT	−0.18462
8	FCT	−0.24329	FCT	−0.28504	FCT	−0.18462
9	FCT	−0.24329	FCT	−0.28504	FCT	−0.18462

Table A.8 Partial charges q_i for the cis- and trans-isomers of HCFO-1233zd and HFO-1336mzz

	trans-1233zd		cis-1233zd		cis-1336mzz		trans-1336mzz	
#	Atom	q_i (e)	Atom	q_i (e)	Atom	q_i (e)	Atom	q_i (e)
1	CM	−0.00108	CM	−0.00427	CT	0.83591	CT	0.81189
2	CM	−0.30674	CM	−0.35183	CM	−0.25029	CM	−0.24485
3	CT	0.78148	CT	0.97116	CM	−0.25029	CM	−0.24485
4	Cl	−0.09402	Cl	−0.07583	CT	0.83591	CT	0.81189
5	$H1$	0.17109	$H1$	0.15783	FCT^h	−0.25389	FCT^h	−0.25015
6	HC	0.19567	HC	0.16832	FCT^h	−0.25389	FCT^h	−0.25015
7	FCT	−0.24880	FCT	−0.28846	FCT^h	−0.25389	FCT^h	−0.25015
8	FCT	−0.24880	FCT	−0.28846	HC	0.17605	HC	0.18341
9	FCT	−0.24880	FCT	−0.28846	HC	0.17605	HC	0.18341
10					FCT^h	−0.25389	FCT^h	−0.25015
11					FCT^h	−0.25389	FCT^h	−0.25015
12					FCT^h	−0.25389	FCT^h	−0.25015

Table A.9 Force field parameters for the new all-atoms force fields for R-32

LJ parameters and partial charges q_i

Atom	ε ($\frac{\text{kJ}}{\text{mol}}$)	σ (Å)	q_i (e)	Comment
C	0.45400	3.15	0.43960	Adjusted to VLE data for R-32
F	0.36600	2.94	−0.26138	Adjusted to VLE data for R-32
$H2$	0.06570	2.2293	0.04158	LJ parameters from AMBER [1]

Force constants

Bond	k_r ($\frac{\text{kJ}}{\text{mol Å}^2}$)	r_0 (Å)	Comment
$C-F$	1544.61	1.369	k_r from HFO model ($CT-FCT$)
$C-H$	1472.89	1.094	From ab initio simulations on R-32
Angle	k_θ ($\frac{\text{kJ}}{\text{mol rad}^2}$)	θ_0 (°)	Comment
$H-C-H$	146.54	113.6	From ab initio simulations on R-32
$F-C-F$	367.61	108.7	k_θ From HFO model ($FCT-CT-FCT$)
$H-C-F$	249.92	108.6	From ab initio simulations on R-32

References

1. Cornell WD, Cieplak P, Bayly CI, Gould IR, Merz KM Jr, Ferguson DM, Spellmeyer DC, Fox T, Caldwell JW, Kollmann PA (1995) A second generation force field for the simulation of proteins, nucleic acids, and organic molecules. J Am Chem Soc 117:5179–5197
2. Kelkar MS, Shiflett MB, Yokozeki A, Maginn EJ (2008) Development of force fields for hydrofluorocarbons. AIChE Annu, Meeting, Philadelphia, USA

Index

G. Raabe, *Molecular Simulation Studies on Thermophysical Properties*,
Molecular Modeling and Simulation, DOI 10.1007/978-981-10-3545-6

Printed by Printforce, the Netherlands